Guardian for Mathematical Physics

But he knows the way that I take; when he has tested me, I will come forth as gold.

(Job 23:10)

Guardian for Mathematical Physics

Hwanbae Park
Kyungpook National University

KNU PRESS

Guardian for Mathematical Physics

초판 1쇄 2016년 12월 15일 | **초판 2쇄** 2020년 2월 27일
지은이 박환배 | **펴낸이** 김상동 | **펴낸 곳** 경북대학교출판부
출판등록 1973년 10월 10일 ㉣97호 | **주소** 대구광역시 북구 대학로 80
전화 053-950-6741~3 | **팩스** 053-953-4692 | **E-mail** press@knu.ac.kr
Homepage http://knupress.com

ISBN 978-89-7180-450-6 93420

정가는 뒤표지에 있습니다.
잘못 만들어진 책은 바꾸어 드립니다.

PREFACE

Throughout my many years of teaching, I have seen many students struggle learning and understanding formulas and their meanings in mathematical physics courses. Learning significant amount of mathematical physics in two semesters is definitely challenging. Students experience a difficult time studying because they are not clear on how and where the contents are being used and applied, which motivated me to write "Guardian for Mathematical Physics".

Through this book, I tried to help students improve their understanding by presenting examples of Classical Mechanics, Electromagnetism, and Quantum Mechanics.

The contents of each Chapter are written so that students can study each Chapter without prior knowledge or without studying previous Chapters. Like the book title, I hope this book encourages students to overcome the obstacles that they had to face and succeed in understanding mathematical physics. I encourage those who want to learn more in-depth to read other mathematical physics books.

Eventually, I hope that students understand the core physics courses better and have productive results.

Dec. 2016

Hwanbae Park

ACKNOWLEDGMENTS

First of all, I would like to thank Stacy and Daniel who inspired and gave motivation to write this book. I also thank Hyunju, Joryang, and Joe who gave great strength and encouragement.

Last but not least, Jihye Kim and Sanghyeon Kim also helped the production by checking formulas and continuity of contents. I am thankful for their contributions.

I would like to thank all of crew members in Kyungpook National University publishers who helped the publication.

CONTENTS

Chapter 1. Introduction 11

1.1 Trigonometric Functions 11

1.2 Double- and Half-angle Formulas 14

1.2.1 Double-angle formulas · 14

1.2.2 Half-angle formulas · 15

1.3 Hyperbolic Functions 16

1.4 Differentiation 17

1.5 Chain Rule 22

1.6 Integration 24

1.7 Comple Numbers 29

1.8 Dirac Delta Function 31

1.9 Equations of Trajectory 32

Chapter 2. Differential Equation 35

2.1 Concept of Differential Equation 35

2.2 First Order Ordinary Differential Equation 37

2.3 Homogeneous Second Order Differential Equation 40

2.4 Inhomogeneous Second Order Differential Equation 45

Chapter 3. Vectors 51

3.1 Coordinate Systems 52

3.2 Vectors in Rectangular Coordinate System 57

3.2.1 Two Vectors Operation · 59

3.2.2 Three or Four Vectors Operation · 64

3.3 Gradient, Divergence and Curl **67**

3.4 Divergence and Stokes' Theorems **77**

3.5 Tensor **86**

Chapter 4. Physical Quantities in Coordinate Systems ——— 93

4.1 Spherical Coordinate System **93**

4.2 Cylindrical Coordinate System **100**

4.3 Curvilinear Coordinate System **102**

Chapter 5. Infinite Series ——— 113

5.1 Infinite Series **113**

5.2 Taylor Series **117**

5.3 Taylor Series for More than One Variable **129**

Chapter 6. Matrices ——— 134

6.1 Matrix Representation **134**

6.2 Operator **145**

6.3 Eigenvalue Equation **151**

6.4 Notation in Quantum Mechanics **156**

Chapter 7. Special Functions ——— 171

7.1 Beta and Gamma Functions **171**

7.2 Legendre Differential Equation **180**

7.2.1 Legendre Polynomials · 180

7.2.2 Properties of Legendre Polynomials · 195

7.3 Solutions for Bessel's Equation **206**

7.3.1 Bessel and Neumann Functions · 206

7.3.2 Other Bessel Functions · 215

Chapter 8. Fourier Transformation ———— 238

8.1 Fourier Series 238

8.2 Fourier Transform 245

Chapter 9. Variational Principle ———— 258

9.1 Lagrange's Equation 258

9.2 Lagrange Multipliers Method 271

Appendix 280

References 288

Index 290

Chapter 01 Introduction

This chapter introduces mathematical functions and formulas which have already been covered in high school and introductory college mathematics courses. These formulas are helpful to learn not only for other Chapters in this book, but also for core undergraduate physics courses, such as Classical Mechanics, Electromagnetism, and Quantum Mechanics.

1.1 Trigonometric Functions

- $\cos(\theta_1 \pm \theta_2) = \cos\theta_1 \cos\theta_2 \mp \sin\theta_1 \sin\theta_2$
- $\sin(\theta_1 \pm \theta_2) = \sin\theta_1 \cos\theta_2 \pm \cos\theta_1 \sin\theta_2$
- $\tan(\theta_1 \pm \theta_2) = \dfrac{\tan\theta_1 \pm \tan\theta_2}{1 \mp \tan\theta_1 \tan\theta_2}$

(a) $\cos(\theta_1 \pm \theta_2) = \cos\theta_1 \cos\theta_2 \mp \sin\theta_1 \sin\theta_2$

This equation is proved in [Ex. 3.2.1.1]. Using this relationship we can look at the following examples.

(i) For $\theta_1 = 0$ and $\theta_2 = \theta$, we get

$$\cos(0 - \theta) = \cos 0 \cos\theta + \sin 0 \sin\theta = \cos\theta$$

Since the left-hand side of the equation is $\cos(0-\theta)=\cos(-\theta)$, we obtain

$$\cos(-\theta)=\cos\theta$$

When a function satisfies $f(-x)=f(x)$, it is called an **even** function.

(ii) For $\theta_1=\theta_2\equiv\theta$, we get

$$\cos(\theta-\theta)=\cos\theta\cos\theta+\sin\theta\sin\theta=\cos^2\theta+\sin^2\theta$$

Since the left-hand side of the equation is $\cos(\theta-\theta)=1$, we obtain

$$1=\cos^2\theta+\sin^2\theta$$

(b) $\sin(\theta_1\pm\theta_2)=\sin\theta_1\cos\theta_2\pm\cos\theta_1\sin\theta_2$

This equation is proved in [Ex. 3.2.1.2]. Consider the following example using this relationship.

For $\theta_1=0$ and $\theta_2=\theta$,

$$\sin(0-\theta)=\sin0\cos\theta-\cos0\sin\theta=-\sin\theta$$

Since the left-hand side of the equation is $\sin(0-\theta)=\sin(-\theta)$, we get

$$\sin(-\theta)=-\sin\theta$$

When a function satisfies $f(-x)=-f(x)$, it is called an **odd** function.

(c) Since we know that $\sin(-\theta)=-\sin\theta$ and $\cos(-\theta)=\cos\theta$, we obtain

$$\tan(-\theta)=\frac{\sin(-\theta)}{\cos(-\theta)}=\frac{-\sin(\theta)}{\cos(\theta)}=-\tan\theta$$

Thus, $\tan\theta$ is the odd function.

Example

1.1.1 Show that

$$\sin A\cos B=\frac{1}{2}[\sin(A+B)+\sin(A-B)] \text{ and}$$

$$\cos A\sin B=\frac{1}{2}[\sin(A+B)-\sin(A-B)].$$

Ans. Since

$$\sin(A+B) = \sin A\cos B + \cos A\sin B$$

and

$$\sin(A-B) = \sin A\cos B - \cos A\sin B,$$

we obtain the following relationships by adding and subtracting the above two equations, respectively:

$$\frac{1}{2}[\sin(A+B) + \sin(A-B)] = \sin A\cos B$$

$$\frac{1}{2}[\sin(A+B) - \sin(A-B)] = \cos A\sin B$$

Similarly, we can easily obtain the following relationships:

- $\sin A\cos B = \frac{1}{2}[\sin(A+B) + \sin(A-B)]$
- $\cos A\sin B = \frac{1}{2}[\sin(A+B) - \sin(A-B)]$
- $\cos A\cos B = \frac{1}{2}[\cos(A+B) + \cos(A-B)]$
- $\sin A\sin B = -\frac{1}{2}[\cos(A+B) - \cos(A-B)]$

1.1.2 Show that $\sin A + \sin B = 2\sin\frac{A+B}{2}\cos\frac{A-B}{2}$.

Ans. We have shown that $\sin A\cos B = \frac{1}{2}[\sin(A+B) + \sin(A-B)]$ in [Ex. 1.1.1]. If we substitute $\frac{A+B}{2}$ instead of A and $\frac{A-B}{2}$ instead of B in the equation, it gives

$$\begin{aligned}\sin\frac{A+B}{2}\cos\frac{A-B}{2} &= \frac{1}{2}[\sin(\frac{A+B}{2} + \frac{A-B}{2}) + \sin(\frac{A+B}{2} - \frac{A-B}{2})] \\ &= \frac{1}{2}(\sin A + \sin B)\end{aligned}$$

Therefore, we obtain

$$\sin A + \sin B = 2\sin\frac{A+B}{2}\cos\frac{A-B}{2}$$

Similarly, we can easily obtain the following relationships:

- $\sin A + \sin B = 2\sin\frac{A+B}{2}\cos\frac{A-B}{2}$
- $\sin A - \sin B = 2\cos\frac{A+B}{2}\sin\frac{A-B}{2}$

$$\cdot\ \cos A+\cos B=2\cos\frac{A+B}{2}\cos\frac{A-B}{2}$$

$$\cdot\ \cos A-\cos B=-2\sin\frac{A+B}{2}\sin\frac{A-B}{2}$$

1.2 Double- and Half-angle Formulas

In Sections 1.2.1 and 1.2.2 we discuss the double-angle and half-angle formalism, respectively, using formulas obtained from the previous section.

1.2.1 Double-angle Formulas

- $\sin 2\theta = 2\sin\theta\cos\theta$
- $\cos 2\theta = 2\cos^2\theta - 1 = 1 - 2\sin^2\theta$
- $\tan 2\theta = \dfrac{2\tan\theta}{1-\tan^2\theta}$

(a) From $\sin(\theta_1+\theta_2)=\sin\theta_1\cos\theta_2+\cos\theta_1\sin\theta_2$ and $\theta_1=\theta_2\equiv\theta$, we obtain

$$\sin 2\theta=\sin\theta\cos\theta+\cos\theta\sin\theta=2\sin\theta\cos\theta$$

(b) From $\cos(\theta_1+\theta_2)=\cos\theta_1\cos\theta_2-\sin\theta_1\sin\theta_2$ and $\theta_1=\theta_2\equiv\theta$, we obtain

$$\cos 2\theta=\cos^2\theta-\sin^2\theta=2\cos^2\theta-1=1-2\sin^2\theta,$$

where $\cos^2\theta+\sin^2\theta=1$ is used.

Therefore, we get

$$\tan 2\theta=\frac{\sin 2\theta}{\cos 2\theta}=\frac{2\sin\theta\cos\theta}{\cos^2\theta-\sin^2\theta}=\frac{2\dfrac{\sin\theta}{\cos\theta}}{1-\dfrac{\sin^2\theta}{\cos^2\theta}}=\frac{2\tan\theta}{1-\tan^2\theta}$$

1.2.2 Half-angle formulas

- $\sin^2\dfrac{\theta}{2}=\dfrac{1}{2}(1-\cos\theta)$
- $\cos^2\dfrac{\theta}{2}=\dfrac{1}{2}(1+\cos\theta)$
- $\tan^2\dfrac{\theta}{2}=\dfrac{1-\cos\theta}{1+\cos\theta}$

(a) If we substitute $\dfrac{\theta}{2}$ instead of θ into the double-angle formula $\cos2\theta=1-2\sin^2\theta$, we obtain

$$\cos\theta=1-2\sin^2\frac{\theta}{2}$$

This equation gives

$$\sin^2\frac{\theta}{2}=\frac{1}{2}(1-\cos\theta)$$

(b) If we substitute $\dfrac{\theta}{2}$ instead of θ into the double-angle formula $\cos2\theta=2\cos^2\theta-1$, we obtain

$$\cos\theta=2\cos^2\frac{\theta}{2}-1$$

This gives

$$\cos^2\frac{\theta}{2}=\frac{1}{2}(1+\cos\theta)$$

(c) From (a) and (b), we obtain

$$\tan^2\frac{\theta}{2}=\frac{\sin^2\frac{\theta}{2}}{\cos^2\frac{\theta}{2}}=\frac{\frac{1}{2}(1-\cos\theta)}{\frac{1}{2}(1+\cos\theta)}=\frac{1-\cos\theta}{1+\cos\theta}$$

1.3 Hyperbolic Functions

- $\cosh x = \dfrac{e^x + e^{-x}}{2}$ (hyperbolic cosine)
- $\sinh x = \dfrac{e^x - e^{-x}}{2}$ (hyperbolic sine)
- $\tanh x = \dfrac{\sinh x}{\cosh x} = \dfrac{e^x - e^{-x}}{e^x + e^{-x}}$ (hyperbolic tangent)
- $e^{i\theta} = \cos\theta + i\sin\theta$ (Euler's formula)

(a) $\cosh^2 x - \sinh^2 x = (\dfrac{e^x + e^{-x}}{2})^2 - (\dfrac{e^x - e^{-x}}{2})^2 = 1$

If we express $e^{i\theta}$ by **Maclaurin series** which is covered in Section 2 of Chapter 5 and compare its real and imaginary parts, we obtain **Euler's formula**

$$e^{i\theta} = \cos\theta + i\sin\theta$$

If we insert $-\theta$ instead of θ, it gives

$$e^{-i\theta} = \cos(-\theta) + i\sin(-\theta) = \cos\theta - i\sin\theta$$

Thus, we get

$$\cos\theta = \frac{e^{i\theta} + e^{-i\theta}}{2} \text{ and } \sin\theta = \frac{e^{i\theta} - e^{-i\theta}}{2i}$$

Since $e^{i\theta} = \cos\theta + i\sin\theta$ from the Euler's formula, we have

$$(e^{i\theta})^n = (\cos\theta + i\sin\theta)^n$$

Therefore, we obtain $e^{in\theta} = (\cos\theta + i\sin\theta)^n$, which is called **De Moivre's formula**.

As an example that uses the De Moivre's formula, we have

$$(e^{i\theta})^2 = e^{i2\theta} = (\cos\theta + i\sin\theta)^2 = \cos^2\theta - \sin^2\theta + 2i\sin\theta\cos\theta$$

We know that the left-hand side of the equation is $e^{i2\theta} = \cos 2\theta + i\sin 2\theta$ from the Euler's formula. By comparing the real and imaginary parts of the left-hand and right-hand sides of the equation, we obtain

$$\cos 2\theta = \cos^2\theta - \sin^2\theta$$
$$\sin 2\theta = 2\sin\theta\cos\theta,$$

which are the equations obtained in the double-angle formulas.

The following relations are obtained by using $\cos\theta = \dfrac{e^{i\theta}+e^{-i\theta}}{2}$ and $\sin\theta = \dfrac{e^{i\theta}-e^{-i\theta}}{2i}$:

- $\cos ix = \dfrac{e^{i(ix)}+e^{-i(ix)}}{2} = \dfrac{e^{-x}+e^{x}}{2} = \cosh x$
- $\sin ix = \dfrac{e^{i(ix)}-e^{-i(ix)}}{2i} = \dfrac{e^{-x}-e^{x}}{2i} = \dfrac{i(e^{x}-e^{-x})}{2} = i\sinh x$
- $\cosh ix = \dfrac{e^{ix}+e^{-ix}}{2} = \cos x$
- $\sinh ix = \dfrac{e^{ix}-e^{-ix}}{2} = i\dfrac{e^{ix}-e^{-ix}}{2i} = i\sin x$

1.4 Differentiation

- $\dfrac{df}{dx} = f'(x) = \lim\limits_{\Delta x \to 0} \dfrac{f(x+\Delta x)-f(x)}{\Delta x}$
- $\dfrac{d}{dx}\sin x = \cos x$
- $\dfrac{d}{dx}\cos x = -\sin x$
- $\dfrac{d}{dx}\tan x = \dfrac{1}{\cos^2 x} = \sec^2 x$
- $\dfrac{d}{dx}\cosh x = \sinh x$
- $\dfrac{d}{dx}\sinh x = \cosh x$
- $\dfrac{d}{dx}\tanh x = \mathrm{sech}^2 x$

There are two types in differentiations: One is the **ordinary** differentiation ($\frac{df}{dx}$) for a function of one variable and the other is the **partial** differentiation ($\frac{\partial f}{\partial x}$) for a function of several variables.

For $f = f(x,y,z)$, the partial derivative $\frac{\partial f}{\partial x}$ means to perform differentiate on function f with respect to variable x while considering y and z as temporarily constants:

$$\frac{df}{dx}\Big|_{y,z=\text{constant}} \equiv \frac{\partial f}{\partial x}$$

The derivative, $\frac{df}{dx} = f'(x)$, is a ratio of the differential of a function with respect to the differential of a variable and is expressed as

$$\frac{df}{dx} = f'(x) \equiv \lim_{\Delta x \to 0} \frac{f(x+\Delta x) - f(x)}{\Delta x},$$

which is the **calculus limit method** of the differentiation. Obtaining the derivative is called **differentiation**.

The derivative corresponds to the slope of the function $f(x)$ between x and $x + \Delta x$, shown in Fig. 1.4.1.

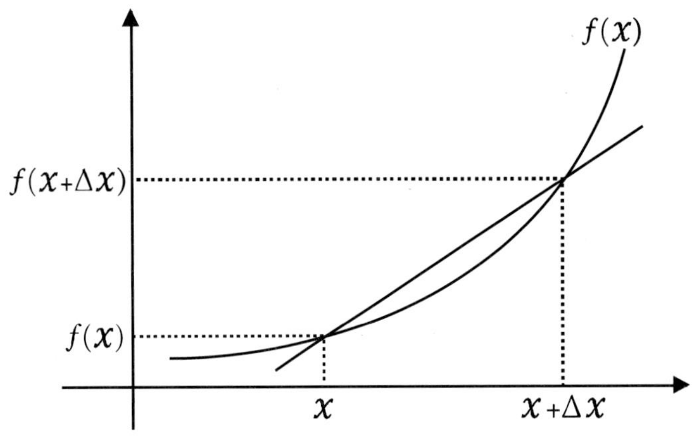

Fig. 1.4.1 $\frac{f(x+\Delta x) - f(x)}{\Delta x}$ is the slope for $\Delta x \to 0$, which is the derivate of a function $f(x)$ with respect to x.

Similarly, the partial differentiation can be written as

$$\frac{\partial f}{\partial x} = \lim_{\Delta x \to 0} \frac{f(x+\Delta x, y, z) - f(x, y, z)}{\Delta x}$$

Example

1.4.1 Calculate $\frac{dy}{dx} = y'$ for $y = f(x) = x^2$ by using the calculus limit method.

Ans. $\lim_{\Delta x \to 0} \frac{(x+\Delta x)^2 - x^2}{\Delta x} = \lim_{\Delta x \to 0} \frac{2x\Delta x + (\Delta x)^2}{\Delta x} = \lim_{\Delta x \to 0} (2x + \Delta x) = 2x$

From this result, we can generalize the differentiation of the function $y = [f(x)]^n$:

$$y' = n\frac{df(x)}{dx}[f(x)]^{n-1}$$

1.4.2 Calculate y' for $y = f(x) = \frac{1}{x^2}$.

Ans. (i) If we use the calculus limit method, we get

$$\begin{aligned} y' &= \lim_{\Delta \to 0} \frac{\frac{1}{(x+\Delta x)^2} - \frac{1}{x^2}}{\Delta x} \\ &= \lim_{\Delta x \to 0} \frac{-2x\Delta x - (\Delta x)^2}{x^2(x+\Delta x)^2 \Delta x} \\ &= \lim_{\Delta x \to 0} \frac{-2x - \Delta x}{x^2(x+\Delta x)^2} \\ &= -\frac{2}{x^3} \end{aligned}$$

(ii) Since $\frac{1}{x^2} = x^{-2}$, the problem is the same as $y' = (x^{-2})'$.

By using the result of [Ex. 1.4.1], $y' = n\frac{df(x)}{dx}[f(x)]^{n-1}$, we obtain

$$y' = -2x^{(-2-1)} = -2x^{-3} = -\frac{2}{x^3}$$

Thus, we can get the same answers for using the calculus limit method and the general differentiation law.

1.4.3 As an example of the partial differentiation, consider the following function, $f(x,y) = x^2 + 2x + 6xy + y^3 + 4$.

Ans. Since the function has two variables x and y, the derivatives of the function with respect to x and y are

$$\frac{df}{dx}\Big|_{y=\text{constant}} = \frac{\partial f}{\partial x} = 2x + 2 + 6y$$

and

$$\frac{\partial f}{\partial y} = 6x + 3y^2,$$

respectively.

1.4.4 Calculate y' for $y = f(x) = \sin x$.

Ans. Using the calculus limit method, we get

$$y' = \lim_{\Delta x \to 0} \frac{\sin(x + \Delta x) - \sin x}{\Delta x}$$

$$= \lim_{\Delta x \to 0} \frac{\sin x \cos \Delta x + \cos x \sin \Delta x - \sin x}{\Delta x},$$

(where $\cos \Delta x = 1$ and $\sin \Delta x = \Delta x$ for $\Delta x \ll 1$ that are proved in Taylor series of Chapter 5),

$$= \lim_{\Delta x \to 0} \frac{\sin x + \Delta x \cos x - \sin x}{\Delta x} = \cos x$$

1.4.5 Calculate y' for $y = f(x) = \cos x$ by using the calculus limit method.

Ans. $$y' = \lim_{\Delta x \to 0} \frac{\cos(x + \Delta x) - \cos x}{\Delta x}$$

$$= \lim_{\Delta x \to 0} \frac{\cos x \cos \Delta x - \sin x \sin \Delta x - \cos x}{\Delta x}$$

$$= \lim_{\Delta x \to 0} \frac{\cos x - \Delta x \sin x - \cos x}{\Delta x} = -\sin x$$

1.4.6 For $y = f(x) = \tan x$, find y' by using the calculus limit method.

Ans. We have

$$y' = \lim_{\Delta x \to 0} \frac{\tan(x+\Delta x) - \tan x}{\Delta x},$$

where the numerator is

$$\tan(x+\Delta x) - \tan x = \frac{\sin x \cos\Delta x + \cos x \sin\Delta x}{\cos x \cos\Delta x - \sin x \sin\Delta x} - \frac{\sin x}{\cos x}$$

$$= \frac{\cos x(\sin x \cos\Delta x + \cos x \sin\Delta x) - \sin x(\cos x \cos\Delta x - \sin x \sin\Delta x)}{\cos x(\cos x \cos\Delta x - \sin x \sin\Delta x)}$$

$$= \frac{(\cos^2 x + \sin^2 x)\sin\Delta x}{\cos x(\cos x \cos\Delta x - \sin x \sin\Delta x)} = \frac{\sin\Delta x}{\cos x(\cos x \cos\Delta x - \sin x \sin\Delta x)}$$

Therefore, we get

$$\begin{aligned} y' &= \lim_{\Delta x \to 0} \frac{\tan(x+\Delta x) - \tan x}{\Delta x} \\ &= \lim_{\Delta x \to 0} \frac{\sin\Delta x}{\cos x(\cos x \cos\Delta x - \sin x \sin\Delta x)\Delta x} \\ &= \lim_{\Delta x \to 0} \frac{\sin\Delta x}{(\cos^2 x \cos\Delta x - \cos x \sin x \sin\Delta x)\Delta x} \\ &= \lim_{\Delta x \to 0} \frac{\Delta x}{(\cos^2 x - \Delta x \cos x \sin x)\Delta x} = \frac{1}{\cos^2 x} \\ &= \sec^2 x \end{aligned}$$

1.4.7 Consider the hyperbolic functions, $\cosh x$, $\sinh x$, $\tanh x$, and find their derivatives.

Ans.

$$\frac{d}{dx}\cosh x = \frac{d}{dx}\left(\frac{e^x + e^{-x}}{2}\right) = \frac{e^x - e^{-x}}{2} = \sinh x$$

$$\frac{d}{dx}\sinh x = \frac{d}{dx}\left(\frac{e^x - e^{-x}}{2}\right) = \frac{e^x + e^{-x}}{2} = \cosh x$$

and

$$\begin{aligned} \frac{d}{dx}\tanh x &= \frac{d}{dx}\left(\frac{\sinh x}{\cosh x}\right) = \frac{\cosh^2 x - \sinh^2 x}{\cosh^2 x} \\ &= \frac{1}{\cosh^2 x} = \operatorname{sech}^2 x \end{aligned}$$

1.5 Chain Rule

Let's consider a function y of the variable u which is itself a function of x. Then, we have

$$\frac{\Delta y}{\Delta x}=\frac{y[u(x+\Delta x)]-y[u(x)]}{\Delta x}$$

$$=\frac{y[u(x+\Delta x)]-y[u(x)]}{u(x+\Delta x)-u(x)}\cdot\frac{u(x+\Delta x)-u(x)}{\Delta x}=\frac{\Delta y}{\Delta u}\cdot\frac{\Delta u}{\Delta x}$$

Since $\Delta u=u(x+\Delta x)-u(x)$, we get $\Delta u\to 0$ as $\Delta x\to 0$. If we take $\lim_{\Delta x\to 0}$ on the both sides of the equation, it gives

$$\frac{dy}{dx}=\frac{dy}{du}\frac{du}{dx}$$

This is known as **chain rule formula** which is to differentiate a function of another function.

If f is a function of two variables x and y which are functions of t, $f=f[x(t),y(t)]$, then we get

$$\frac{df}{dt}=\frac{\partial f}{\partial x}\Big|_y\frac{dx}{dt}+\frac{\partial f}{\partial y}\Big|_x\frac{dy}{dt},$$

where $\frac{\partial f}{\partial x}\Big|_y$ and $\frac{\partial f}{\partial y}\Big|_x$ are the partial derivatives considering y and x be constants, respectively.

If f is explicitly a function of time, it is expressed by $f=f[x(t),y(t),t]$ and then we have

$$\frac{df}{dt}=\frac{\partial f}{\partial t}\Big|_{x,y}+\frac{\partial f}{\partial x}\Big|_{t,y}\frac{dx}{dt}+\frac{\partial f}{\partial y}\Big|_{t,x}\frac{dy}{dt},$$

where $\frac{\partial f}{\partial t}\Big|_{x,y}$ is the partial derivatives considering x and y be constants.

Example

1.5.1 Consider $y = ln(\sin 2x)$ and find $\frac{dy}{dx}$.

Ans. Let $u = \sin 2x$, then the original equation is $y = ln(u)$.
Thus, we obtain

$$\frac{dy}{dx} = \frac{dy}{du}\frac{du}{dx} = \frac{1}{u}(2\cos 2x) = 2\frac{\cos 2x}{\sin 2x} = 2\cot 2x$$

1.5.2 For $y = e^{x^3}$, calculate $\frac{dy}{dx}$.

Ans. Let $u = x^3$, and then the original equation is $y = e^u$.
Thus, we obtain

$$\frac{dy}{dx} = \frac{dy}{du}\frac{du}{dx} = e^u(3x^2) = 3x^2e^{x^3}$$

1.5.3 In classical Mechanics, the radial component of an orbit of a particle with mass m moving in a central field $f(r)$ is given as follows:

$$m(\ddot{r} - r\dot{\theta}^2) = f(r)$$

Express the equation in the variable u as a function of θ when $r \equiv \frac{1}{u}$ and $\dot{\theta} = \frac{d\theta}{dt} \equiv \ell u^2$, where ℓ is the constant. The derivation of these equations is given in [Ex. 4.1.2].

Ans. We have

$$\dot{r} = \frac{dr}{dt} = \frac{dr}{du}\frac{du}{dt} = (-\frac{1}{u^2})\frac{du}{d\theta}\frac{d\theta}{dt} = -\ell\frac{du}{d\theta}$$

and

$$\begin{aligned}\ddot{r} &= \frac{d^2r}{dt^2} = \frac{d}{dt}(\frac{dr}{dt}) = -\ell\frac{d}{dt}(\frac{du}{d\theta}) = -\ell\frac{d\theta}{dt}\frac{d}{d\theta}(\frac{du}{d\theta}) \\ &= -\ell(\ell u^2)\frac{d^2u}{d\theta^2} = -\ell^2u^2\frac{d^2u}{d\theta^2}\end{aligned}$$

If we substitute these into the original equation, we get

$$\begin{aligned}\ddot{r} - r\dot{\theta}^2 = \frac{f(r)}{m} &\Rightarrow \ell^2u^2\frac{d^2u}{d\theta^2} + \frac{1}{u}\ell^2u^4 = -\frac{f(u^{-1})}{m} \\ &\Rightarrow \frac{d^2u}{d\theta^2} + u = -\frac{1}{m\ell^2u^2}f(u^{-1}) \qquad (1.5.3.1)\end{aligned}$$

For the gravitational field $f(r) = -\dfrac{k}{r^2}$ and a constant k, we have

$$f(u^{-1}) = -ku^2$$

Eq. (1.5.3.1) is then

$$\frac{d^2u}{d\theta^2} + u = -\frac{1}{m\ell^2u^2} f(u^{-1}) = \frac{k}{m\ell^2}$$

This is the second order ordinary differential equation with a constant on the right-hand side of the equation, which can be solvable.

1.5.4 Find $\dfrac{df}{dt}$ for $f = xy^2 + 2y$, in here $x = 2t$ and $y = t^3$.

Ans. We have

$$\begin{aligned}\frac{df}{dt} &= \frac{\partial f}{\partial x}\frac{dx}{dt} + \frac{\partial f}{\partial y}\frac{dy}{dt} \\ &= y^2(2) + (2xy+2)3t^2 \\ &= 2y^2 + 6xyt^2 + 6t^2 \\ &\quad \text{(by substituting } x = 2t \text{ and } y = t^3\text{)} \\ &= 6t^2 + 14t^6\end{aligned}$$

1.6 Integration

When a function f is defined over the entire interval, an accumulation function is given by

$$F(x) = \int f(x)dx,$$

which is an **anti-derivative** and $\dfrac{dF}{dx} = f(x)$. This integral is called the **indefinite integral**. The terminology of the integration is the process of solving the anti-derivative.

The meaning of the integral is

$$\lim_{\Delta x \to 0} \sum_{k=1}^{n} f_k(x)\Delta x = \int f(x)dx$$

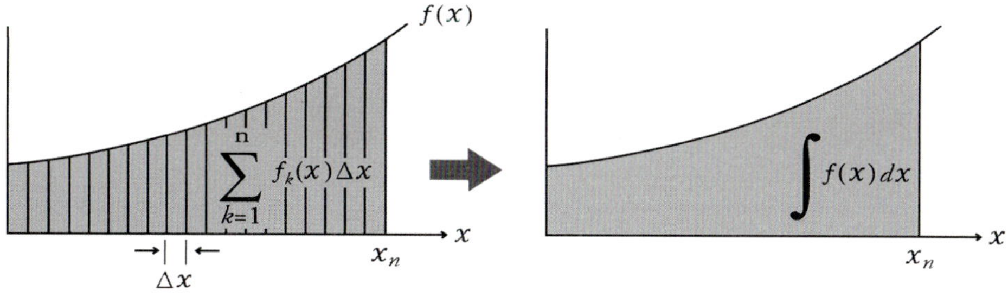

Fig. 1.6.1 A sum of area $f_k(x)\Delta x$ is changed to an integral for $\Delta x = \frac{x_n}{n} \rightarrow 0$ as $n \rightarrow \infty$.

In the limit $\Delta x \rightarrow 0$, $f(x)$ has an integral form and is called the **integrand** of the integral.

When the integral has an interval between a and b, the equation is

$$\int_a^b f(x)dx = \int_a^b \frac{dF}{dx}dx$$
$$= F(x)|_{x=a}^{x=b} = F(b) - F(a)$$

This integral stands in contrast to the indefinite integral, and it is known as the **definite integral** because it has a definite numerical value.

The validity of the solution can be proved as to whether the derivative of the solution gives the integrand or not. Sometimes it is not easy to think the anti-derivative and then trial guesses may a good approach in that case. It may lead to a better trial anti-derivative by differentiating the trial guessing of the anti-derivative. Some useful techniques to solve integrals are **integration by substitution** and **integration by parts**.

Example

1.6.1 As an example of the integration by substitution, consider $\int \sin 2\theta d\theta$.

Ans. We already know that $\frac{d\cos\theta}{d\theta} = -\sin\theta$. Let $x = 2\theta$ and the original integral is then

$$\int \sin 2\theta d\theta = \int \sin x\left(\frac{1}{2}dx\right)$$
$$= \frac{1}{2}\int \sin x dx = -\frac{1}{2}\int \frac{d\cos x}{dx}dx$$
$$= -\frac{1}{2}\cos x + C = -\frac{1}{2}\cos 2\theta + C,$$

where C is the integration constant.

If we differentiate the right-hand side of the equation with respect to θ, we obtain

$$\frac{d}{d\theta}\left(-\frac{1}{2}\cos 2\theta + C\right) = -\frac{1}{2}\frac{d\cos 2\theta}{d\theta} = -\frac{1}{2}(-2\sin 2\theta)$$
$$= \sin 2\theta$$

As expected we obtain

$$\frac{d}{d\theta}\int \sin 2\theta d\theta = \sin 2\theta$$

1.6.2 As another example of the integration by substitution, consider the form of $\int \frac{dx}{\sqrt{ax^2+bx+c}}$ with $a<0$. This equation is seen in Classical Mechanics to find equation for an orbit from energy equation.

Ans. We can express the equation by

$$\int \frac{dx}{\sqrt{ax^2+bx+c}} = \frac{1}{\sqrt{-a}}\int \frac{dx}{\sqrt{-\left(x+\frac{b}{2a}\right)^2 + \frac{b^2}{4a^2} - \frac{c}{a}}}$$

When we define $\sqrt{\frac{b^2}{4a^2} - \frac{c}{a}} = k$, the equation gives

$$\int \frac{dx}{\sqrt{ax^2+bx+c}} = \frac{1}{\sqrt{-a}}\frac{1}{k}\int \frac{dx}{\sqrt{1-\left\{\frac{x+\frac{b}{2a}}{k}\right\}^2}}$$

Let's define $\frac{x+\frac{b}{2a}}{k} = \cos\theta$, which then gives $dx = -k\sin\theta d\theta$.

The equation is then

$$\int \frac{dx}{\sqrt{ax^2+bx+c}} = \frac{1}{\sqrt{-a}}\int \frac{-\sin\theta d\theta}{\sin\theta} = -\frac{\theta}{\sqrt{-a}},$$

where

$$-\theta = -\cos^{-1}\left(\frac{x+\frac{b}{2a}}{k}\right) = \cos^{-1}\left(\frac{x+\frac{b}{2a}}{k}\right)$$

Therefore, we have

$$\int \frac{dx}{\sqrt{ax^2+bx+c}} = \frac{1}{\sqrt{-a}}\cos^{-1}\left(\frac{x+\dfrac{b}{2a}}{k}\right)$$

$$= \frac{1}{\sqrt{-a}}\cos^{-1}\left(\frac{x+\dfrac{b}{2a}}{\sqrt{\dfrac{b^2}{4a^2}-\dfrac{c}{a}}}\right)$$

1.6.3 Consider $\int x\cos x dx$ as an example of the integration by parts.

Ans. We can express it by

$$\int x\cos x dx = \int x(\sin x)' dx,$$

where $(\sin x)' = \frac{d}{dx}\sin x$.

Since $\frac{d}{dx}(x\sin x) = \sin x\frac{d}{dx}x + x\frac{d\sin x}{dx} = \sin x + x(\sin x)'$, we get

$$x(\sin x)' = \frac{d}{dx}(x\sin x) - \sin x$$

If we substitute this equation into the original integral, we obtain

$$\begin{aligned}\int x\cos x dx &= \int x(\sin x)' dx \\ &= \int \left[\frac{d}{dx}(x\sin x) - \sin x\right]dx \\ &= \int \frac{d}{dx}(x\sin x)dx - \int \sin x dx \\ &= x\sin x + \cos x + C\end{aligned}$$

1.6.4 Consider $\int \frac{dx}{\sqrt{a^2-a^2x^2}}$ with constant a as an example of the integration by substitution.

Ans. The equation is

$$\int \frac{dx}{\sqrt{a^2-a^2x^2}} = \frac{1}{a}\int \frac{dx}{\sqrt{1-x^2}}$$

If we let $x = \sin\theta$, we get $dx = \cos\theta d\theta$.

By substituting these into the integral, we have

$$\frac{1}{a}\int \frac{\cos\theta d\theta}{\sqrt{1-\sin^2\theta}} = \frac{1}{a}\int d\theta = \frac{\theta}{a} + C,$$

where C is the integration constant.

Since we define $x = \sin\theta$, we have $\theta = \sin^{-1}x$.

Thus, we obtain

$$\int \frac{dx}{\sqrt{a^2 - a^2x^2}} = \frac{\theta}{a} + C = \frac{\sin^{-1}x}{a} + C$$

1.6.5 Consider another example of the integration by substitution, $\int \frac{dx}{a^2x^2 + a^2}$ with constant a.

Ans. If we let $x = \tan\theta$, we have $dx = \frac{d\theta}{\cos^2\theta}$.

The original problem is then

$$\int \frac{dx}{a^2x^2 + a^2} = \frac{1}{a^2}\int \frac{d\theta}{\cos^2\theta} \frac{1}{(\tan^2\theta + 1)} = \frac{1}{a^2}\int d\theta$$
$$= \frac{1}{a^2} \tan^{-1}x + C$$

1.6.6 As an example for the double integral, let's calculate the area of a triangle which has base a and height b.

Ans. An area of a triangle is expressed by

$$A = \int dA = \int_0^a dx \int_0^{\frac{b}{a}x} dy$$
$$= \int_0^a \frac{b}{a} x dx$$
$$= \frac{b}{a}\frac{1}{2}x^2 \Big|_{x=0}^{x=a} = \frac{1}{2}ab$$

1.6.7 Calculate the area of a circle with radius a.

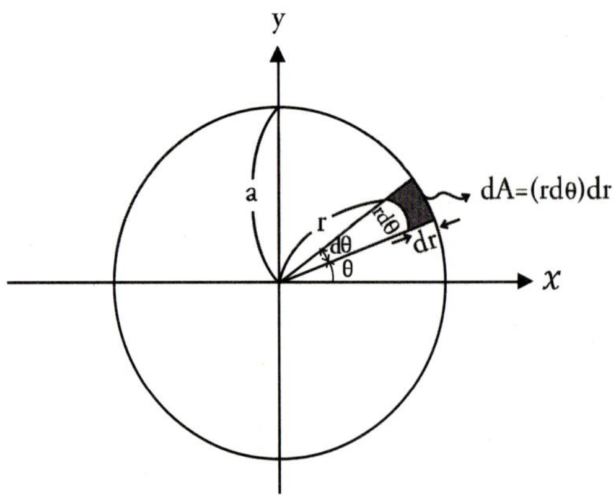

Ans. An area of a circle is expressed by

$$A = \int dA = \int_{r=0}^{a} \int_{\theta=0}^{2\pi} r dr d\theta$$

$$= 2\pi \int_0^a r dr \ = 2\pi \frac{1}{2} r^2 |_{r=0}^{r=a} \ = \pi a^2$$

1.7 Complex Numbers

A complex number z is represented by $z = x + iy$, where x and y are the real numbers, and $i \equiv \sqrt{-1}$. Sometimes $x = Re(z)$ and $y = Im(z)$ notation are used. This complex number is plotted in Fig. 1.7.1 on an **Argand diagram** where $x-$axis and $y-$axis represent the real and imaginary parts of the complex number.

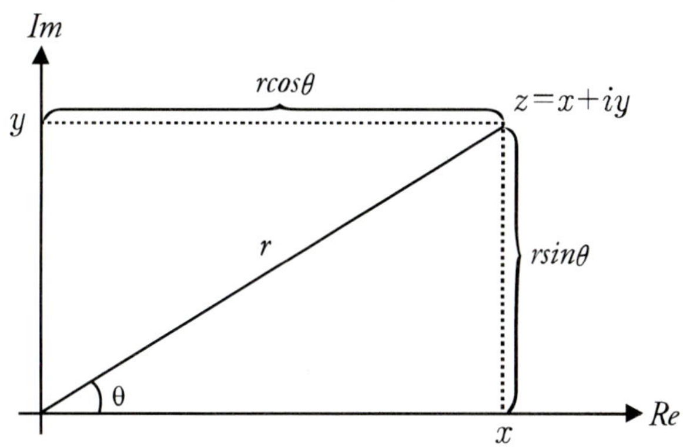

Fig. 1.7.1 A complex number $z = x + iy$ is plotted in the complex plane.

The **complex conjugate** z^* of z is defined as

$$z^* = x - iy$$

Then, we have

$$|z|^2 = zz^* = (x + iy)(x - iy) = x^2 + y^2$$

Its absolute value gives

$$|z| = \sqrt{x^2 + y^2}$$

and is called the **modulus** of the complex number z.

Sometimes z is expressed in the polar form, $z = re^{i\theta}$, where $r = \sqrt{x^2 + y^2}$ (r is the modulus) and $\theta = \tan^{-1}\frac{y}{x}$ or **arctan**$(\frac{y}{x})$ (θ is called the **argument** or **phase**). This polar form is convenient in calculations of multiplications and divisions of the complex numbers.

Example

1.7.1 A function $e^{i\theta}$ can be expressed in infinite series which is proved in Chapter 5:

$$\begin{aligned} e^{i\theta} &= 1 + i\theta + \frac{(i\theta)^2}{2!} + \frac{(i\theta)^3}{3!} + \frac{(i\theta)^4}{4!} + \cdots\cdots \\ &= 1 - \frac{\theta^2}{2!} + \frac{\theta^4}{4!} - \cdots\cdots + i(\theta - \frac{\theta^3}{3!} + \frac{\theta^5}{5!} - \cdots\cdots) \\ &= \cos\theta + i\sin\theta \end{aligned}$$

This is the Euler's formula which was previously mentioned.

1.7.2 Calculate $z_1 z_2$ for the complex number z_1 and z_2.

Ans. We have

$$z_1 z_2 = r_1 e^{i\theta_1} r_2 e^{i\theta_2} = r_1 r_2 e^{i(\theta_1 + \theta_2)}$$

This gives

$$r_1 r_2 (\cos\theta_1 + i\sin\theta_1)(\cos\theta_2 + i\sin\theta_2) = r_1 r_2 [\cos(\theta_1 + \theta_2) + i\sin(\theta_1 + \theta_2)]$$

The left-hand side of the above equation is

$$(\cos\theta_1\cos\theta_2 - \sin\theta_1\sin\theta_2) + i(\cos\theta_1\sin\theta_2 + \sin\theta_1\cos\theta_2)$$

By comparing the left-hand side of the equation with the right-hand side of the equation, we obtain

$$\cos(\theta_1 + \theta_2) = \cos\theta_1\cos\theta_2 - \sin\theta_1\sin\theta_2$$

and

$$\sin(\theta_1 + \theta_2) = \cos\theta_1\sin\theta_2 + \sin\theta_1\cos\theta_2$$

This is the same result obtained in Section 1.1.

1.7.3 For $z=\cos\theta+i\sin\theta$, calculate $\dfrac{dz}{d\theta}$.

Ans. The differentiation of the complex function is

$$\frac{dz(\theta)}{d\theta}=\frac{dx(\theta)}{d\theta}+i\frac{dy(\theta)}{d\theta}$$

and then the equation is

$$\frac{dz}{d\theta}=\frac{d}{d\theta}(\cos\theta+i\sin\theta)=-\sin\theta+i\cos\theta$$

This is also calculated as

$$\frac{dz}{d\theta}=\frac{d}{d\theta}e^{i\theta}=ie^{i\theta}=i(\cos\theta+i\sin\theta)=-\sin\theta+i\cos\theta$$

1.8 Dirac Delta Function

The values of **Dirac delta function** $\delta(x)$ are zero and ∞ for $x\neq 0$ and $x=0$, respectively, so that $\int_{-\infty}^{\infty}\delta(x)dx=1$,
which means the delta function is infinitely peaked at $x=0$ and its total area is 1.

Let's look at some of the important properties of the delta function.

(a) $\int_{-\infty}^{\infty}\delta(x)f(x)dx=0$

Since $\delta(x)=0$ except $x=0$, the integrand is then $\delta(x)f(x)=\delta(x)f(0)$. Therefore, $f(0)$ is not dependent on x and can be moved outside of the integral,

$$\int_{-\infty}^{\infty}\delta(x)f(x)dx=f(0)\int_{-\infty}^{\infty}\delta(x)dx=f(0)$$

(b) $\int_{-\infty}^{\infty}\delta(x-x')f(x)dx=f(x')$

(c) $\delta(-x)=\delta(x)$

If we let $y=-x$, $\int_{-\infty}^{\infty}\delta(-x)f(x)dx$ gives

$$-\int_{\infty}^{-\infty}\delta(y)f(-y)dy$$

$$=\int_{-\infty}^{\infty}\delta(y)f(-y)dy$$

$$=f(0)$$

$$=\int_{-\infty}^{\infty}\delta(x)f(x)dx$$

Thus, we obtain $\delta(-x)=\delta(x)$.

(d) $\delta(ax)=\frac{1}{|a|}\delta(x)$ for any real number a.

If we let $y=ax$, we have

$$\int_{-\infty}^{\infty}\delta(ax)f(x)dx=\frac{1}{a}\int_{-\infty}^{\infty}\delta(y)f(\frac{y}{a})dy=\frac{1}{a}f(0)$$

$$=\frac{1}{a}\int_{-\infty}^{\infty}f(x)\delta(x)dx=\int_{-\infty}^{\infty}f(x)[\frac{1}{a}\delta(x)]dx$$

or

$$=\frac{1}{a}\int_{-\infty}^{\infty}f(x)\delta(-x)dx=\int_{-\infty}^{\infty}f(x)[\frac{1}{a}\delta(-x)]dx$$

By comparing the left- and right-hand sides of the equation, we obtain

$$\delta(ax)=\frac{1}{|a|}\delta(x)$$

1.9 Equations of Trajectory

Equations for circle, ellipse, parabola and hyperbola are as follows:
(a) $(x-x_0)^2+(y-y_0)^2=a^2$ is the equation for the circle centered at $(x_{0,}\ y_0)$ with radius a.

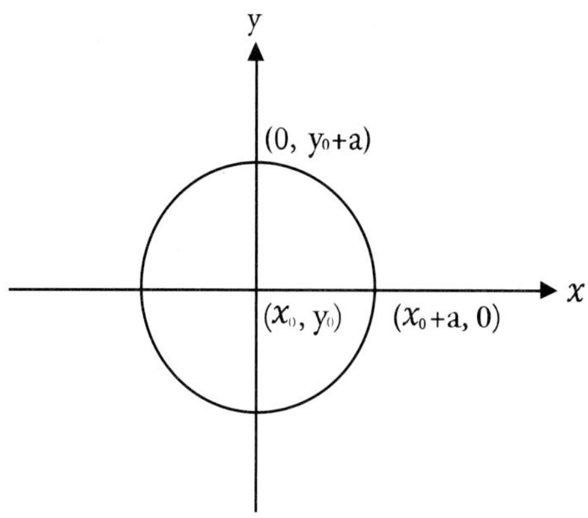

(b) $\frac{x^2}{a^2}+\frac{y^2}{b^2}=1$ is the equation for the ellipse of foci $(\pm\sqrt{a^2-b^2},\ 0)$ with a semi-major and semi-minor axis a and b, respectively. It is called a horizontal ellipse when $a>b$ as shown below while it is called the vertical ellipse for $a<b$.

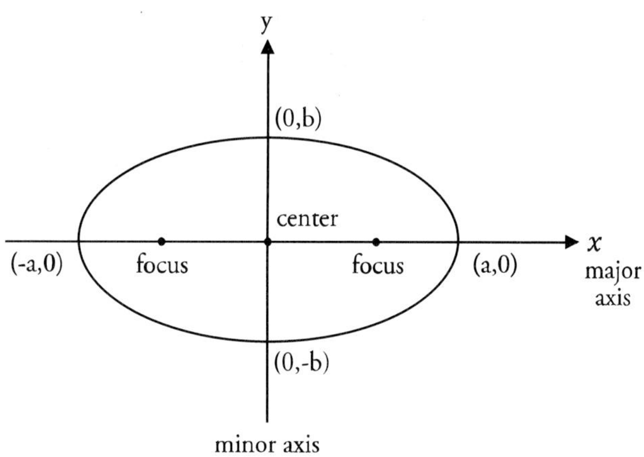

(c) $y=a(x-x_0)^2+b$ and $x=a(y-y_0)^2+b$ are the equations for the vertical and horizontal parabolas with the vertices $(x_{0,}\ b)$ and $(b,\ y_0)$, respectively.

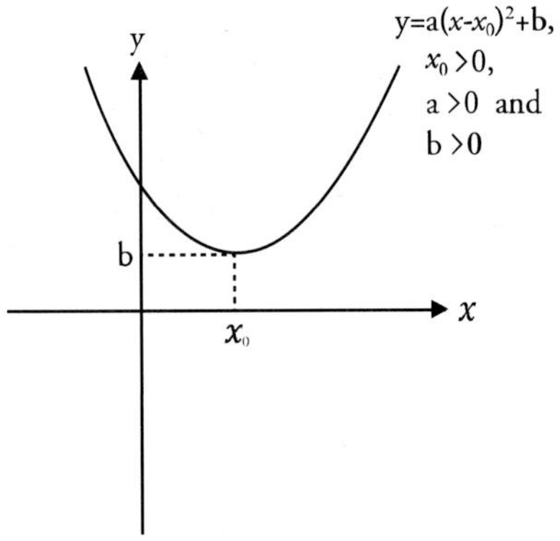

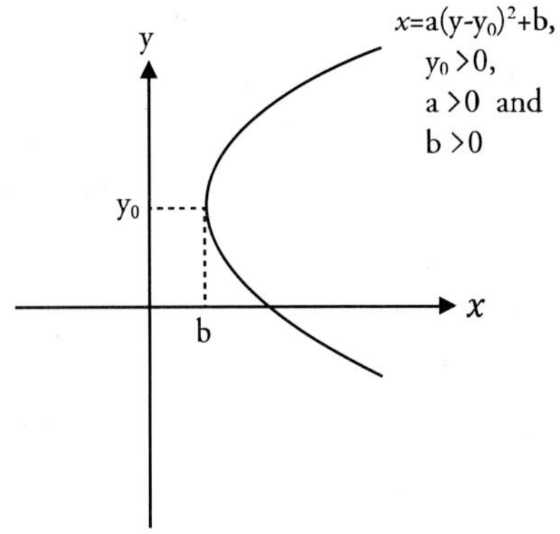

(d) $\dfrac{x^2}{a^2} - \dfrac{y^2}{b^2} = 1$ is the equation for the hyperbola with a center $(0,0)$, vertices $(\pm a,0)$, and foci $(\pm\sqrt{a^2+b^2},0)$. When x^2 (y^2) comes first in the equation, it is called the horizontal (vertical) hyperbola.

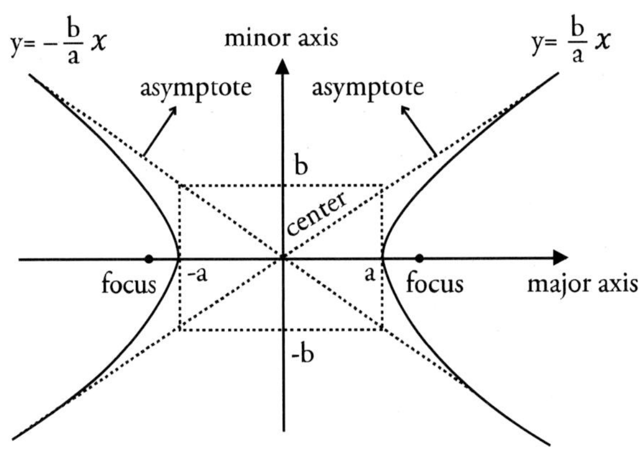

We can determine whether a general quadratic equation for the conic section $ax^2 + bxy + cy^2 + dx + ey = f$ represents a parabolic, ellipse or hyperbola depending on whether the discriminant $b^2 - 4ac$ is negative, zero, or positive, respectively. This statement is proved in [Ex. 6.4.7].

Chapter 02 Differential Equation

Equations containing derivatives are called differential equations. An order of a differential equation is determined by the order of the highest derivative in the equation.

In Sections 2.1~2.3 we study how to solve first and second order differential equations, which appear in undergraduate physics courses. Solving inhomogeneous second order differential equations is described in Section 2.4.

2.1 Concept of Differential Equation

· linear ⇔ nonlinear
· ordinary ⇔ partial
· homogeneous ⇔ inhomogeneous

When a function and its derivatives in the differential equation are not a product of a function and its derivative, we call it a **linear** differential equation. Otherwise, it is a **nonlinear** differential equation.

When all functions including derivatives in the differential equation are functions of only one variable, the equation is called an **ordinary** differential equation. In contrast, a **partial** differential equation contains functions with multiple variables.

When a function on the right-hand side of the differential equation is zero (non-zero), it is called a **homogeneous** (inhomogeneous) differential equation.

For instance, $a\frac{d^2x}{dt^2}+b\frac{dx}{dt}+cx=f(t)$, in here a, b, and c are constants, is the second order linear ordinary homogeneous (inhomogeneous) differential equation for $f(t)=0$ $(f(t)\neq 0)$.

On the other hand, $a(x)\frac{d^2x}{dt^2}+b(x)\frac{dx}{dt}+cx=f(t)$ is the nonlinear differential equation.

Example

2.1.1 Consider the first order nonlinear ordinary differential equation $x\frac{dy}{dx}=y+1$ and solve the differential equation.

Ans. This equation can be solved by a **method of separation of variables** which rearranges the equation so that each side has only one variable.

The original equation can be expressed by

$$\frac{dy}{y+1}=\frac{dx}{x},$$

where the left- and right-hand sides of the equation only have variable y and x, respectively.

If we integrate this equation, we have

$$\begin{aligned} ln(y+1) &= lnx + C \ (C \text{ is the integration constant}) \\ &= ln(ax) \ (ln(a) = C \text{ is another constant}), \end{aligned}$$

which gives

$$y+1=ax$$

We obtain the general solution

$$y(x)=ax-1$$

2.1.2 Consider the second order linear ordinary differential equation $m\frac{d^2x}{dt^2}=mg$, where g is the gravitational constant, and solve the differential equation.

Ans. If we integrate the original equation, we have

$$\int\frac{d}{dt}(\frac{dx}{dt})dt=g\int dt \Rightarrow \frac{dx}{dt}=gt+C$$

As an intial condition, let's consider the velocity and position at $t=0$ are $v(t=0)=v_0$ and $x(t=0)=x_0$, respectively.

We then get

$$v(t)=\frac{dx}{dt}=gt+v_0$$

If we integrate this equation, we have

$$\int(\frac{dx}{dt})dt=g\int tdt+v_0\int dt$$

$$\Rightarrow x(t)=\frac{1}{2}gt^2+v_0t+C$$

Then, we obtain the solution

$$x(t)=\frac{1}{2}gt^2+v_0t+x_0$$

2.2 First Order Ordinary Differential Equation

· For the first order differential equation $y'+P(x)y=Q(x)$, a solution is

$$y(x)=e^{-I}\int Q(x)e^{I}dx+Ce^{-I},$$

where P and Q are the functions of x, and $I(x)=\int P(x)dx$

Let's find a way to obtain a general solution for the first order ordinary differential equation,

$$y'+P(x)y=Q(x), \tag{2.2.1}$$

where P and Q are the functions of a variable x, and y' is $\frac{dy}{dx}$.

First, consider $Q=0$ corresponding to the homogeneous differential equation and then Eq. (2.2.1) is

$$\frac{dy}{dx}+Py=0 \tag{2.2.2}$$

Using the method of separation of variables, it gives

$$\frac{dy}{y}=-Pdx,$$

which yields

$$lny=-\int Pdx+C \quad (C \text{ is the integration constant})$$

This gives

$$y=Ae^{-\int Pdx},$$

where $A=e^{C}$ is the constant.

If we define $I=\int Pdx$, we have

$$\frac{dI}{dx}=P$$

Since we have

$$\begin{aligned}\frac{d}{dx}(ye^{I})&=\frac{dy}{dx}e^{I}+ye^{I}\frac{dI}{dx}\\&=y'e^{I}+ye^{I}P\\&=(y'+yP)e^{I}\end{aligned}$$

and the original differential equation is $y'+Py=Q$, this equation can be expressed by

$$\frac{d}{dx}(ye^{I})=Qe^{I} \tag{2.2.3}$$

If we integrate this equation, we get

$$ye^{I}=\int Qe^{I}dx+C \,(C \text{ is the integration constant})$$

We obtain the general solution for the first order ordinary differential equation, $y'+Py=Q$,

$$y = e^{-I}\int Qe^{I}dx + Ce^{-I}, \tag{2.2.4}$$

where $I = \int Pdx$.

Example

2.2.1 Let's apply the described method to get a general solution for the first order differential equation in [Ex. 2.1.1], $x\frac{dy}{dx} = y + 1$.

Ans. Since the original equation can be written as

$$y' - \frac{1}{x}y = \frac{1}{x} \tag{2.2.1.1}$$

By comparing Eqs. (2.2.1.1) and (2.2.1), we get

$$P(x) = -\frac{1}{x} \text{ and } Q(x) = \frac{1}{x}$$

Since $I = \int P(x)dx = -\int \frac{1}{x}dx = -lnx$, we get

$$I = -lnx$$

By inserting these into Eq. (2.2.4) for the general solution, we have

$$y = e^{-I}\int Qe^{I}dx + Ce^{-I}$$

$$= e^{lnx}\int \frac{1}{x}e^{-lnx}dx + Ce^{lnx}$$

To calculate e^{lnx} in this equation, let's define $e^{lnx} = f$.
This gives

$$lnf = lnx \Rightarrow f = x$$

Therefore, we get $e^{lnx} = x$. Similarly, we can obtain $e^{-lnx} = \frac{1}{x}$.

The result for the general solution is then

$$y = x\int \frac{1}{x}(\frac{1}{x})dx + Cx$$

$$= x\frac{-1}{x} + Cx$$

$$= Cx - 1$$

This is the same answer as obtained in [Ex. 2.1.1].

2.2.2 Find a general solution for $xy' + 2y = x^2 - x + 1$.

Ans. The original equation can be expressed as $y' + \frac{2}{x}y = x - 1 + \frac{1}{x}$.
This corresponds to $P(x) = \frac{2}{x}$ and $Q(x) = x - 1 + \frac{1}{x}$ in Eq. (2.2.1).
Since $I = \int P(x)dx = \int \frac{2}{x}dx = 2lnx$, the general solution is

$$\begin{aligned}
y &= e^{-I}\int Qe^{I}dx + Ce^{-I} \\
&= e^{-2lnx}\int (x - 1 + \frac{1}{x})e^{2lnx}dx + Ce^{-2lnx} \\
&= \frac{1}{x^2}\int (x - 1 + \frac{1}{x})x^2 dx + C\frac{1}{x^2} \\
&= \frac{1}{x^2}\int (x^3 - x^2 + x)dx + C\frac{1}{x^2} \\
&= \frac{1}{x^2}(\frac{1}{4}x^4 - \frac{1}{3}x^3 + \frac{1}{2}x^2) + C\frac{1}{x^2}
\end{aligned}$$

2.3 Homogeneous Second Order Differential Equation

· For the second order differential equation $ay'' + by' + cy = 0$, where a, b and c are constants, a solution is

$$y(x) = A_1 e^{D_1 x} + A_2 e^{D_2 x},$$

where $D_{1,2} = \frac{-b \pm \sqrt{b^2 - 4ac}}{2a}$

Consider the second order ordinary differential equation with constant coefficients,

$$a\frac{d^2y(x)}{dx^2} + b\frac{dy(x)}{dx} + cy(x) = f(x) \tag{2.3.1}$$

It is called homogeneous second order ordinary differential equation for $f(x)=0$ and inhomogeneous second order ordinary differential equation for $f(x)\neq 0$.

Let's find a way to obtain a general solution for the homogeneous second order ordinary differential equation by introducing the **differential operator**, $\frac{d}{dx}\equiv D$.

Then, we have

$$\frac{d^2y}{dx^2}=\frac{d}{dx}(\frac{d}{dx})y=\frac{d}{dx}Dy=D^2y$$

Eq. (2.3.1) for the homogeneous differential equation can be expressed as

$$(aD^2+bD+c)y(x)=0, \tag{2.3.2}$$

which is called the **auxiliary equation**.

If D_1 and D_2 are two roots for $aD^2+bD+c=0$, the equation is

$$(D-D_1)(D-D_2)y(x)=0,$$

where $D_{1,2}=\frac{-b\pm\sqrt{b^2-4ac}}{2a}$.

Therefore, we get

$$(D-D_1)y(x)=0 \text{ or } (D-D_2)y(x)=0$$

(a) For $(D-D_1)y(x)=0$,

it is $\frac{dy}{dx}=D_1y$, which gives $\frac{dy}{y}=D_1dx$.

If we integrate this equation, we get

$$lny=D_1x+C$$

Thus, we obtain the solution

$$y(x)=A_1e^{D_1x}$$

(b) For $(D-D_2)y(x)=0$,

it is $\frac{dy}{dx} = D_2 y$, which gives $\frac{dy}{y} = D_2 dx$.

Similarly as (a), we obtain the solution

$$y(x) = A_2 e^{D_2 x}$$

Since $y(x) = A_1 e^{D_1 x}$ and $y(x) = A_2 e^{D_2 x}$ are the solutions for Eq. (2.3.2), their linear combination is also the solution.

Thus, the general solution is

$$y(x) = A_1 e^{D_1 x} + A_2 e^{D_2 x}$$

(c) If two roots of the auxiliary equation are equal, $D_1 = D_2$, we have

$$(D - D_1)(D - D_1) y(x) = 0$$

If we let $(D - D_1) y(x) = u(x)$, then it is written as

$$(D - D_1) u(x) = 0,$$

which is $\frac{du(x)}{dx} = D_1 u(x)$.

This gives $\frac{du}{u} = D_1 dx$ and its solution is $u(x) = Ae^{D_1 x}$.

Since $(D - D_1) y(x) = u(x)$, we get $(D - D_1) y(x) = Ae^{D_1 x}$,

which is

$$\frac{dy}{dx} - D_1 y = Ae^{D_1 x}$$

By comparing this equation with Eq. (2.2.1), we know that $P(x) = -D_1$ and $Q(x) = Ae^{D_1 x}$.

The general solution for the first order differential equation is given by

$$y(x) = e^{-I} \int Qe^{I} dx + Ce^{-I} \text{ and } I = \int P dx,$$

where we get

$$I = -\int D_1 dx = -D_1 x$$

The general solution is then

$$y(x) = e^{D_1 x} \int Ae^{D_1 x} e^{-D_1 x} dx + Ce^{D_1 x}$$
$$= Axe^{D_1 x} + Ce^{D_1 x} = (Ax + C) e^{D_1 x}$$

Example

2.3.1 Show that the equation of motion for a **simple harmonic oscillator** is $\frac{d^2x(t)}{dt^2}+\omega_0^2x(t)=0$, where ω_0^2 is the constant, and solve the differential equation.

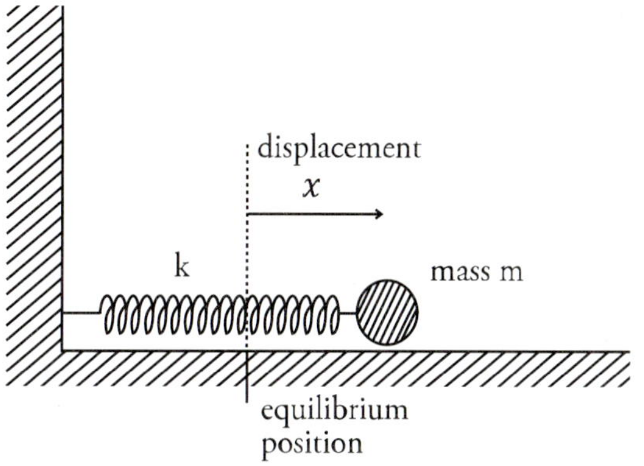

Simple harmonic oscillator: A spring obeys Hook's law, $F=-kx$, where F is the force exerted by the spring and x is the displacement of the spring from its equilibrium position.

Ans. Since the oscillator follows Hook's law $F=-kx$, it satisfies

$$F=m\ddot{x}\ =-kx$$

from Newton's second law of motion.

This equation gives

$$m\frac{d^2x}{dt^2}+kx=0,$$

which is expressed by

$$\frac{d^2x(t)}{dt^2}+\omega_0^2x(t)=0$$

assuming $\omega_0=\sqrt{\frac{k}{m}}$.

This homogeneous second order differential equation corresponds to case of $a=1$, $b=0$ and $c=\omega_0^2$ in Eq. (2.3.2) and gives

$$D_{1,2}=\frac{-b\pm\sqrt{b^2-4ac}}{2a}=\frac{\pm\sqrt{-4\omega_0^2}}{2}=\pm\omega_0 i,$$

where $\sqrt{-1}\equiv i$

The general solution is then given by

$$x(t) = A_1 e^{D_1 t} + A_2 e^{D_2 t} = A_1 e^{i\omega_0 t} + A_2 e^{-i\omega_0 t} \text{ or}$$

$$x(t) = B_1 \cos\omega_0 t + B_2 \sin\omega_0 t \text{ or}$$

$$x(t) = C\cos(\omega_0 t + \phi),$$

where $\omega_0 t + \phi$ is called the **phase** and ϕ is called the **phase angle**.

2.3.2 As another example, let's consider $\frac{d^2x(t)}{dt^2} + 2\gamma\frac{dx(t)}{dt} + \omega_0^2 x(t) = 0$, which is the equation for a harmonic oscillator with a damping (or frictional force) term (is called **damped harmonic oscillator** in contrast to the simple harmonic oscillator). In here $\omega_0^2 = \frac{k}{m}$ is the constant.

Ans. First, let's show how the original differential equation is obtained. In general the damping term is proportional to a velocity and its direction is opposite to a motion so that it can be written as $F_{damp} = -b\dot{x}$, where b is the proportional constant called **damping coefficient**.

Then, from Newton's second law of motion, we get

$$F_{tot} = m\ddot{x} \Rightarrow -kx - b\dot{x} = m\ddot{x}$$

Assuming $\frac{b}{m} = 2\gamma$, the equation is then expressed as

$$\frac{d^2x(t)}{dt^2} + 2\gamma\frac{dx(t)}{dt} + \omega_0^2 x(t) = 0$$

In this case, we get

$$D_{1,2} = \frac{-b \pm \sqrt{b^2 - 4ac}}{2a} = -\gamma \pm \sqrt{\gamma^2 - \omega_0^2}$$

The general solution is then as follows:

(i) For $\gamma^2 > \omega_0^2$, we obtain

$$x(t) = A_1 e^{D_1 t} + A_2 e^{D_2 t} = [A_1 e^{\sqrt{\gamma^2 - \omega_0^2}} + A_2 e^{-\sqrt{\gamma^2 - \omega_0^2}}]e^{-\gamma t},$$

which is called an **overdamped** oscillator.

(ii) For $\gamma^2 < \omega_0^2$, we obtain

$$x(t) = A_1 e^{D_1 t} + A_2 e^{D_2 t} = [A_1 e^{i\sqrt{\omega_0^2 - \gamma^2}} + A_2 e^{-i\sqrt{\omega_0^2 - \gamma^2}}]e^{-\gamma t} \text{ or}$$

$$x(t) = e^{-\gamma t}[B_1 \cos\sqrt{(\omega_0^2 - \gamma^2)}\,t + B_2 \sin\sqrt{(\omega_0^2 - \gamma^2)}\,t] \text{ or}$$

$$x(t) = Ce^{-\gamma t}\cos(\sqrt{(\omega_0^2 - \gamma^2)}\,t + \phi),$$

which is called an **underdamped** oscillator.

(iii) For $\gamma^2 = \omega_0^2$,

it is the case of $D_1 = D_2 = -\gamma$ and the general solution is then given by

$$x(t) = (At + C)e^{D_1 t} = (At + C)e^{-\gamma t},$$

which is called a **critical damping**.

Figure shows three different cases, overdamped, underdamped, and critical damping. The solutions for all three cases tend to go to zero as $t \to \infty$ and are often called the **transient** solutions.

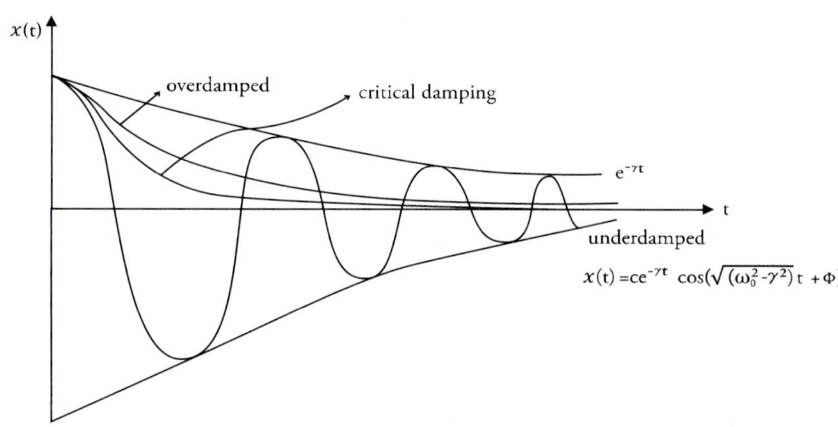

Damped harmonic oscillator: A damping force such as frictional force exists and it is proportional to a velocity of an object. Depending on the damping coefficient and angular frequency, its motion has overdamped, underdamped, or critical damping.

2.4 Inhomogeneous Second Order Differential Equation

Consider the second order differential equation with constant coefficients and a non-zero in the right-hand side,

$$a\frac{d^2y}{dx^2} + b\frac{dy}{dx} + cy = f(x), \qquad (2.4.1)$$

where $f(x) \neq 0$

Three steps are involved to obtain a general solution of this inhomogeneous second order differential equation:

(1) Find the general solution of the related homogeneous equation, $a\frac{d^2y}{dx^2}+b\frac{dy}{dx}+cy=0$, which is discussed in Section 2.3. This solution is referred as a **complementary solution** (y_c) in contrast to the solution of Eq. (2.4.1), a **particular solution** (y_p). The constant coefficients in the complementary solution are obtained by boundary (or initial) conditions.

(2) Find the particular solution by using an appropriate trial solution. The constant coefficients in a trial solution are obtained by substituting a trial solution into Eq. (2.4.1).

(3) Since $ay_c''+by_c'+cy_c=0$ and $ay_p''+by_p'+cy_p=f(x)$, these equations can be expressed by $(aD^2+bD+c)y_c=0$ and $(aD^2+bD+c)y_p=f(x)$ using the differential operator D. If we add these two equations, it gives

$$(aD^2+bD+c)(y_c+y_p)=f(x)$$

Therefore, the sum of the complementary and particular solutions, y_c+y_p, is also the solution and is called a **general solution** of Eq. (2.4.1). Thus, the complete general solution is given by

$$y(x)=y_p(x)+y_c(x)$$

Example

2.4.1 Consider $\frac{d^2x(t)}{dt^2}+\omega_0^2x(t)=F_0\cos\omega t$, where ω_0^2 and F_0 are the constants. This is the equation for the **forced** (or **driven**) **oscillator without a damping term**.

Ans. Let's follow the three steps described earlier to obtain the general solution.
(i) We already know that the complementary solution of the equation is

$$x_c(t)=B_1\cos\omega_0 t+B_2\sin\omega_0 t$$

(ii) As a trial solution, let's take

$$x_p(t) = C_1\cos\omega t + C_2\sin\omega t$$

and substitute it into the original equation $\dfrac{d^2x_p(t)}{dt^2} + \omega_0^2 x_p(t) = F_0\cos\omega t$.

Then, we have

$$(-\omega^2 C_1\cos\omega t - \omega^2 C_2\sin\omega t) + (\omega_0^2 C_1\cos\omega t + \omega_0^2 C_2\sin\omega t) = F_0\cos\omega t$$

$$\Rightarrow (\omega_0^2 - \omega^2)C_1\cos\omega t + (\omega_0^2 - \omega^2)C_2\sin\omega t = F_0\cos\omega t$$

This equation satisfies for all t so that it gives

$$(\omega_0^2 - \omega^2)C_2 = 0$$

and

$$(\omega_0^2 - \omega^2)C_1 = F_0$$

These relations give $C_2 = 0$ and $C_1 = \dfrac{F_0}{\omega_0^2 - \omega^2}$ for $\omega_0^2 \neq \omega^2$.

Therefore, the particular solution is

$$x_p(t) = \frac{F_0}{\omega_0^2 - \omega^2}\cos\omega t$$

(iii) Since the general solution is the sum of $x_c(t)$ and $x_p(t)$, we obtain

$$x(t) = x_p(t) + x_c(t) = \frac{F_0}{\omega_0^2 - \omega^2}\cos\omega t + B_1\cos\omega_0 t + B_2\sin\omega_0 t, \quad (2.4.1.1)$$

where B_1 and B_2 are constants which are determined by boundary conditions.

For the boundary condition $x(0) = 0$ and $\dfrac{dx(t)}{dt}\Big|_{t=0} = \dot{x}(0) = 0$, Eq. (2.4.1.1) is

$$x(0) = \frac{F_0}{\omega_0^2 - \omega^2} + B_1 = 0 \quad (2.4.1.2)$$

Since $\dot{x} = \dfrac{-\omega F_0}{\omega_0^2 - \omega^2}\sin\omega t - \omega_0 B_1\sin\omega_0 t + \omega_0 B_2\cos\omega_0 t$, it gives

$$\dot{x}(0) = \omega_0 B_2 = 0 \quad (2.4.1.3)$$

From Eqs. (2.4.1.2) and (2.4.1.3), B_1 and B_2 are determined to be

$$B_1 = -\frac{F_0}{\omega_0^2 - \omega^2} \text{ and } B_2 = 0$$

The complete general solution is then

$$x(t) = \frac{F_0}{\omega_0^2 - \omega^2}(\cos\omega t - \cos\omega_0 t)$$

2.4.2 As another example, let's consider $\frac{d^2x(t)}{dt^2}+2\gamma\frac{dx(t)}{dt}+\omega_0^2x(t)=F_0\cos\omega t$, which is the equation for the **forced** (or **driven**) **oscillator with damping term**.

Ans. Let's follow the three steps described earlier to obtain the general solution.
(i) There are three different cases (overdamped, underdamped, and critical damping) depending on the magnitude of γ^2 compared to ω_0^2 as discussed in [Ex. 2.3.2] and we already know the complementary solution for each case.
(ii) Let's take a trial solution $X=X_R+iX_I$, where X_R and X_I are the real and imaginary parts of $X=C(\omega)e^{i\omega t}$.
Since $F_0\cos\omega t=Re(F_0e^{i\omega t})$, the real part of X is the particular solution of the original equation.
Now let's solve the equation

$$\frac{d^2X(t)}{dt^2}+2\gamma\frac{dX(t)}{dt}+\omega_0^2X(t)=F_0e^{i\omega t} \tag{2.4.2.1}$$

By substituting the trial solution into Eq. (2.4.2.1), we get

$$(-\omega^2+2i\gamma\omega+\omega_0^2)C=F_0$$

This gives

$$C=\frac{F_0}{\omega_0^2-\omega^2+2i\gamma\omega}$$
$$=\frac{F_0}{\sqrt{(\omega_0^2-\omega^2)^2+4\gamma^2\omega^2}}\left(\frac{\omega_0^2-\omega^2}{\sqrt{(\omega_0^2-\omega^2)^2+4\gamma^2\omega^2}}-\frac{2i\gamma\omega}{\sqrt{(\omega_0^2-\omega^2)^2+4\gamma^2\omega^2}}\right)$$

If we let $A=\sqrt{(\omega_0^2-\omega^2)^2+4\gamma^2\omega^2}$, $\cos\phi=\frac{\omega_0^2-\omega^2}{\sqrt{(\omega_0^2-\omega^2)^2+4\gamma^2\omega^2}}$,
and $\sin\phi=\frac{2i\gamma\omega}{\sqrt{(\omega_0^2-\omega^2)^2+4\gamma^2\omega^2}}$,
we get $C=\frac{F_0}{A}(\cos\phi-i\sin\phi)=\frac{F_0}{A}e^{-i\phi}$

Then, we have

$$X(t)=C(\omega)e^{i\omega t}=\frac{F_0}{A}e^{-i\phi}e^{i\omega t}=\frac{F_0}{A}e^{i(\omega t-\phi)}$$

The real part of the $X(t)$ is then

$$x_p(t)=X_R(t)=\frac{F_0}{A}\cos(\omega t-\phi)$$

(iii) Finally, by summing the complementary and particular solutions, we can obtain the complete general solution for the forced (or driven) oscillator with damping term.

2.4.3 Find the equation of motion for a particle with mass m moving in an inverse square repulsive force, $f(r) = -\dfrac{k}{r^2}$ (k is a positive constant).

Ans. From Newton's second law $f(r)\hat{r} = m\vec{a}$, we obtain

$$m\ddot{r} - mr\dot{\theta}^2 = f(r)$$

with $\dot{\theta} = \dfrac{\ell}{r^2}$ (ℓ is the angular momentum per unit mass) for the radial component in the spherical coordinates which is covered in detail in Chapter 4.

If we define $u(r) = \dfrac{1}{r}$, we get

$$\dot{r} = -\ell\frac{du}{d\theta} \text{ and } \ddot{r} = -\ell\frac{d}{dt}\left(\frac{du}{d\theta}\right) = -\ell^2 u^2 \frac{d^2u}{d\theta^2}$$

Therefore, the equation is then expressed as

$$\frac{d^2u}{d\theta^2} + u = \frac{k}{m\ell^2},$$

where the right-hand side of the equation is the constant.

This gives the general solution

$$u(\theta) = A\cos(\theta - \theta_0) + \frac{k}{m\ell^2}$$

Since $u(r) = \dfrac{1}{r}$, this equation gives

$$r(\theta) = \frac{1}{A\cos(\theta - \theta_0) + \dfrac{k}{m\ell^2}}$$

Table 2.4.1 is the trial solutions for $f(x)$ in the right-hand side of the inhomogeneous second order differential equation.

$f(x)$	trial solution for x_p
arbitrary constant C	C
$\cos nx$, n is an arbitrary constant	$Re[Ce^{inx}]$ or $c_1\cos nx + c_2\sin nx$
$\sin nx$, n is an arbitrary constant	$Im[Ce^{inx}]$ or $c_1\cos nx + c_2\sin nx$
e^{nx}, n is an arbitrary constant	Ce^{nx}
$x^n - x$, n is an integer	$c_n x^n + c_{n-1}x^{n-1} + \cdots\cdots + c_1 x + c_0$

Table 2.4.1 Trial solutions for given $f(x)$ in the inhomogeneous second order differential equation.

2.4.4 Let's take a look at one of important examples,

$$\frac{d^2x}{dt^2}+\frac{dx}{dt}-2x=e^t \tag{2.4.4.1}$$

Ans. This equation is expressed as

$$(D^2+D-2)x=e^t$$

by introducing the differential operator, $D=\dfrac{dx}{dt}$.

The original equation is then written as

$$(D+2)(D-1)x=e^t$$

(i) The complementary solution is then $x_c(t)=Ae^t+Be^{-2t}$.

We notice that one of roots for the auxiliary equation, Ae^t, is equal to the function e^t on the right-hand side of Eq. (2.4.4.1).

(ii) In this case, the particular solution has a form of $x_p=Cte^t$ as the similar manner discussed earlier for $D_1=D_2$ of the second order differential equation in Section 2.3.

Since the particular solution is $x_p=Cte^t$, the first and second derivatives of it are

$$\dot{x}_p=Ce^t+Cte^t$$

and

$$\ddot{x}_p=2Ce^t+Cte^t$$

respectively.

If we substitute these into the original equation, we get

$$3C=1,$$

which gives

$$C=\frac{1}{3}$$

The particular solution is then

$$x_p=\frac{1}{3}te^t$$

Thus, we obtain the general solution

$$x(t)=x_c(t)+x_p(t)=Ae^t+Be^{-2t}+\frac{1}{3}te^t$$

Chapter 03 Vectors

When a position of a point in a space is described uniquely and completely by coordinates, a system consisted of the coordinates is called a coordinate system. Depending on the problem, the coordinate system, which provides simple and convenient ways for calculations, is chosen.

Basic concepts for the rectangular (or Cartesian), plane polar, spherical, and cylindrical coordinate systems are discussed in Section 3.1. Quantities which have both magnitude and direction, vectors, and their operations in the rectangular coordinate system are described in Section 3.2.

The differential vector operator $\overrightarrow{\nabla}$ is defined and its operations on another vector or scalar are described in Section 3.3. In Section 3.4 the divergence theorem and Stokes' theorem, which are very useful in Electromagnetism are discussed. The basic concept of a tensor is briefly discussed with examples from Classical Mechanics and Electromagnetism are presented in Section 3.5.

3.1 Coordinate Systems

- (x, y, z) in rectangular coordinate system

  $-\infty < x < \infty$, $-\infty < y < \infty$, and $-\infty < z < \infty$

- (r, θ) for $x-y$ plane in polar coordinate system

  $0 \le r < \infty$ and $0 \le \theta \le 2\pi$

- (r, θ, ϕ) in spherical coordinate system

  $0 \le r < \infty$, $0 \le \theta \le \pi$, and $0 \le \phi \le 2\pi$

- (ρ, ϕ, z) in cylindrical coordinate system

  $0 \le \rho < \infty$, $0 \le \phi \le 2\pi$, and $-\infty < z < \infty$

We consider four coordinate systems in this and following Chapters: Rectangular (or Cartesian), Plane Polar, Spherical and Cylindrical coordinate systems.

(1) **Rectangular coordinate system** in a three dimensional space is represented by three axes of $x-$axis, $y-$axis, and $z-$axis which are perpendicular to each other as shown in Fig. 3.1.1.

By a notation convention for vectors we follow a **right-handed coordinate system** (or **right-hand rule**), in which a right thumb points along a positive $z-$direction when fingers curl from the $x-$axis to the $y-$axis. If this movement is viewed from the $z-$axis, it is counterclockwise so that we defines a positive rotation is counterclockwise about a rotation axis.

A position vector $\vec{r}$ is the distance from the origin to observation point P. In order to cover the whole three-dimensional space, the ranges of x, y and z are

$$-\infty < x < \infty, \quad -\infty < y < \infty, \quad \text{and} \quad -\infty < z < \infty$$

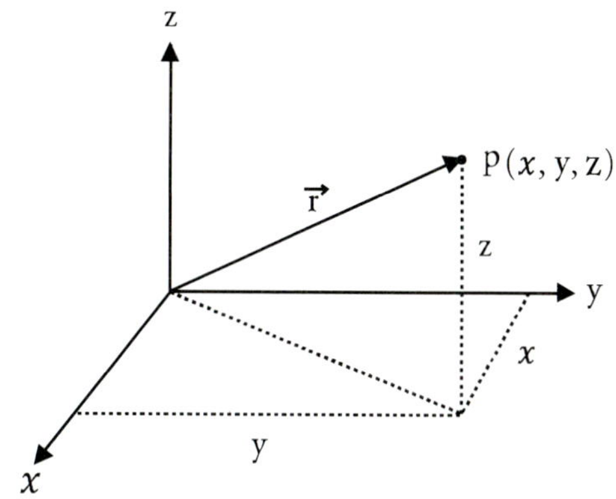

Fig. 3.1.1 Rectangular (Cartesian) coordinate system, (x, y, z).
A position vector from the origin to the point P is $\vec{r} = x\hat{x} + y\hat{y} + z\hat{z}$.

(2) A point P in the $x-y$ plane can be represented by r and θ in the **plane polar coordinate system**, shown in Fig. 3.1.2, where θ is the angle from the $x-$axis and r is the radial distance from the origin to the point P. The quantities, x and y, in the plane polar coordinate system are given by

$$x = r\cos\theta \text{ and } y = r\sin\theta \tag{3.1.1}$$

In order to cover the whole $x-y$ plane, the ranges of r and θ are

$$0 \leq r < \infty \text{ and } 0 \leq \theta \leq 2\pi$$

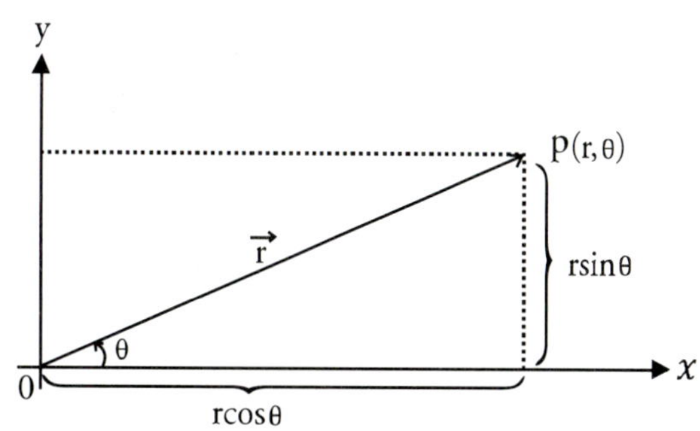

Fig. 3.1.2 Plane polar coordinate system, (r, θ). A position vector from the origin to the point P is $\vec{r} = r\hat{r} = r\cos\theta\hat{x} + r\sin\theta\hat{y}$.

(3) **Spherical coordinate system** is represented by three variables, r, θ and ϕ, which are perpendicular to each other as shown in Fig. 3.1.3, where θ and ϕ are the polar angle measured from the $z-$axis and the azimuthal

angle measured from the x−axis in the $x-y$ plane, respectively, and r is the radial distance from the origin to the observation point P.
The quantities, x, y and z, are then given by

$$x = r\sin\theta\cos\phi,\quad y = r\sin\theta\sin\phi,\ \text{and}\ z = r\cos\theta$$

In order to cover the whole three-dimensional space, the ranges of r, θ, and ϕ are

$$0 \le r < \infty,\ 0 \le \theta \le \pi,\ \text{and}\ 0 \le \phi \le 2\pi$$

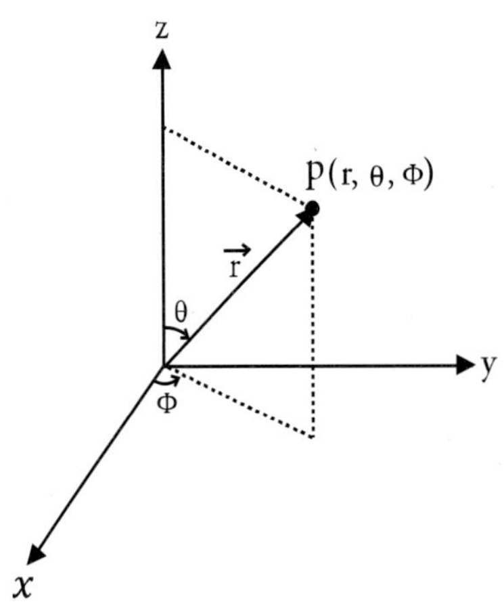

Fig. 3.1.3 Spherical coordinate system, (r,θ,ϕ). A position vector from the origin to the point P is $\vec{r} = r\hat{r} = r\sin\theta\cos\phi\hat{x} + r\sin\theta\sin\phi\hat{y} + r\cos\theta\hat{z}$, where θ is the polar angle and ϕ is the azimuthal angle.

(4) **Cylindrical coordinate system** is represented by three variables, ρ, ϕ and z, which are perpendicular to each other as shown in Fig. 3.1.4. The radial distance ρ in the $x-y$ plane and the azimuthal angle ϕ correspond to the radial distance and angle in the plane polar coordinate system, and z is the height from the $x-y$ plane.
Since x and y in the plane polar coordinates are

$$x = \rho\cos\phi \ \text{and}\ y = \rho\sin\phi,$$

the ranges of ρ, ϕ, and z are in order to cover the whole three-dimensional space

$$0 \le \rho < \infty,\ 0 \le \phi \le 2\pi,\ \text{and}\ -\infty < z < \infty$$

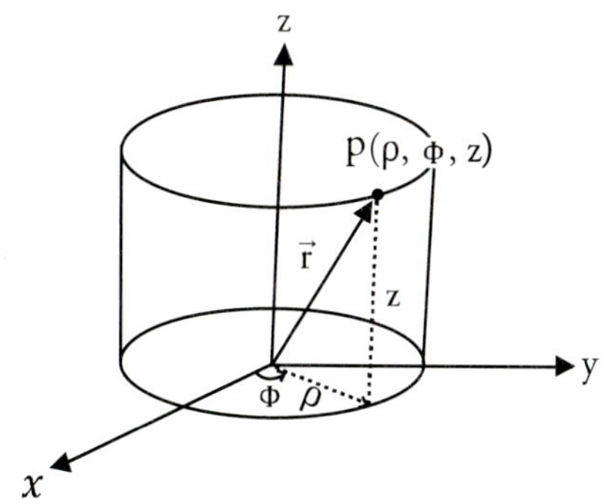

Fig. 3.1.4 Cylindrical coordinate system, (ρ,ϕ,z). A position vector from the origin to the point P is $\vec{r}=\rho\cos\phi\hat{x}+\rho\sin\phi\hat{y}+z\hat{z}$, where ρ is the radial distance in the $x-y$ plane, ϕ is the azimuthal angle, and z is the distance to the point P in $z-$direction from the $x-y$ plane.

We focus on calculations of physical quantities in the rectangular coordinate system in this chapter and calculations in another coordinate systems are covered in Chapter 4.

Example

3.1.1 Calculate $\int_{-\infty}^{\infty} e^{-x^2}dx$.

Ans. $[\int_{-\infty}^{\infty} e^{-x^2}dx]^2 = \int_{-\infty}^{\infty} e^{-x^2}dx \int_{-\infty}^{\infty} e^{-y^2}dy = \int_{-\infty}^{\infty}\int_{-\infty}^{\infty} e^{-(x^2+y^2)}dxdy$

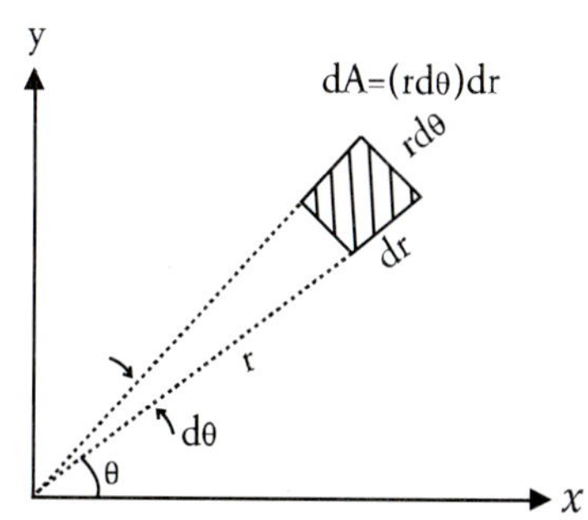

In the plane polar coordinates, $x=r\cos\theta,\ y=r\sin\theta$ and $dA=dxdy=rd\theta dr$.

By considering this problem in the plane polar coordinate system, we get

$$x^2+y^2=r^2\cos^2\theta+r^2\sin^2\theta=r^2 \text{ and } dxdy=dA=rd\theta dr$$

By inserting these into the integral, we have

$$\int_{-\infty}^{\infty} e^{-x^2}dx \int_{-\infty}^{\infty} e^{-y^2}dy = \int\int e^{-r^2} r d\theta dr$$

$$= \int_0^{\infty} e^{-r^2} r dr \int_0^{2\pi} d\theta = 2\pi \int_0^{\infty} e^{-r^2} r dr$$

$$= 2\pi \int_0^{\infty} e^{-r^2} \frac{1}{2} d(r^2) = \pi \int_0^{\infty} e^{-r^2} d(r^2)$$

$$= -\pi e^{-r^2}\Big|_0^{\infty} = -\pi(0-1) = \pi$$

Thus, we obtain

$$\int_{-\infty}^{\infty} e^{-x^2}dx = \sqrt{\pi}$$

Since $\int_{-\infty}^{\infty} e^{-x^2}dx = 2\int_0^{\infty} e^{-x^2}dx$, we get

$$\int_0^{\infty} e^{-x^2}dx = \frac{\sqrt{\pi}}{2}$$

This integration result is very often used in many physics courses. One of examples is given in the following [Ex. 3.1.2].

▶ $\int_{-\infty}^{\infty} e^{-x^2}dx = \sqrt{\pi}$ or $\int_0^{\infty} e^{-x^2}dx = \frac{\sqrt{\pi}}{2}$

3.1.2 The **probability density function** (PDF) of the **normal** (or **Gaussian**) **distribution** is given as

$$f(x) = \frac{1}{\sqrt{2\pi}\,\sigma} e^{-\frac{(x-\mu)^2}{2\sigma^2}},$$

where μ and σ are the mean and standard deviation, respectively. Calculate the integral of the function $f(x)$.

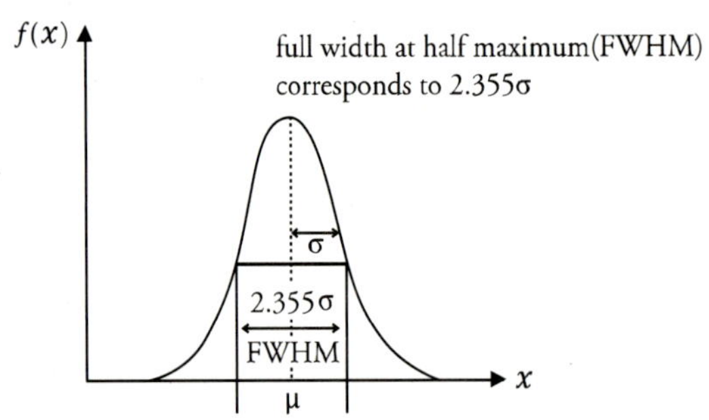

Normal(Gaussian) probability density distribution function.

Ans. If we define $\dfrac{x-\mu}{\sqrt{2}\,\sigma}=y$, we get $dx=\sqrt{2}\,\sigma dy$.

Then, the integral of the function is

$$\int_{-\infty}^{\infty} f(x)dx = \frac{1}{\sqrt{2\pi}\,\sigma}\int_{-\infty}^{\infty} e^{-\frac{(x-\mu)^2}{2\sigma^2}}dx = \frac{1}{\sqrt{2\pi}\,\sigma}\int_{-\infty}^{\infty}\sqrt{2}\,\sigma e^{-y^2}dy$$

$$= \frac{1}{\sqrt{\pi}}\int_{-\infty}^{\infty} e^{-y^2}dy = 1 \quad \text{(by using the result of [Ex. 3.1.1])}$$

Thus, we obtain

$$\int_{-\infty}^{\infty}\frac{1}{\sqrt{2\pi}\,\sigma}e^{-\frac{(x-\mu)^2}{2\sigma^2}}dx = 1,$$

which shows $\dfrac{1}{\sqrt{2\pi}\,\sigma}$ in the PDF is the normalization constant.

3.2 Vectors in Rectangular Coordinate System

A physical quantity is classified as either a **scalar** or **vector**. The scalar quantities such as speed, time, and mass, have only magnitudes of them. On the other hand, the vector quantities such as velocity, acceleration, and force, have both magnitudes and directions.

The vector is represented by an arrow and labeled by a boldface letter. The length and direction of an arrow of the vector in geometrical representation denote its magnitude and direction, respectively. A vector with magnitude A is expressed by $\vec{A}$, in which the magnitude of $\vec{A}$ is given by $|\vec{A}|$ or A.

An **unit vector** is a vector of magnitude 1. The unit vectors representing coordinates of vectors in the rectangular, spherical or cylindrical coordinate system are $(\hat{x},\hat{y},\hat{z})$, $(\hat{r},\hat{\theta},\hat{\phi})$ or $(\hat{\rho},\hat{\phi},\hat{z})$, respectively. Therefore, we have

$$|\hat{x}|=|\hat{y}|=|\hat{z}|=1,\ \ |\hat{r}|=|\hat{\theta}|=|\hat{\phi}|=1,\ \text{and}\ \ |\hat{\rho}|=|\hat{\phi}|=|\hat{z}|=1$$

A_x, A_y, and A_z are the projections of a vector $\vec{A}$ on the $x-$axis, $y-$axis, and $z-$axis, respectively, shown in Fig. 3.2.1. They are referred to the components of $\vec{A}$ on the respective axes. The vector $\vec{A}$ in the rectangular coordinate system is expressed in terms of its components:

$$\vec{A} = A_x\hat{x} + A_y\hat{y} + A_z\hat{z} \text{ or } (A_x, A_y, A_z)$$

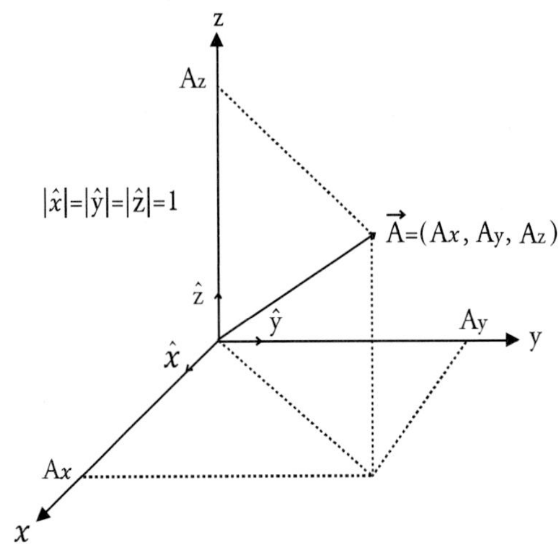

Fig. 3.2.1 A vector $\vec{A} = (A_x, A_y, A_z)$ in the rectangular coordinate system. Each component of the vector is the projection of the vector on the corresponding axis. The vector can be expressed by algebraically in terms of the unit vectors, $\hat{x}$, $\hat{y}$ and $\hat{z}$.

The addition and subtraction of the vectors can be calculated geometrically or algebraically. The geometrical method is well covered in general physics and will not be discussed in this book. When the components of vectors are used in vector calculations, it is called the **algebraical method**. Only same components of the vectors are added or subtracted in the addition or subtraction. The vectors obey the following properties:

- Vector addition: $\vec{A}+\vec{B} = (A_x + B_x)\hat{x} + (A_y + B_y)\hat{y} + (A_z + B_z)\hat{z}$
- Vector subtraction: $\vec{A}-\vec{B} = (A_x - B_x)\hat{x} + (A_y - B_y)\hat{y} + (A_z - B_z)\hat{z}$
- Vector equality: $\vec{A}=\vec{B} \Rightarrow A_x = B_x$, $A_y = B_y$, and $A_z = B_z$
- Scalar multiplication with a constant c: $c\vec{A} = cA_x\hat{x} + cA_y\hat{y} + cA_z\hat{z}$

3.2.1 Two Vectors Operation

- For $\vec{A} = A_x\hat{x} + A_y\hat{y} + A_z\hat{z}$,

  its magnitude is $A = |\vec{A}| = \sqrt{A_x^2 + A_y^2 + A_z^2}$
- $\vec{A} \cdot \vec{B} = |A||B|\cos\theta = A_xB_x + A_yB_y + A_zB_z$,

  where θ is the included angle between $\vec{A}$ and $\vec{B}$.
- $\vec{A} \times \vec{B} = \begin{vmatrix} \hat{x} & \hat{y} & \hat{z} \\ A_x & A_y & A_z \\ B_x & B_y & B_z \end{vmatrix}$

  $= (A_yB_z - A_zB_y)\hat{x} + (A_zB_x - A_xB_z)\hat{y} + (A_xB_y - A_yB_x)\hat{z}$

An operation of any two vectors, let's say $\vec{A}$ and $\vec{B}$, results in either a scalar or vector:

(a) A result of **scalar** (or **dot**, **inner**) **product** is a scalar:

$$\vec{A} \cdot \vec{B} = |A||B|\cos\theta,$$

where θ is the included angle between $\vec{A}$ and $\vec{B}$.
Since the scalar does not have direction, we have $\vec{A} \cdot \vec{B} = \vec{B} \cdot \vec{A}$. This scalar multiplication is called **commutative**.
If $\vec{A} \cdot \vec{B} = 0$ for $\vec{A} \neq 0$ and $\vec{B} \neq 0$, we get $\cos\theta = 0$, which means two vectors, $\vec{A}$ and $\vec{B}$, are **perpendicular** (or **orthogonal**) to each other. In that sense $\hat{x}$, $\hat{y}$, and $\hat{z}$ are mutually orthogonal.

(b) A result of **vector** (or **cross**, **outer**) **product** is a vector:

$$\vec{A} \times \vec{B} = \vec{C} = |\vec{A}||\vec{B}|\sin\theta\hat{n},$$

where the magnitude and direction of the produced vector $\vec{C}$ is $|\vec{A}||\vec{B}|\sin\theta$ and is perpendicularly outward to the plane of two vectors, $\vec{A}$ and $\vec{B}$, formed by following the right-hand rule. Since the direction of the vector follows the right-hand rule, it gives $\vec{A} \times \vec{B} = -\vec{B} \times \vec{A}$. Therefore, the vector multiplication is not commutative (or **anti-commutative**).

For the scalar product of the unit vector $\hat{x}$ with another unit vector $\hat{y}$, we have

$$\hat{x} \cdot \hat{y} = |\hat{x}||\hat{y}|\cos\frac{\pi}{2} = 0$$

Similarly, we have

$$\hat{y} \cdot \hat{z} = \hat{z} \cdot \hat{x} = 0 = \hat{y} \cdot \hat{x} = \hat{z} \cdot \hat{y} = \hat{x} \cdot \hat{z}$$

On the other hand, we obtain

$$\hat{x} \cdot \hat{x} = |\hat{x}||\hat{x}|\cos 0 = 1 = \hat{y} \cdot \hat{y} = \hat{z} \cdot \hat{z}$$

From the results of the scalar product for the unit vectors, the scalar product of $\vec{A}$ and $\vec{B}$ is expressed by

$$\begin{aligned}
\vec{A} \cdot \vec{B} &= (A_x\hat{x} + A_y\hat{y} + A_z\hat{z}) \cdot (B_x\hat{x} + B_y\hat{y} + B_z\hat{z}) \\
&= A_xB_x\hat{x} \cdot \hat{x} + A_xB_y\hat{x} \cdot \hat{y} + A_xB_z\hat{x} \cdot \hat{z} \\
&\quad + A_yB_x\hat{y} \cdot \hat{x} + A_yB_y\hat{y} \cdot \hat{y} + A_yB_z\hat{y} \cdot \hat{z} \\
&\quad + A_zB_x\hat{z} \cdot \hat{x} + A_zB_y\hat{z} \cdot \hat{y} + A_zB_z\hat{z} \cdot \hat{z} \\
&= A_xB_x + A_yB_y + A_zB_z
\end{aligned}$$

Thus, we obtain

$$\vec{A} \cdot \vec{B} = A_xB_x + A_yB_y + A_zB_z$$

Since $\vec{A} \cdot \vec{A} = |\vec{A}||\vec{A}|\cos 0 = A^2$ and $\vec{A} \cdot \vec{A} = A_xA_x + A_yA_y + A_zA_z$, the magnitude of the vector $\vec{A}$ is given by

$$|\vec{A}| = A = \sqrt{A_x^2 + A_y^2 + A_z^2}$$

For the vector product of the unit vectors, we have

$$\hat{x} \times \hat{y} = |\hat{x}||\hat{y}|\sin\frac{\pi}{2}\hat{z} = \hat{z},$$

because the direction of the produced vector follows the right-hand rule. Similarly, we have

$$\hat{y} \times \hat{z} = \hat{x}, \quad \text{and} \quad \hat{z} \times \hat{x} = \hat{y}$$

On the other hand, we get

$$|\hat{x} \times \hat{x}| = |\hat{x}||\hat{x}|\sin 0 = 0 = |\hat{y} \times \hat{y}| = |\hat{z} \times \hat{z}|$$

Since the vector product is the anti-commutative, we have

$$\hat{y}\times\hat{x}=-(\hat{x}\times\hat{y})=-\hat{z},\quad \hat{z}\times\hat{y}=-(\hat{y}\times\hat{z})=-\hat{x},\ \text{and}\ \hat{x}\times\hat{z}=-(\hat{z}\times\hat{x})=-\hat{y}$$

For the vector product of $\vec{A}$ and $\vec{B}$, we get

$$\begin{aligned}\vec{A}\times\vec{B} &= (A_x\hat{x} + A_y\hat{y} + A_z\hat{z})\times(B_x\hat{x} + B_y\hat{y} + B_z\hat{z}) \\ &= A_xB_x\hat{x}\times\hat{x}+A_xB_y\hat{x}\times\hat{y}+A_xB_z\hat{x}\times\hat{z} \\ &\quad +A_yB_x\hat{y}\times\hat{x}+A_yB_y\hat{y}\times\hat{y}+A_yB_z\hat{y}\times\hat{z} \\ &\quad +A_zB_x\hat{z}\times\hat{x}+A_zB_y\hat{z}\times\hat{y}+A_zB_z\hat{z}\times\hat{z} \\ &= A_xB_y\hat{z}-A_xB_z\hat{y}-A_yB_x\hat{z}+A_yB_z\hat{x}+A_zB_x\hat{y}-A_zB_y\hat{x} \\ &= (A_yB_z-A_zB_y)\hat{x}+(A_zB_x-A_xB_z)\hat{y}+(A_xB_y-A_yB_x)\hat{z}\end{aligned}$$

This result can be expressed in compact form by introducing **determinant**. The determinant is calculated by multiplying elements at opposite corners and then subtracting the products:

$$\begin{aligned}\vec{A}\times\vec{B} &= \begin{vmatrix} \hat{x} & \hat{y} & \hat{z} \\ A_x & A_y & A_z \\ B_x & B_y & B_z \end{vmatrix} \\ &= (A_yB_z-A_zB_y)\hat{x}+(A_zB_x-A_xB_z)\hat{y}+(A_xB_y-A_yB_{x)}\hat{z}\end{aligned}$$

Example

3.2.1.1 Prove the relationship

$$\cos(\theta_1 \pm \theta_2) = \cos\theta_1\cos\theta_2 \mp \sin\theta_1\sin\theta_2$$

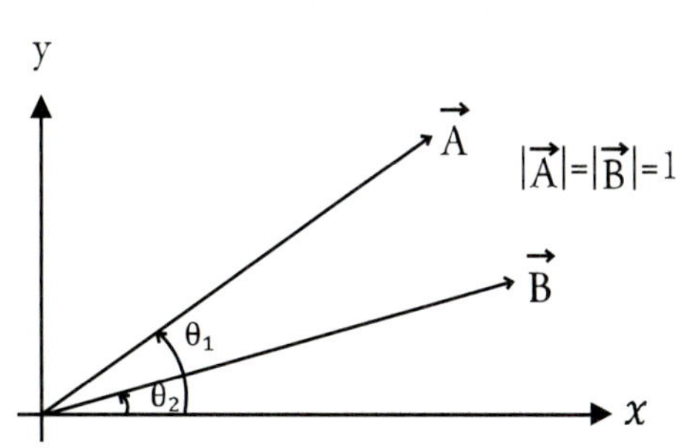

Two vectors $\vec{A}$ and $\vec{B}$ with the magnitudes $|\vec{A}|=|\vec{B}|=1$ in the plane polar coordinate system.

Ans. Let's assume that $\vec{A}$ and $\vec{B}$ are vectors with unit length (or magnitude) in the $x-y$ plane. Then, $\vec{A}$ and $\vec{B}$ are given by

$$\vec{A} = \hat{x}\cos\theta_1 + \hat{y}\sin\theta_1 \text{ and } \vec{B} = \hat{x}\cos\theta_2 + \hat{y}\sin\theta_2$$

Their scalar product gives

$$\vec{A} \cdot \vec{B} = \cos\theta_1\cos\theta_2 + \sin\theta_1\sin\theta_2$$

Since the included angle between $\vec{A}$ and $\vec{B}$ is $\theta_1 - \theta_2$, the left-hand side of the equation is

$$\vec{A} \cdot \vec{B} = AB\cos(\theta_1 - \theta_2) = \cos(\theta_1 - \theta_2)$$

because $|\vec{A}| = |\vec{B}| = 1$.

Thus, we obtain

$$\cos(\theta_1 - \theta_2) = \cos\theta_1\cos\theta_2 + \sin\theta_1\sin\theta_2 \tag{3.2.1.1.1}$$

If we substitute $-\theta_2$ instead of θ_2 in Eq. (3.2.1.1.1), we have

$$\begin{aligned}\cos(\theta_1 + \theta_2) &= \cos\theta_1\cos(-\theta_2) + \sin\theta_1\sin(-\theta_2) \\ &= \cos\theta_1\cos\theta_2 - \sin\theta_1\sin\theta_2\end{aligned}$$

3.2.1.2 Prove the relationship,

$$\sin(\theta_1 \pm \theta_2) = \sin\theta_1\cos\theta_2 \pm \cos\theta_1\sin\theta_2,$$

assuming $\vec{A}$ and $\vec{B}$ are vectors with unit length in the $x-y$ plane as shown in Figure of [Ex. 3.2.1.1].

Ans. The $\vec{A}$ and $\vec{B}$ are given by

$$\vec{A} = \hat{x}\cos\theta_1 + \hat{y}\sin\theta_1 \text{ and } \vec{B} = \hat{x}\cos\theta_2 + \hat{y}\sin\theta_2$$

The vector product of two vectors is then

$$\vec{A}\times\vec{B} = (\hat{x}\times\hat{y})\cos\theta_1\sin\theta_2 + (\hat{y}\times\hat{x})\sin\theta_1\cos\theta_2$$

because $\hat{x}\times\hat{x} = \hat{y}\times\hat{y} = 0$.

Since $\hat{x}\times\hat{y} = \hat{z}$ by following the right-hand rule, the equation is then

$$\begin{aligned}\vec{A}\times\vec{B} &= \cos\theta_1\sin\theta_2(\hat{z}) + \sin\theta_1\cos\theta_2(-\hat{z}) \\ &= (\cos\theta_1\sin\theta_2 - \sin\theta_1\cos\theta_2)\hat{z}\end{aligned}$$

The left-hand side of the equation is

$$\begin{aligned}\vec{A}\times\vec{B} &= AB\sin(\theta_1 - \theta_2)(-\hat{z}) \\ &= -\sin(\theta_1 - \theta_2)\hat{z}\end{aligned}$$

because the direction of a movement from the $x-$axis to the $y-$axis is opposite to that of a movement from $\vec{A}$ to $\vec{B}$.

Thus, we obtain

$$\sin(\theta_1 - \theta_2) = \sin\theta_1\cos\theta_2 - \cos\theta_1\sin\theta_2 \qquad (3.2.1.2.1)$$

If we substitute $-\theta_2$ instead of θ_2 in Eq. (3.2.1.2.1), we have

$$\begin{aligned}\sin(\theta_1 + \theta_2) &= \sin\theta_1\cos(-\theta_2) - \cos\theta_1\sin(-\theta_2) \\ &= \sin\theta_1\cos\theta_2 + \cos\theta_1\sin\theta_2\end{aligned}$$

3.2.1.3 An angular momentum vector $\vec{L}$ is given by $\vec{L} = \vec{r} \times \vec{P}$, where $\vec{r} = x\hat{x} + y\hat{y} + z\hat{z}$ is the position vector and $\vec{P} = P_x\hat{x} + P_y\hat{y} + P_z\hat{z}$ is the linear momentum. Calculate the angular momentum in the rectangular coordinate system.

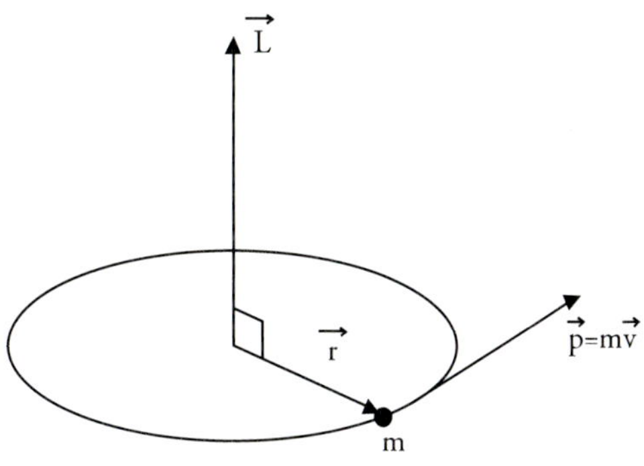

Ans. Since the angular momentum is the vector product of the position and linear momentum vectors, in general, it can be expressed by the determinant:

$$\begin{aligned}\vec{L} = \vec{r} \times \vec{P} &= \begin{vmatrix} \hat{x} & \hat{y} & \hat{z} \\ x & y & z \\ P_x & P_y & P_z \end{vmatrix} \\ &= (yP_z - zP_y)\hat{x} + (zP_x - xP_z)\hat{y} + (xP_y - yP_x)\hat{z}\end{aligned}$$

Since the angular momentum vector in the left-hand side of the equation is expressed by $\vec{L} = L_x\hat{x} + L_y\hat{y} + L_z\hat{z}$, we obtain

$$L_x = yP_z - zP_y,\ L_y = zP_x - xP_z \text{ and } L_z = xP_y - yP_x$$

from the property of the vector equality.

In case of Figure, $\vec{r} = x\hat{x} + y\hat{y}$ and $\vec{P} = P_x\hat{x} + P_y\hat{y}$ so that

$$\vec{L} = \begin{vmatrix} \hat{x} & \hat{y} & \hat{z} \\ x & y & 0 \\ P_x & P_y & 0 \end{vmatrix} = (xP_y - yP_x)\hat{z}$$

Therefore, the direction of the angular momentum is z-axis.

3.2.2 Three or Four Vectors Operation

- $\sum_k \epsilon_{ijk}\epsilon_{k\ell m} = \delta_{i\ell}\delta_{jm} - \delta_{im}\delta_{j\ell}$
- $\vec{A}\times(\vec{B}\times\vec{C}) = \vec{B}(\vec{A}\cdot\vec{C}) - \vec{C}(\vec{A}\cdot\vec{B})$ ("BAC-CAB"rule)
- $\vec{A}\cdot(\vec{B}\times\vec{C}) = \vec{A}\times\vec{B}\cdot\vec{C} = \overrightarrow{ABC}$ ("ABC"rule)
- $(\vec{A}\times\vec{B})\cdot(\vec{C}\times\vec{D}) = (\vec{A}\cdot\vec{C})(\vec{B}\cdot\vec{D}) - (\vec{B}\cdot\vec{C})(\vec{A}\cdot\vec{D})$

A triple product is an operation of three vectors. It is convenient to know the **Levi-Civita density** ϵ_{ijk} (or **permutation symbol**) and **Kronecker delta symbol** δ_{ij} for the triple product of the vectors. Here, i, j and k run over 1, 2, and 3, corresponding to x, y, and z components of the vectors, respectively.

i The Levi-Civita density is defined to be

- $\epsilon_{ijk} = 0$ if any index is equal to another index
- $\epsilon_{ijk} = 1$ if any index forms an even permutation of 1,2,3

  for example, $\epsilon_{123} = \epsilon_{231} = \epsilon_{312} = 1$
- $\epsilon_{ijk} = -1$ if any index forms an odd permutation of 1,2,3

  for example, $\epsilon_{132} = \epsilon_{213} = \epsilon_{321} = -1$

ii The Kronecker delta symbol is defined to be

- $\delta_{ij} = 0$ if $i \neq j$
- $\delta_{ij} = 1$ if $i = j$

For the vector product $\vec{A}\times\vec{B} = \vec{C}$, a component i of the vector $\vec{C}$ is expressed by

$$C_i = \sum_{j,k} \epsilon_{ijk} A_j B_k$$

using the Levi-Civita density.

For instance, we get for $i = 1,\ 2,\ 3$

$$C_1 = C_x = \epsilon_{123}A_2B_3 + \epsilon_{132}A_3B_2 = A_2B_3 - A_3B_2 = A_yB_z - A_zB_y,$$
$$C_2 = C_y = \epsilon_{213}A_1B_3 + \epsilon_{231}A_3B_1 = -A_1B_3 + A_3B_1 = A_zB_x - A_xB_z,$$
$$C_3 = C_z = \epsilon_{312}A_1B_2 + \epsilon_{321}A_2B_1 = A_1B_2 - A_2B_1 = A_xB_y - A_yB_x$$

This is what we expected for the vector product $\vec{A}\times\vec{B} = \vec{C}$.

Example

3.2.2.1 Simplify the triple vector product, $\vec{A}\times(\vec{B}\times\vec{C})$, by using the Levi-Civita density.

Ans. The component i of its vector product is

$$[\vec{A}\times(\vec{B}\times\vec{C})]_i = \sum_{j,k}\epsilon_{ijk}A_j(\vec{B}\times\vec{C})_k$$
$$= \sum_{j,k}\epsilon_{ijk}A_j\sum_{\ell,m}\epsilon_{k\ell m}B_\ell C_m$$
$$= \sum_{j,k}\sum_{\ell,m}\epsilon_{ijk}\epsilon_{k\ell m}A_jB_\ell C_m$$
$$= \sum_{j,\ell,m}(\delta_{i\ell}\delta_{jm} - \delta_{im}\delta_{j\ell})A_jB_\ell C_m,$$

(by using $\sum_k \epsilon_{ijk}\epsilon_{k\ell m} = \delta_{i\ell}\delta_{jm} - \delta_{im}\delta_{j\ell}$[1])

$$= \sum_{j,\ell,m}\delta_{i\ell}\delta_{jm}A_jB_\ell C_m - \sum_{j,\ell,m}\delta_{im}\delta_{j\ell}A_jB_\ell C_m$$
$$= B_i\sum_j A_jC_j - C_i\sum_j A_jB_j$$
$$= B_i(\vec{A}\cdot\vec{C}) - C_i(\vec{A}\cdot\vec{B}),$$

which is

$$[\vec{A}\times(\vec{B}\times\vec{C})]_x = B_x(\vec{A}\cdot\vec{C}) - C_x(\vec{A}\cdot\vec{B}) \text{ for } i = 1.$$

Similarly, we get

$$[\vec{A}\times(\vec{B}\times\vec{C})]_y = B_y(\vec{A}\cdot\vec{C}) - C_y(\vec{A}\cdot\vec{B})$$

and

$$[\vec{A}\times(\vec{B}\times\vec{C})]_z = B_z(\vec{A}\cdot\vec{C}) - C_z(\vec{A}\cdot\vec{B}),$$

respectively, for $i = 2$ and 3.

[1] It is proved in Appendix 3-A.

Therefore, we have

$$\vec{A}\times(\vec{B}\times\vec{C}) = [\vec{A}\times(\vec{B}\times\vec{C})]_x\hat{x} + [\vec{A}\times(\vec{B}\times\vec{C})]_y\hat{y} + [\vec{A}\times(\vec{B}\times\vec{C})]_z\hat{z}$$
$$= (B_x\hat{x} + B_y\hat{y} + B_z\hat{z})(\vec{A}\cdot\vec{C}) - (C_x\hat{x} + C_y\hat{y} + C_z\hat{z})(\vec{A}\cdot\vec{B})$$

Thus, we obtain

$$\vec{A}\times(\vec{B}\times\vec{C}) = \vec{B}(\vec{A}\cdot\vec{C}) - \vec{C}(\vec{A}\cdot\vec{B}),$$

which is called **"BAC-CAB" rule**.

3.2.2.2 Simplify the triple scalar product, $\vec{A}\cdot(\vec{B}\times\vec{C})$, where the result of this operation of three vectors is the scalar because $\vec{B}\times\vec{C}$ results in a vector.

Ans. The given triple scalar product can be expressed by

$$\vec{A}\cdot(\vec{B}\times\vec{C}) = \sum_i A_i(\vec{B}\times\vec{C})_i = \sum_i\sum_{j,k}\epsilon_{ijk}A_iB_jC_k \qquad (3.2.2.2.1)$$

for its component i $(i = 1,2,3)$.

On the other hand, we have

$$\vec{A}\times\vec{B}\cdot\vec{C} = \sum_k(\vec{A}\times\vec{B})_kC_k = \sum_k\sum_{\ell,m}\epsilon_{k\ell m}A_\ell B_mC_k$$
$$= \sum_k\sum_{\ell,m}\epsilon_{\ell mk}A_\ell B_mC_k \qquad (3.2.2.2.2)$$

If we substitute i and j for ℓ and m, respectively, in Eq. (3.2.2.2.2), we know that it is the same as Eq. (3.2.2.2.1) and then we get

$$\vec{A}\cdot(\vec{B}\times\vec{C}) = \vec{A}\times\vec{B}\cdot\vec{C}$$

Therefore, we notice that "dot" and "cross" can be exchanged in the triple scalar product.

From Eq. (3.2.2.2.1) we have

$$\vec{A}\cdot(\vec{B}\times\vec{C}) = \sum_i\sum_{j,k}\epsilon_{ijk}A_iB_jC_k$$
$$= \epsilon_{123}A_1B_2C_3 + \epsilon_{132}A_1B_3C_2 + \epsilon_{213}A_2B_1C_3$$
$$+ \epsilon_{231}A_2B_3C_1 + \epsilon_{312}A_3B_1C_2 + \epsilon_{321}A_3B_2C_1$$
$$= A_1(B_2C_3 - B_3C_2) + A_2(B_3C_1 - B_1C_3) + A_3(B_1C_2 - B_2C_1)$$
$$= A_x(B_yC_z - B_zC_y) + A_y(B_zC_x - B_xC_z) + A_z(B_xC_y - B_yC_x)$$

Thus, this triple scalar product can be simply expressed by $\vec{A}\vec{B}\vec{C}$ or by the determinant:

$$\vec{A}\cdot(\vec{B}\times\vec{C}) = \vec{A}\times\vec{B}\cdot\vec{C} = \begin{vmatrix} A_x & A_y & A_z \\ B_x & B_y & B_z \\ C_x & C_y & C_z \end{vmatrix} \equiv \vec{A}\vec{B}\vec{C}$$

Keep in mind that $(\vec{A}\cdot\vec{B})\times\vec{C}$ is not allowed product because $\vec{A}\cdot\vec{B}$ produces a scalar quantity, and the vector product of a scalar and a vector is not allowed.

3.2.2.3 Consider the combination of four vectors, $(\vec{A}\times\vec{B})\cdot(\vec{C}\times\vec{D})$ and simplify this product.

$$
\begin{aligned}
(\vec{A}\times\vec{B})\cdot(\vec{C}\times\vec{D}) &= \sum_i(\vec{A}\times\vec{B})_i(\vec{C}\times\vec{D})_i \\
&= \sum_i\sum_{j,k}\epsilon_{ijk}A_jB_k\sum_{\ell,m}\epsilon_{i\ell m}C_\ell D_m \\
&= \sum_i\sum_{j,k}\epsilon_{jki}\sum_{\ell,m}\epsilon_{i\ell m}A_jB_kC_\ell D_m \\
&= \sum_{j,k}\sum_{\ell,m}(\delta_{j\ell}\delta_{km}-\delta_{jm}\delta_{k\ell})A_jB_kC_\ell D_m \\
&= \sum_j A_jC_j\sum_k B_kD_k-\sum_k B_kC_k\sum_j A_jD_j \\
&= (\vec{A}\cdot\vec{C})(\vec{B}\cdot\vec{D})-(\vec{B}\cdot\vec{C})(\vec{A}\cdot\vec{D})
\end{aligned}
$$

3.3 Gradient, Divergence and Curl

- $\vec{\nabla}=\hat{x}\frac{\partial}{\partial x}+\hat{y}\frac{\partial}{\partial y}+\hat{z}\frac{\partial}{\partial z}$ ("del" vector)
- $\vec{\nabla}f=\frac{df}{dr}\hat{r}$ for $f=f(r)$ (Gradient of function f)
- $\vec{\nabla}\cdot\vec{F}=\frac{\partial F_x}{\partial x}+\frac{\partial F_y}{\partial y}+\frac{\partial F_z}{\partial z}$ (Divergent of vector F)
- $\vec{\nabla}\times\vec{F}=(\frac{\partial F_z}{\partial y}-\frac{\partial F_y}{\partial z})\hat{x}+(\frac{\partial F_x}{\partial z}-\frac{\partial F_z}{\partial x})\hat{y}+(\frac{\partial F_y}{\partial x}-\frac{\partial F_x}{\partial y})\hat{z}$ (Curl of vector F)
- $\vec{\nabla}\cdot\vec{\nabla}\phi=\nabla^2\phi=\frac{\partial^2\phi}{\partial x^2}+\frac{\partial^2\phi}{\partial y^2}+\frac{\partial^2\phi}{\partial z^2}$
- $\vec{\nabla}\times(\vec{\nabla}\times\vec{A})=\vec{\nabla}(\vec{\nabla}\cdot\vec{A})-\nabla^2\vec{A}$
- $\vec{\nabla}\times(\vec{A}\times\vec{B})=(\vec{B}\cdot\vec{\nabla})\vec{A}+\vec{A}(\vec{\nabla}\cdot\vec{B})-\vec{B}(\vec{\nabla}\cdot\vec{A})-(\vec{A}\cdot\vec{\nabla})\vec{B}$
- $\vec{\nabla}\cdot(\phi\vec{A})=\vec{A}\cdot\vec{\nabla}\phi+\phi(\vec{\nabla}\cdot\vec{A})$
- $\vec{\nabla}(\vec{A}\cdot\vec{B})=\vec{A}\times(\vec{\nabla}\times\vec{B})+\vec{B}\times(\vec{\nabla}\times\vec{A})+(\vec{A}\cdot\vec{\nabla})\vec{B}+(\vec{B}\cdot\vec{\nabla})\vec{A}$

(a) **Gradient** of any function points along the direction of the greatest rate of increase of the function so that a gradient operator is the directional derivative. Its magnitude is the slope of a function in that direction. Let's define the gradient as $\overrightarrow{\nabla}=\hat{x}\frac{\partial}{\partial x}+\hat{y}\frac{\partial}{\partial y}+\hat{z}\frac{\partial}{\partial z}$, called "**del**" vector operator.

The gradient of a scalar function $\phi(x,y,z)$ is then

$$grad\ \phi=\overrightarrow{\nabla}\phi=\frac{\partial\phi}{\partial x}\hat{x}+\frac{\partial\phi}{\partial y}\hat{y}+\frac{\partial\phi}{\partial z}\hat{z}$$

Consider a function $f=f(r)$ and $r=\sqrt{x^2+y^2+z^2}$. The gradient of the function f is then

$$\begin{aligned}\overrightarrow{\nabla}f&=\frac{\partial f}{\partial x}\hat{x}+\frac{\partial f}{\partial y}\hat{y}+\frac{\partial f}{\partial z}\hat{z}\\&=\frac{\partial f}{\partial r}\frac{\partial r}{\partial x}\hat{x}+\frac{\partial f}{\partial r}\frac{\partial r}{\partial y}\hat{y}+\frac{\partial f}{\partial r}\frac{\partial r}{\partial z}\hat{z},\end{aligned}\tag{3.3.1}$$

where $\frac{\partial r}{\partial x}=\frac{\partial}{\partial x}(x^2+y^2+z^2)^{\frac{1}{2}}=\frac{1}{2}2x(x^2+y^2+z^2)^{-\frac{1}{2}}=\frac{x}{r}$ and similarly, $\frac{\partial r}{\partial y}=\frac{y}{r}$ and $\frac{\partial r}{\partial z}=\frac{z}{r}$.

Then, the equation is

$$\overrightarrow{\nabla}f=\frac{\partial f}{\partial r}\frac{x}{r}\hat{x}+\frac{\partial f}{\partial r}\frac{y}{r}\hat{y}+\frac{\partial f}{\partial r}\frac{z}{r}\hat{z}$$

Since f is the function of r only, it gives $\frac{\partial f}{\partial r}=\frac{df}{dr}$.

Thus, we obtain

$$\overrightarrow{\nabla}f(r)=\frac{df}{dr}\frac{x}{r}\hat{x}+\frac{df}{dr}\frac{y}{r}\hat{y}+\frac{df}{dr}\frac{z}{r}\hat{z}=\frac{df}{dr}\frac{\vec{r}}{r}=\frac{df}{dr}\hat{r}$$

(b) **Divergence** of any vector function $\vec{F}(x,y,z)$ is defined as the scalar product of the del vector $\overrightarrow{\nabla}$ and vector $\vec{F}$, giving a scalar function as result. The physical meaning of the divergence is that it represents quantity of flowing of the vector field outward from an infinitesimal volume around a given position in the vector field.

$\vec{\nabla}\cdot$ is called "**divergence**" operator.

$$div\ \vec{F}=\vec{\nabla}\cdot\vec{F}=(\frac{\partial}{\partial x}\hat{x}+\frac{\partial}{\partial y}\hat{y}+\frac{\partial}{\partial z}\hat{z})\cdot(F_x\hat{x}+F_y\hat{y}+F_z\hat{z})$$
$$=\frac{\partial F_x}{\partial x}+\frac{\partial F_y}{\partial y}+\frac{\partial F_z}{\partial z}$$

(c) **Curl** of any vector function $\vec{F}(x,y,z)$ is in the form of the vector product of the del vector $\vec{\nabla}$ and vector $\vec{F}$, giving a vector function as result. The physical meaning of the curl is that it represents the rotation at a given point in the vector field. The direction and magnitude of the curl are the rotational axis, following the right-hand rule, and rotational magnitude, respectively.

$\vec{\nabla}\times$ is called "**curl**" operator.

$$curl\ \vec{F}=\vec{\nabla}\times\vec{F}=\begin{vmatrix}\hat{x} & \hat{y} & \hat{z}\\ \frac{\partial}{\partial x} & \frac{\partial}{\partial y} & \frac{\partial}{\partial z}\\ F_x & F_y & F_z\end{vmatrix}$$
$$=(\frac{\partial F_z}{\partial y}-\frac{\partial F_y}{\partial z})\hat{x}+(\frac{\partial F_x}{\partial z}-\frac{\partial F_z}{\partial x})\hat{y}+(\frac{\partial F_y}{\partial x}-\frac{\partial F_x}{\partial y})\hat{z}$$

(d) The gradient of any scalar function $\phi(x,y,z)$ produces the vector function and then the divergence of this vector can be represented by $\vec{\nabla}\cdot(\vec{\nabla}\phi)$:

$$\vec{\nabla}\cdot\vec{\nabla}\phi=(\hat{x}\frac{\partial}{\partial x}+\hat{y}\frac{\partial}{\partial y}+\hat{z}\frac{\partial}{\partial z})\cdot(\frac{\partial\phi}{\partial x}\hat{x}+\frac{\partial\phi}{\partial y}\hat{y}+\frac{\partial\phi}{\partial y}\hat{y})$$
$$=\frac{\partial^2\phi}{\partial x^2}+\frac{\partial^2\phi}{\partial y^2}+\frac{\partial^2\phi}{\partial z^2},$$

where $\vec{\nabla}\cdot\vec{\nabla}=\nabla^2$ and is called **Laplacian**.

We call **Laplace**'s equation and **Poisson**'s equation for $\nabla^2\phi=0$ and $\nabla^2\phi=-\frac{\rho}{\epsilon_0}$, respectively, in Electromagnetism[2].

---

[2] This is the reason why we call it Laplacian.

Example

3.3.1 Calculate the divergence of a vector function $\vec{r}f(r)$, where $\vec{r}=x\hat{x}+y\hat{y}+z\hat{z}$.

Ans. $\vec{\nabla}\cdot\vec{r}f(r)=(\frac{\partial}{\partial x}\hat{x}+\frac{\partial}{\partial y}\hat{y}+\frac{\partial}{\partial z}\hat{z})\cdot[f(r)x\hat{x}+f(r)y\hat{y}+f(r)z\hat{z}]$

$$=\frac{\partial}{\partial x}[xf(r)]+\frac{\partial}{\partial y}[yf(r)]+\frac{\partial}{\partial z}[zf(r)], \quad (3.3.1.1)$$

where

$$\frac{\partial}{\partial x}[xf(r)]=f(r)+x\frac{\partial f(r)}{\partial x}=f(r)+x\frac{df(r)}{dr}\frac{\partial r}{\partial x}$$

$$=f(r)+\frac{df(r)}{dr}\frac{x^2}{r}$$

Similarly, we get

$$\frac{\partial}{\partial y}[yf(r)]=f(r)+\frac{df(r)}{dr}\frac{y^2}{r} \text{ and } \frac{\partial}{\partial z}[zf(r)]=f(r)+\frac{df(r)}{dr}\frac{z^2}{r}$$

By inserting these into Eq. (3.3.1.1), we obtain

$$\vec{\nabla}\cdot\vec{r}f(r)=3f(r)+\frac{df(r)}{dr}\frac{1}{r}(x^2+y^2+z^2)=3f(r)+r\frac{df(r)}{dr} \quad (3.3.1.2)$$

For $f(r)=r^{n-1}$, the right-hand side of Eq. (3.3.1.2) is

$$3f(r)+r\frac{df(r)}{dr}=3r^{n-1}+r\frac{d}{dr}r^{n-1}$$

$$=3r^{n-1}+(n-1)r^{n-1}=(n+2)r^{n-1} \quad (3.3.1.3)$$

As an application of this, consider the electric field which is expressed by

$$\vec{E}(r)=\frac{q}{r^2}\hat{r}=\frac{q}{r^3}\vec{r}$$

This corresponds to $n=-2$ in the $f(r)=r^{n-1}$

Then, from Eq. (3.3.1.3), we get the right-hand side of Eq. (3.3.1.2) is zero.

Thus, we obtain

$$\vec{\nabla}\cdot\vec{E}=0,$$

which is one of the Maxwell's equations.

3.3.2 A vector potential $\vec{A}$ is given by $\vec{A}=\frac{1}{2}(\vec{B}\times\vec{r})$ in Electromagnetism, in which $\vec{B}$ is the magnetic field vector and $\vec{r}$ is the position vector at which the vector potential is measured. Find the curl of the vector potential $\vec{A}$.

Ans. The curl of the vector potential is

$$\vec{\nabla}\times\vec{A}=\frac{1}{2}\vec{\nabla}\times(\vec{B}\times\vec{r})$$

The i component of the triple vector product is then

$$\begin{aligned}
[\vec{\nabla}\times(\vec{B}\times\vec{r})]_i &= \sum_{j,k}\epsilon_{ijk}\partial_j(\vec{B}\times\vec{r})_k, \quad \text{where } \frac{\partial}{\partial x_j}\equiv\partial_j, \\
&= \sum_{j,k}\sum_{\ell,m}\epsilon_{ijk}\partial_j\epsilon_{k\ell m}B_\ell r_m = \sum_{j,k}\sum_{\ell,m}\epsilon_{ijk}\epsilon_{k\ell m}\partial_j B_\ell r_m \\
&= \sum_{j,k}\sum_{\ell,m}\epsilon_{ijk}\epsilon_{k\ell m}(B_\ell\partial_j r_m + r_m\partial_j B_\ell) \\
&= \sum_{j}\sum_{\ell,m}(\delta_{i\ell}\delta_{jm}-\delta_{im}\delta_{j\ell})(B_\ell\partial_j r_m + r_m\partial_j B_\ell) \\
&= \sum_{j}\sum_{\ell,m}\delta_{i\ell}\delta_{jm}B_\ell\partial_j r_m + \sum_{j}\sum_{\ell,m}\delta_{i\ell}\delta_{jm}r_m\partial_j B_\ell \\
&\quad - \sum_{j}\sum_{\ell,m}\delta_{im}\delta_{j\ell}B_\ell\partial_j r_m - \sum_{j}\sum_{\ell,m}\delta_{im}\delta_{j\ell}r_m\partial_j B_\ell
\end{aligned}$$

The first term of the right-hand side of the equation is

$$\sum_{j}\sum_{\ell,m}\delta_{i\ell}\delta_{jm}B_\ell\partial_j r_m = \sum_{j}B_i\partial_j r_j = [\sum_{j}\partial_j r_j]B_i = (\vec{\nabla}\cdot\vec{r})B_i$$

Since $\sum_{j}\sum_{\ell,m}\delta_{i\ell}\delta_{jm}\partial_j r_m B_\ell = \sum_{j}\sum_{\ell,m}\delta_{i\ell}\delta_{jm}(r_m\partial_j B_\ell + B_\ell\partial_j r_m)$, the second term of the right-hand side of the equation is

$$\begin{aligned}
\sum_{j}\sum_{\ell,m}\delta_{i\ell}\delta_{jm}r_m\partial_j B_\ell &= \sum_{j}\sum_{\ell,m}\delta_{i\ell}\delta_{jm}\partial_j r_m B_\ell - \sum_{j}\sum_{\ell,m}\delta_{i\ell}\delta_{jm}B_\ell\partial_j r_m \\
&= \sum_{j}\partial_j r_j B_i - \sum_{j}B_i\partial_j r_j = 0
\end{aligned}$$

The third term of the right-hand side of the equation is

$$\sum_{j}\sum_{\ell,m}\delta_{im}\delta_{j\ell}B_\ell\partial_j r_m = [\sum_{j}B_j\partial_j]r_i = (\vec{B}\cdot\vec{\nabla})r_i$$

The fourth term of the right-hand side of the equation is

$$\sum_{j}\sum_{\ell,m}\delta_{im}\delta_{j\ell}r_m\partial_j B_\ell = \sum_{j}r_i\partial_j B_j = [\sum_{j}\partial_j B_j]r_i = (\vec{\nabla}\cdot\vec{B})r_i = 0$$

because $\vec{\nabla}\cdot\vec{B}=0$ from the Maxwell equation.

Therefore, we get

$$[\vec{\nabla}\times(\vec{B}\times\vec{r})]_i = (\vec{\nabla}\cdot\vec{r})B_i - (\vec{B}\cdot\vec{\nabla})r_i,$$

which gives

$$\vec{\nabla}\times(\vec{B}\times\vec{r}) = (\vec{\nabla}\cdot\vec{r})\vec{B} - (\vec{B}\cdot\vec{\nabla})\vec{r}$$

Thus, we obtain

$$\vec{\nabla}\times\vec{A}=\frac{1}{2}\vec{\nabla}\times(\vec{B}\times\vec{r})=\frac{1}{2}[(\vec{\nabla}\cdot\vec{r})\vec{B}-(\vec{B}\cdot\vec{\nabla})\vec{r}] \qquad (3.3.2.1)$$

In this equation, we have

$$\begin{aligned}(\vec{B}\cdot\vec{\nabla})\vec{r} &= (B_x\frac{\partial}{\partial x}+B_y\frac{\partial}{\partial y}+B_z\frac{\partial}{\partial z})(x\hat{x}+y\hat{y}+z\hat{z})\\ &= B_x\hat{x}+B_y\hat{y}+B_z\hat{z} = \vec{B}\end{aligned}$$

and $\vec{\nabla}\cdot\vec{r}$ corresponds to $n=1$ in the $\vec{\nabla}\cdot\vec{r}f(r)$ with $f(r)=r^{n-1}$, giving

$$\vec{\nabla}\cdot\vec{r}=(n+2)r^{n-1}=3$$

By substituting these into Eq. (3.3.2.1), we obtain

$$\vec{\nabla}\times\vec{A}=\frac{1}{2}(3\vec{B}-\vec{B})=\vec{B}$$

Thus, the curl of the vector potential $\vec{A}$ is the magnetic field $\vec{B}$.

3.3.3 Show that $\vec{B}=\vec{\nabla}\times\vec{A}$ in [Ex. 3.3.2] is the solution of the Maxwell equation $\vec{\nabla}\cdot\vec{B}=0$.

Ans. $\vec{\nabla}\cdot\vec{B}=\vec{\nabla}\cdot(\vec{\nabla}\times\vec{A})$

The right-hand side of the equation is then

$$\begin{aligned}\vec{\nabla}\cdot(\vec{\nabla}\times\vec{A}) &= \sum_i\partial_i(\vec{\nabla}\times\vec{A})_i=\sum_i\partial_i(\sum_{j,k}\epsilon_{ijk}\partial_jA_k)=\sum_i\sum_{j,k}\epsilon_{ijk}\partial_i\partial_jA_k\\ &= \sum_k[\sum_{i,j}\epsilon_{kij}\partial_i\partial_j]A_k\\ &= \sum_k(\vec{\nabla}\times\vec{\nabla})_kA_k\\ &= (\vec{\nabla}\times\vec{\nabla})\cdot\vec{A},\end{aligned}$$

where

$$\begin{aligned}\vec{\nabla}\times\vec{\nabla}=\begin{vmatrix}\hat{x} & \hat{y} & \hat{z}\\ \frac{\partial}{\partial x} & \frac{\partial}{\partial y} & \frac{\partial}{\partial z}\\ \frac{\partial}{\partial x} & \frac{\partial}{\partial y} & \frac{\partial}{\partial z}\end{vmatrix} &= \left(\frac{\partial^2}{\partial y\partial z}-\frac{\partial^2}{\partial z\partial y}\right)\hat{x}\\ &+\left(\frac{\partial^2}{\partial z\partial x}-\frac{\partial^2}{\partial x\partial z}\right)\hat{y}\\ &+\left(\frac{\partial^2}{\partial x\partial y}-\frac{\partial^2}{\partial y\partial x}\right)\hat{z}=0\end{aligned}$$

Thus, we obtain

$$\vec{\nabla} \cdot (\vec{\nabla} \times \vec{A}) = (\vec{\nabla} \times \vec{\nabla}) \cdot \vec{A} = 0$$

which means that the divergence of the curl is zero.

Therefore, $\vec{B} = \vec{\nabla} \times \vec{A}$ is the solution of $\vec{\nabla} \cdot \vec{B} = 0$.

3.3.4 Show that the curl of the gradient of a scalar function $\phi(x, y, z)$ is zero.

Ans.
$$\vec{\nabla} \times \vec{\nabla}\phi = \begin{vmatrix} \hat{x} & \hat{y} & \hat{z} \\ \frac{\partial}{\partial x} & \frac{\partial}{\partial y} & \frac{\partial}{\partial z} \\ \frac{\partial \phi}{\partial x} & \frac{\partial \phi}{\partial y} & \frac{\partial \phi}{\partial z} \end{vmatrix}$$

$$= \left(\frac{\partial^2 \phi}{\partial y \partial z} - \frac{\partial^2 \phi}{\partial z \partial y}\right)\hat{x} + \left(\frac{\partial^2 \phi}{\partial z \partial x} - \frac{\partial^2 \phi}{\partial x \partial z}\right)\hat{y} + \left(\frac{\partial^2 \phi}{\partial x \partial y} - \frac{\partial^2 \phi}{\partial y \partial x}\right)\hat{z} = 0$$

because $\frac{\partial^2 \phi}{\partial y \partial z} = \frac{\partial^2 \phi}{\partial z \partial y}$, ⋯⋯ for the continuous function $\phi(x, y, z)$.

3.3.5 A mechanical energy in Classical Mechanics is the sum of the kinetic energy and potential energy, $H = K.E. + P.E. = \frac{p^2}{2m} + U(r)$.

If we express this equation by replacing momentum $\vec{p}$ with the momentum operator $\vec{p}_{op} = \frac{\hbar}{i}\vec{\nabla}$, it gives $H = -\frac{\hbar^2}{2m}\nabla^2 + U(r)$ in which H becomes from the mechanical energy to **Hamiltonian operator**.

Ans. From $H\Psi(\vec{r}) = E\Psi(\vec{r})$ for time-independent wavefuntion $\Psi(\vec{r})$, we get

$$\left[-\frac{\hbar^2}{2m}\nabla^2 + U(r)\right]\Psi(\vec{r}) = E\Psi(\vec{r}), \tag{3.3.5.1}$$

which is called **time-independent Schrödinger equation** in Quantum Mechanics. For simplicity, consider a free particle (which is the case for $U(r) = 0$) case in one-dimension. The equation is then expressed as

$$-\frac{\hbar^2}{2m}\frac{d^2\Psi(x)}{dx^2} = E\Psi(x),$$

which gives

$$\frac{d^2\Psi}{dx^2} + \frac{2mE}{\hbar^2}\Psi = 0 \tag{3.3.5.2}$$

This is the homogeneous ordinary second order differential equation.

By defining the constant $\frac{2mE}{\hbar^2} = k^2$ for the positive energy E, we get

$$\frac{d^2\Psi}{dx^2} + k^2\Psi = 0$$

The general solution for this equation is

$$\Psi(x) = A\cos kx + B\sin kx,$$

where constants A and B are determined by boundary conditions.

3.3.6 For time-dependent wavefuntion $\Psi(\vec{r},t)$, we replace E with $i\hbar\frac{\partial}{\partial t}$ in the equation $H\Psi(\vec{r},t) = E\Psi(\vec{r},t)$ and then we get

$$[-\frac{\hbar^2}{2m}\nabla^2 + U(r)]\Psi(\vec{r},t) = i\hbar\frac{\partial}{\partial t}\Psi(\vec{r},t),$$

which is called the **time-dependent Schrödinger equation** in Quantum Mechanics.

Ans. Consider one-dimensional and a free particle case for simplicity. The original equation is then written by

$$-\frac{\hbar^2}{2m}\frac{\partial^2}{\partial x^2}\Psi(x,t) = i\hbar\frac{\partial}{\partial t}\Psi(x,t) \quad (3.3.6.1)$$

If we let $\Psi(x,t) = u(x)T(t)$, the equation gives

$$-\frac{\hbar^2}{2m}\frac{1}{u(x)}\frac{d^2u(x)}{dx^2} = i\hbar\frac{1}{T(t)}\frac{dT(t)}{dt}$$

Since x and t are independent variables, the left- and right-hand sides of the equation should be a constant, let's say this constant is E, to satisfy the equality of the equation for all x and t.

Then, we have two independent equations,

$$\frac{d^2u(x)}{dx^2} + \frac{2mE}{\hbar^2}u(x) = 0 \quad (3.3.6.2)$$

$$\frac{dT(t)}{dt} = -\frac{i}{\hbar}ET(t) \quad (3.3.6.3)$$

The general solution of Eq. (3.3.6.2) is

$$u(x) = A\cos kx + B\sin kx, \quad (3.3.6.4)$$

where $k = \frac{\sqrt{2mE}}{\hbar}$ with the positive energy E.

For Eq. (3.3.6.3), we have

$$\frac{dT}{T} = -\frac{i}{\hbar}Edt,$$

which gives the general solution

$$T(t) = Ce^{-\frac{i}{\hbar}Et}, \quad (3.3.6.5)$$

where C is the integration constant.

Since the time-dependent wavefuntion is given by $\Psi(x,t) = u(x)T(t)$, we obtain the general solution

$$\Psi(x,t) = (A_1 \cos kx + A_2 \sin kx) e^{-\frac{i}{\hbar}Et}$$

from Eqs. (3.3.6.4) and (3.3.6.5).

3.3.7 Prove the following relation which appears in Electromagnetism,

$$\overrightarrow{\nabla} \times (\overrightarrow{A} \times \overrightarrow{B}) = (\overrightarrow{B} \cdot \overrightarrow{\nabla})\overrightarrow{A} + \overrightarrow{A}(\overrightarrow{\nabla} \cdot \overrightarrow{B}) - \overrightarrow{B}(\overrightarrow{\nabla} \cdot \overrightarrow{A}) - (\overrightarrow{A} \cdot \overrightarrow{\nabla})\overrightarrow{B}$$

Ans. The i component of the vector $\overrightarrow{\nabla} \times (\overrightarrow{A} \times \overrightarrow{B})$ is

$$\begin{aligned}
[\overrightarrow{\nabla} \times (\overrightarrow{A} \times \overrightarrow{B})]_i &= \sum_{j,k} \epsilon_{ijk} \partial_j (\overrightarrow{A} \times \overrightarrow{B})_k \\
&= \sum_{j,k} \epsilon_{ijk} \partial_j (\sum_{\ell,m} \epsilon_{k\ell m} A_\ell B_m) \\
&= \sum_{j,k} \sum_{\ell,m} \epsilon_{ijk} \epsilon_{k\ell m} \partial_j (A_\ell B_m) = \sum_{j,k} \sum_{\ell,m} \epsilon_{ijk} \epsilon_{k\ell m} (B_m \partial_j A_\ell + A_\ell \partial_j B_m) \\
&= \sum_{j} \sum_{\ell,m} (\delta_{i\ell} \delta_{jm} - \delta_{im} \delta_{j\ell})(B_m \partial_j A_\ell + A_\ell \partial_j B_m) \\
&= \sum_{j} \sum_{l,m} (\delta_{i\ell} \delta_{jm} B_m \partial_j A_\ell + \delta_{i\ell} \delta_{jm} A_\ell \partial_j B_m - \delta_{im} \delta_{j\ell} B_m \partial_j A_\ell - \delta_{im} \delta_{j\ell} A_\ell \partial_j B_m) \\
&= (\overrightarrow{B} \cdot \overrightarrow{\nabla}) A_i + A_i (\overrightarrow{\nabla} \cdot \overrightarrow{B}) - B_i (\overrightarrow{\nabla} \cdot \overrightarrow{A}) - (\overrightarrow{A} \cdot \overrightarrow{\nabla}) B_i
\end{aligned}$$

Thus, we obtain

$$\overrightarrow{\nabla} \times (\overrightarrow{A} \times \overrightarrow{B}) = (\overrightarrow{B} \cdot \overrightarrow{\nabla})\overrightarrow{A} + \overrightarrow{A}(\overrightarrow{\nabla} \cdot \overrightarrow{B}) - \overrightarrow{B}(\overrightarrow{\nabla} \cdot \overrightarrow{A}) - (\overrightarrow{A} \cdot \overrightarrow{\nabla})\overrightarrow{B}$$

3.3.8 Prove the following relation in Electromagnetism,

$$\overrightarrow{\nabla} \times (\overrightarrow{\nabla} \times \overrightarrow{A}) = \overrightarrow{\nabla}(\overrightarrow{\nabla} \cdot \overrightarrow{A}) - \nabla^2 \overrightarrow{A}$$

Ans. The i component of the vector $\overrightarrow{\nabla} \times (\overrightarrow{\nabla} \times \overrightarrow{A})$ is

$$\begin{aligned}
[\overrightarrow{\nabla} \times (\overrightarrow{\nabla} \times \overrightarrow{A})]_i &= \sum_{j,k} \epsilon_{ijk} \partial_j (\overrightarrow{\nabla} \times \overrightarrow{A})_k \\
&= \sum_{j,k} \sum_{\ell,m} \epsilon_{ijk} \partial_j (\epsilon_{k\ell m} \partial_\ell A_m) = \sum_{j,k} \sum_{\ell,m} \epsilon_{ijk} \epsilon_{k\ell m} \partial_j (\partial_\ell A_m) \\
&= \sum_{j} \sum_{\ell,m} (\delta_{i\ell} \delta_{jm} - \delta_{im} \delta_{j\ell}) \partial_j (\partial_\ell A_m) \\
&= \sum_{j} \sum_{\ell,m} \delta_{i\ell} \delta_{jm} \partial_j (\partial_\ell A_m) - \sum_{j} \sum_{\ell,m} \delta_{im} \delta_{j\ell} \partial_j (\partial_\ell A_m) \\
&= \sum_{j} \partial_i (\partial_j A_j) - \sum_{j} (\partial_j \partial_j) A_i \\
&= \partial_i (\overrightarrow{\nabla} \cdot \overrightarrow{A}) - (\overrightarrow{\nabla} \cdot \overrightarrow{\nabla}) A_i = \partial_i (\overrightarrow{\nabla} \cdot \overrightarrow{A}) - \nabla^2 A_i
\end{aligned}$$

We then obtain

$$\overrightarrow{\nabla} \times (\overrightarrow{\nabla} \times \overrightarrow{A}) = \overrightarrow{\nabla}(\overrightarrow{\nabla} \cdot \overrightarrow{A}) - \nabla^2 \overrightarrow{A}$$

3.3.9 Prove the following relation which appears in Electromagnetism,

$$\vec{\nabla}\cdot(\phi\vec{A})=\vec{A}\cdot\vec{\nabla}\phi+\phi(\vec{\nabla}\cdot\vec{A})$$

Ans.

$$\begin{aligned}\vec{\nabla}\cdot(\phi\vec{A})&=\sum_i\partial_i(\phi\vec{A})_i=\sum_i\partial_i(\phi A_i)=\sum_i A_i(\partial_i\phi)+\sum_i\phi(\partial_i A_i)\\&=\sum_i A_i(\partial\phi)_i+\sum_i\phi(\partial_i A_i)=\vec{A}\cdot\vec{\nabla}\phi+\phi(\vec{\nabla}\cdot\vec{A})\end{aligned}$$

3.3.10 Prove the following relation which appears in Electromagnetism,

$$\vec{\nabla}(\vec{A}\cdot\vec{B})=\vec{A}\times(\vec{\nabla}\times\vec{B})+\vec{B}\times(\vec{\nabla}\times\vec{A})+(\vec{A}\cdot\vec{\nabla})\vec{B}+(\vec{B}\cdot\vec{\nabla})\vec{A}$$

Ans. The i component of the vector $\vec{A}\times(\vec{\nabla}\times\vec{B})$ is

$$\begin{aligned}[\vec{A}\times(\vec{\nabla}\times\vec{B})]_i&=\sum_{j,k}\epsilon_{ijk}A_j(\vec{\nabla}\times\vec{B})_k\\&=\sum_{j,k}\sum_{\ell,m}\epsilon_{ijk}\epsilon_{k\ell m}A_j\partial_\ell B_m=\sum_j\sum_{\ell,m}(\delta_{i\ell}\delta_{jm}-\delta_{im}\delta_{j\ell})A_j\partial_\ell B_m\\&=\sum_j\sum_{\ell,m}\delta_{i\ell}\delta_{jm}A_j\partial_\ell B_m-\sum_j\sum_{\ell,m}\delta_{im}\delta_{j\ell}A_j\partial_\ell B_m\\&=\sum_j\sum_{\ell,m}\delta_{i\ell}\delta_{jm}A_j\partial_\ell B_m-(\vec{A}\cdot\vec{\nabla})B_i\end{aligned}$$

Similarly, we have

$$[\vec{B}\times(\vec{\nabla}\times\vec{A})]_i=\sum_j\sum_{\ell,m}\delta_{i\ell}\delta_{jm}B_j\partial_\ell A_m-(\vec{B}\cdot\vec{\nabla})A_i$$

By adding these two equations, we get

$$\begin{aligned}&[\vec{A}\times(\vec{\nabla}\times\vec{B})]_i+[\vec{B}\times(\vec{\nabla}\times\vec{A})]_i\\&=\sum_j\sum_{\ell,m}\delta_{i\ell}\delta_{jm}(A_j\partial_\ell B_m+B_j\partial_\ell A_m)-(\vec{A}\cdot\vec{\nabla})B_i-(\vec{B}\cdot\vec{\nabla})A_i\\&=\sum_{j,\ell}\delta_{i\ell}(A_j\partial_\ell B_j+B_j\partial_\ell A_j)-(\vec{A}\cdot\vec{\nabla})B_i-(\vec{B}\cdot\vec{\nabla})A_i\\&=\sum_{j,\ell}\delta_{i\ell}\partial_\ell(A_jB_j)-(\vec{A}\cdot\vec{\nabla})B_i-(\vec{B}\cdot\vec{\nabla})A_i\\&=\sum_j\partial_i(A_jB_j)-(\vec{A}\cdot\vec{\nabla})B_i-(\vec{B}\cdot\vec{\nabla})A_i\\&=\partial_i(\vec{A}\cdot\vec{B})-(\vec{A}\cdot\vec{\nabla})B_i-(\vec{B}\cdot\vec{\nabla})A_i\end{aligned}$$

Thus, we get

$$\vec{A}\times(\vec{\nabla}\times\vec{B})+\vec{B}\times(\vec{\nabla}\times\vec{A})=\vec{\nabla}(\vec{A}\cdot\vec{B})-(\vec{A}\cdot\vec{\nabla})\vec{B}-(\vec{B}\cdot\vec{\nabla})\vec{A},$$

which gives

$$\vec{\nabla}(\vec{A}\cdot\vec{B})=\vec{A}\times(\vec{\nabla}\times\vec{B})+\vec{B}\times(\vec{\nabla}\times\vec{A})+(\vec{A}\cdot\vec{\nabla})\vec{B}+(\vec{B}\cdot\vec{\nabla})\vec{A}$$

3.4 Divergence and Stokes' Theorems

- $\int_V \vec{\nabla} \cdot \vec{F}\, dV = \int_A \vec{F} \cdot d\vec{A}$ (Divergence theorem)
- $\int_A \vec{\nabla} \times \vec{F} \cdot d\vec{A} = \oint_C \vec{F} \cdot d\vec{\ell}$ (Stokes' theorem)

(a) Let's prove **divergence** (or **Gauss**) **theorem**,

$$\int_V \vec{\nabla} \cdot \vec{F}\, dV = \int_A \vec{F} \cdot d\vec{A},$$

where the volume V has the boundary surface A oriented outward.

[proof] Consider a rectangular object with $x_1 \le x \le x_2$, $y_1 \le y \le y_2$ and $z_1 \le z \le z_2$, shown in Fig. 3.4.1, which has two faces A_1 and A_2 perpendicular to the x-axis.

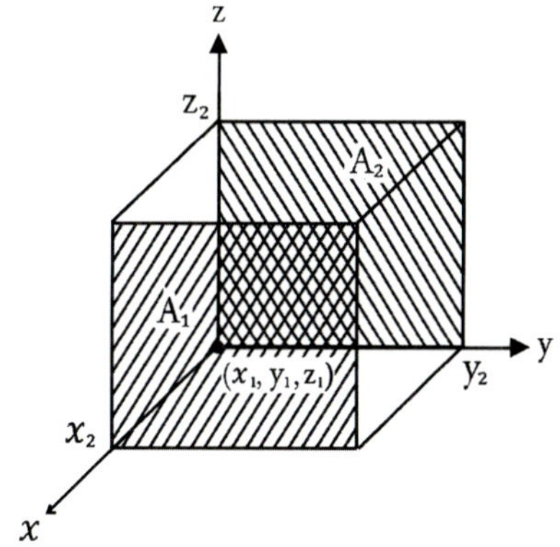

Fig 3.4.1 A rectangular object has $x_1 \le x \le x_2$, $y_1 \le y \le y_2$ and $z_1 \le z \le z_2$ for width, length and height, respectively. Two faces A_1 and A_2 are perpendicular to the x-axis.

If we calculate $\int_{A_1} \vec{F} \cdot d\vec{A} + \int_{A_2} \vec{F} \cdot d\vec{A}$ assuming the flux of vector field $\vec{F}$[3] be passing through from the faces A_2 to A_1, it gives

[3] A vector field $F(x, y, z)$ is assumed to be a smooth function.

$$-\int_{z=z_1}^{z_2}\int_{y=y_1}^{y_2} F(x_1,y,z)dydz + \int_{z=z_1}^{z_2}\int_{y=y_1}^{y_2} F(x_2,y,z)dydz$$

$$=\int_{z=z_1}^{z_2}\int_{y=y_1}^{y_2} [F(x_2,y,z)-F(x_1,y,z)]dydz$$

$$=\int_{z=z_1}^{z_2}\int_{y=y_1}^{y_2} [F(x_1+\Delta x,y,z)-F(x_1,y,z)]dydz \text{ (by using } x_2=x_1+\Delta x)$$

$$=\int_{z=z_1}^{z_2}\int_{y=y_1}^{y_2} [\frac{\partial F}{\partial x}\Delta x]dydz = \int_{z=z_1}^{z_2}\int_{y=y_1}^{y_2} [\frac{\partial F}{\partial x}\int_{x=x_1}^{x_1+\Delta x} dx]dydz$$

$$=\int_{z=z_1}^{z_2}\int_{y=y_1}^{y_2} [\int_{x=x_1}^{x_2} \frac{\partial F}{\partial x}dx]dydz$$

$$=\int\int\int \frac{\partial F}{\partial x}dV$$

In a similar way, we can obtain the flux of vector field $\vec{F}$ passes through for the other two faces perpendicular to the $y-$ and $z-$axis, respectively, is

$$\int\int\int \frac{\partial F}{\partial y}dV \text{ and } \int\int\int \frac{\partial F}{\partial z}dV$$

By adding all these together, we get

$$\int_A \vec{F}\cdot d\vec{A} = \int\int\int \frac{\partial F}{\partial x}dV + \int\int\int \frac{\partial F}{\partial y}dV + \int\int\int \frac{\partial F}{\partial z}dV$$

$$=\int_V (\frac{\partial F}{\partial x}+\frac{\partial F}{\partial y}+\frac{\partial F}{\partial z})dV$$

$$=\int_V \vec{\nabla}\cdot\vec{F}\,dV$$

Thus, we obtain the divergence theorem

$$\int_V \vec{\nabla}\cdot\vec{F}\,dV = \int_A \vec{F}\cdot d\vec{A}$$

(b) Let's prove **Stokes' theorem**

$$\int_A \vec{\nabla}\times\vec{F}\cdot d\vec{A} = \oint_C \vec{F}\cdot d\vec{\ell},$$

where the surface A oriented outward following the right-hand rule is bounded by the curve C.

[proof] Consider the surface A bounded by the curve C, shown in Fig. 3.4.2.

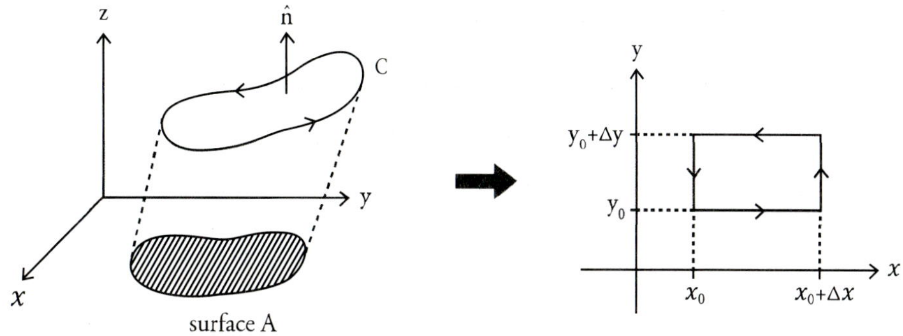

Fig 3.4.2 The surface A in the $x-y$ plane is bounded by the curve C and its direction follows the right-hand rule.

If we calculate $\oint \vec{F} \cdot d\vec{\ell}$ in the $x-y$ plane, it gives

$$\oint \vec{F} \cdot d\vec{\ell} = \int_{x_0}^{x_0+\Delta x} F_x(x,y_0)dx + \int_{y_0}^{y_0+\Delta y} F_y(x_0+\Delta x,y)dy$$

$$+\int_{x_0+\Delta x}^{x_0} F_x(x,y_0+\Delta y)dx + \int_{y_0+\Delta y}^{y_0} F_y(x_0,y)dy$$

$$= \int_{x_0}^{x_0+\Delta x} [F_x(x,y_0) - F_x(x,y_0+\Delta y)]dx$$

$$+\int_{y_0}^{y_0+\Delta y} [F_y(x_0+\Delta x,y) - F_y(x_0,y)]dy$$

$$= \int_{x_0}^{x_0+\Delta x} [-\frac{\partial F_x}{\partial y}\Delta y]dx + \int_{y_0}^{y_0+\Delta y} [\frac{\partial F_y}{\partial x}\Delta x]dy$$

$$= \int_{x_0}^{x_0+\Delta x}\int_{y_0}^{y_0+\Delta y} (-\frac{\partial F_x}{\partial y})dxdy + \int_{x_0}^{x_0+\Delta x}\int_{y_0}^{y_0+\Delta y} (\frac{\partial F_y}{\partial x})dxdy$$

$$= \int_{x_0}^{x_0+\Delta x}\int_{y_0}^{y_0+\Delta y} (\frac{\partial F_y}{\partial x} - \frac{\partial F_x}{\partial y})dxdy$$

$$= \int_{x_0}^{x_0+\Delta x}\int_{y_0}^{y_0+\Delta y} (\overrightarrow{\nabla} \times \vec{F})_z dA_z$$

Similarly, we can obtain

$$\int_{y_0}^{y_0+\Delta y}\int_{z_0}^{z_0+\Delta z}(\vec{\nabla}\times\vec{F})_x dA_x \text{ and } \int_{z_0}^{z_0+\Delta z}\int_{x_0}^{x_0+\Delta x}(\vec{\nabla}\times\vec{F})_y dA_y$$

for the $y-z$ plane and the $z-x$ plane, respectively.

If we add all these together, it gives

$$\begin{aligned}\oint_C \vec{F}\cdot d\vec{\ell} &= \int_A(\vec{\nabla}\times\vec{F})_x dA_x + \int_A(\vec{\nabla}\times\vec{F})_y dA_y + \int_A(\vec{\nabla}\times\vec{F})_z dA_z \\ &= \int_A \vec{\nabla}\times\vec{F}\cdot d\vec{A}\end{aligned}$$

Thus, we obtain Stokes' theorem

$$\int_A \vec{\nabla}\times\vec{F}\cdot d\vec{A} = \oint_C \vec{F}\cdot d\vec{\ell}$$

Example

3.4.1 Many of physical quantities such as force and electric field are proportional to the inverse square of the distance. Consider a function of $\vec{F} = \frac{1}{r^2}\hat{r}$.

Ans. The divergence of the function $\vec{F}$ is then expressed as

$$\begin{aligned}\vec{\nabla}\cdot\vec{F} &= \frac{1}{r^2}\frac{\partial}{\partial r}(r^2 F_r)^{[4]} \\ &= \frac{1}{r^2}\frac{\partial}{\partial r}(r^2\frac{1}{r^2}) = \frac{1}{r^2}\frac{\partial}{\partial r}(1) \\ &= 0\end{aligned}$$

This gives

$$\int \vec{\nabla}\cdot\vec{F}\, dV = 0 \qquad (3.4.1.1)$$

Using the divergence theorem, this equation gives

$$\int \vec{F}\cdot d\vec{A} = 0 \qquad (3.4.1.2)$$

If we integrate over a sphere with radius a centered at the origin, we have

[4] See Section 4.3.

$$\int \overrightarrow{\nabla} \cdot \overrightarrow{F}\, dV = \int \overrightarrow{F} \cdot d\overrightarrow{A} = \int_{\phi=0}^{2\pi} \int_{\theta=0}^{\pi} (\frac{1}{a^2}\hat{r}) \cdot (a^2 \sin\theta d\theta d\phi \hat{r})$$

$$= 4\pi \tag{3.4.1.3}$$

This result contradicts with Eq. (3.4.1.1). This trouble can be avoided by letting $\overrightarrow{\nabla} \cdot \overrightarrow{F} = 0$ for any place except the origin. This is expressed mathematically using the Dirac delta function, which is defined as $\delta(x) = 0$ for $x \neq 0$ and $\int_{-\infty}^{\infty} \delta(x) dx = 1$ in one-dimension:

$$\overrightarrow{\nabla} \cdot \overrightarrow{F} = \overrightarrow{\nabla} \cdot \frac{\hat{r}}{r^2}$$

$$= 4\pi\delta^3(\overrightarrow{r})$$

$\delta(r)$, $\delta^2(\overrightarrow{r})$ and $\delta^3(\overrightarrow{r})$ are one-, two- and three-dimensional delta functions, respectively.

3.4.2 Calculate $\nabla^2 \frac{1}{r}$.

Ans. First, calculate the gradient of the function

$$\overrightarrow{\nabla} \frac{1}{r} = \frac{\partial}{\partial r}(\frac{1}{r})\hat{r} = -\frac{1}{r^2}\hat{r}$$

The Laplacian of the function is then given as

$$\overrightarrow{\nabla} \cdot \overrightarrow{\nabla} \frac{1}{r} = -\overrightarrow{\nabla} \cdot \frac{\hat{r}}{r^2}$$

$$= -4\pi\delta^3(\overrightarrow{r})$$

from [Ex. 3.4.1].

3.4.3 The electric field due to continuous charge distribution is given by $\overrightarrow{E} = \frac{1}{4\pi\epsilon_0} \int \frac{\hat{r}}{r^2} \rho(\overrightarrow{r}) dV$, where $\rho(\overrightarrow{r})$ is the volume charge density. Calculate the divergence of the electric field.

Ans. $$\overrightarrow{\nabla} \cdot \overrightarrow{E} = \frac{1}{4\pi\epsilon_0} \int (\overrightarrow{\nabla} \cdot \frac{\hat{r}}{r^2}) \rho dV$$

$$= \frac{1}{4\pi\epsilon_0} \int 4\pi\delta^3(\overrightarrow{r}) \rho dV$$

$$= \frac{1}{\epsilon_0} \int \delta^3(\overrightarrow{r}) \rho dV$$

$$= \frac{\rho}{\epsilon_0}$$

Thus, we obtain one of the Maxwell equations

$$\vec{\nabla} \cdot \vec{E} = \frac{\rho}{\epsilon_0}$$

3.4.4 Express the differential Maxwell equation, $\vec{\nabla} \cdot \vec{E} = \frac{1}{\epsilon_0}\rho$, into the integral form.

Ans. If we take the integration of the original equation, we have

$$\int_V \vec{\nabla} \cdot \vec{E}\, dV = \frac{1}{\epsilon_0}\int_V \rho dV = \frac{1}{\epsilon_0}Q,$$

where Q is the total charge in the volume V.

The left-hand side of the equation is expressed as

$$\int_V \vec{\nabla} \cdot \vec{E}\, dV = \int_A \vec{E} \cdot d\vec{A}$$

by using the divergence theorem.

Thus, we obtain the Gauss theorem in the integral form

$$\int_A \vec{E} \cdot d\vec{A} = \frac{Q}{\epsilon_0}$$

3.4.5 As an example of Stokes' theorem, consider one of the problems in Electromagnetism. The electric field due to a point charge q (which is discrete charge distribution) is given by

$$\vec{E} = \frac{1}{4\pi\epsilon_0}\frac{q}{r^2}\hat{r}$$

Ans. If we calculate $\oint \vec{E} \cdot d\vec{r}$, it gives a result in the spherical coordinate system

$$\begin{aligned}\oint \vec{E} \cdot d\vec{r} &= \frac{q}{4\pi\epsilon_0}\oint \frac{\hat{r}}{r^2} \cdot (dr\hat{r} + rd\theta\hat{\theta} + r\sin\theta d\phi\hat{\phi})^{[5]} \\ &= \frac{q}{4\pi\epsilon_0}\oint \frac{1}{r^2}dr \\ &= -\frac{q}{4\pi\epsilon_0}\left[\frac{1}{r}\right]_{r=r_0}^{r=r_0} = -\frac{q}{4\pi\epsilon_0}\left(\frac{1}{r_0} - \frac{1}{r_0}\right) \\ &= 0\end{aligned}$$

Since $\oint \vec{E} \cdot d\vec{r} = \int (\vec{\nabla} \times \vec{E}) \cdot d\vec{A}$ from Stokes' theorem, we obtain

$$\int (\vec{\nabla} \times \vec{E}) \cdot d\vec{A} = 0,$$

---

[5] From Eqs. (4.1.7) and (4.1.8), we obtain $\vec{r} = dr\hat{r} + rd\theta\hat{\theta} + r\sin\theta d\phi\hat{\phi}$.

which gives

$$\overrightarrow{\nabla}\times\overrightarrow{E} = 0$$

Since the curl of the gradient of any function is zero as proved in [Ex. 3.3.4], the electric field can be defined as the gradient of a scalar potential ϕ,

$$\overrightarrow{E} = -\overrightarrow{\nabla}\phi$$

3.4.6 From [Ex. 3.4.5] and [Ex. 3.4.4], we have

$$\overrightarrow{\nabla}\cdot(-\overrightarrow{\nabla}\phi) = \frac{\rho}{\epsilon_0}$$

Ans. This gives

$$\nabla^2\phi = -\frac{\rho}{\epsilon_0},$$

which is the Poisson's equation.
If there is no charge density in a space, the Laplace equation is obtained

$$\nabla^2\phi = 0$$

3.4.7 Show that the electromagnetic force per unit volume, $\overrightarrow{f}$, given by $\overrightarrow{f} = \rho\overrightarrow{E} + \overrightarrow{J}\times\overrightarrow{B}$ is also expressed by

$$\overrightarrow{f} = \epsilon_0[(\overrightarrow{\nabla}\cdot\overrightarrow{E})\overrightarrow{E} + (\overrightarrow{E}\cdot\overrightarrow{\nabla})\overrightarrow{E}] + \frac{1}{\mu_0}[(\overrightarrow{\nabla}\cdot\overrightarrow{B})\overrightarrow{B} + (\overrightarrow{B}\cdot\overrightarrow{\nabla})\overrightarrow{B}]$$
$$-\frac{1}{2}\overrightarrow{\nabla}(\epsilon_0 E^2 + \frac{1}{\mu_0}B^2) - \epsilon_0\frac{\partial}{\partial t}(\overrightarrow{E}\times\overrightarrow{B})$$

Ans. (i) The Lorentz force acting on a point charge q due to electromagnetic field is

$$\overrightarrow{F} = q(\overrightarrow{E} + \overrightarrow{v}\times\overrightarrow{B})$$

Then, the total electromagnetic force acting on charges in a volume is given by

$$\overrightarrow{F} = \int(\overrightarrow{E} + \overrightarrow{v}\times\overrightarrow{B})\rho dV = \int(\rho\overrightarrow{E} + \rho\overrightarrow{v}\times\overrightarrow{B})dV = \int(\rho\overrightarrow{E} + \overrightarrow{J}\times\overrightarrow{B})dV,$$

where $\overrightarrow{J}$ is the current density.
Since $\overrightarrow{F} = \int\overrightarrow{f}dV$, the electromagnetic force per unit volume is

$$\overrightarrow{f} = \rho\overrightarrow{E} + \overrightarrow{J}\times\overrightarrow{B} \qquad (3.4.7.1)$$

(ii) From the Maxwell's equations, $\overrightarrow{\nabla}\cdot\overrightarrow{E} = \frac{\rho}{\epsilon_0}$ and $\overrightarrow{\nabla}\times\overrightarrow{B} = \mu_0\overrightarrow{J} + \mu_0\epsilon_0\frac{\partial\overrightarrow{E}}{\partial t}$, we have

$$\rho = \epsilon_0\overrightarrow{\nabla}\cdot\overrightarrow{E} \text{ and } \overrightarrow{J} = \frac{1}{\mu_0}\overrightarrow{\nabla}\times\overrightarrow{B} - \epsilon_0\frac{\partial\overrightarrow{E}}{\partial t},$$

respectively. If we substitute these into Eq. (3.4.7.1), we get

$$\begin{aligned}\vec{f} &= \epsilon_0(\vec{\nabla}\cdot\vec{E})\vec{E} + (\frac{1}{\mu_0}\vec{\nabla}\times\vec{B} - \epsilon_0\frac{\partial\vec{E}}{\partial t})\times\vec{B} \\ &= \epsilon_0(\vec{\nabla}\cdot\vec{E})\vec{E} + \frac{1}{\mu_0}(\vec{\nabla}\times\vec{B})\times\vec{B} - \epsilon_0\frac{\partial\vec{E}}{\partial t}\times\vec{B} \qquad (3.4.7.2)\end{aligned}$$

Since we have $\frac{\partial}{\partial t}(\vec{E}\times\vec{B}) = \frac{\partial\vec{E}}{\partial t}\times\vec{B} + \vec{E}\times\frac{\partial\vec{B}}{\partial t} = \frac{\partial\vec{E}}{\partial t}\times\vec{B} + \vec{E}\times(-\vec{\nabla}\times\vec{E})$[6], we get

$$\frac{\partial\vec{E}}{\partial t}\times\vec{B} = \frac{\partial}{\partial t}(\vec{E}\times\vec{B}) + \vec{E}\times(\vec{\nabla}\times\vec{E})$$

By substituting this into the third term on the right-hand side of Eq. (3.4.7.2), we obtain

$$\begin{aligned}\vec{f} &= \epsilon_0[(\vec{\nabla}\cdot\vec{E})\vec{E} - \vec{E}\times(\vec{\nabla}\times\vec{E})] \qquad (3.4.7.3) \\ &\quad - \frac{1}{\mu_0}[\vec{B}\times(\vec{\nabla}\times\vec{B})] - \epsilon_0\frac{\partial}{\partial t}(\vec{E}\times\vec{B})\end{aligned}$$

Let's calculate $\vec{E}\times(\vec{\nabla}\times\vec{E})$ and $\vec{B}\times(\vec{\nabla}\times\vec{B})$ in Eq. (3.4.7.3) by using the Levi-Civita density:

$$\begin{aligned}[\vec{E}\times(\vec{\nabla}\times\vec{E})]_i &= \sum_{j,k}\epsilon_{ijk}E_j(\vec{\nabla}\times\vec{E})_k = \sum_{j,k}\sum_{\ell,m}\epsilon_{ijk}E_j\epsilon_{k\ell m}\partial_\ell E_m \\ &= \sum_{j,k}\sum_{\ell,m}\epsilon_{ijk}\epsilon_{k\ell m}E_j\partial_\ell E_m = \sum_{j}\sum_{\ell,m}(\delta_{i\ell}\delta_{jm} - \delta_{im}\delta_{j\ell})E_j\partial_\ell E_m \\ &= \sum_j E_j\partial_i E_j - \sum_j E_j\partial_j E_i,\end{aligned}$$

where $\partial_i(E_jE_j) = 2E_j\partial_iE_j \Rightarrow E_j\partial_iE_j = \frac{1}{2}\partial_i(E_jE_j)$.

Therefore, we have

$$[\vec{E}\times(\vec{\nabla}\times\vec{E})]_i = \frac{1}{2}\partial_i(\sum_j E_jE_j) - (\sum_j E_j\partial_j)E_i = \frac{1}{2}\partial_iE^2 - (\vec{E}\cdot\vec{\nabla})E_i,$$

which gives

$$\vec{E}\times(\vec{\nabla}\times\vec{E}) = \frac{1}{2}\vec{\nabla}E^2 - (\vec{E}\cdot\vec{\nabla})\vec{E}$$

Similarly, we have

$$\vec{B}\times(\vec{\nabla}\times\vec{B}) = \frac{1}{2}\vec{\nabla}B^2 - (\vec{B}\cdot\vec{\nabla})\vec{B}$$

If we insert these two equations into Eq. (3.4.7.3), we obtain

[6] $\vec{\nabla}\times\vec{E} = -\frac{\partial\vec{B}}{\partial t}$ from the Maxwell equation.

$$\vec{f}=\epsilon_0[(\vec{\nabla}\cdot\vec{E})\vec{E}+(\vec{E}\cdot\vec{\nabla})\vec{E}]+\frac{1}{\mu_0}[(\vec{\nabla}\cdot\vec{B})\vec{B}+(\vec{B}\cdot\vec{\nabla})\vec{B}]$$
$$-\frac{1}{2}\vec{\nabla}(\epsilon_0 E^2+\frac{1}{\mu_0}B^2)-\epsilon_0\frac{\partial}{\partial t}(\vec{E}\times\vec{B}) \qquad (3.4.7.4)$$

In this equation, we can add $(\vec{\nabla}\cdot\vec{B})\vec{B}$ term because it does not affect result of the equation due to $\vec{\nabla}\cdot\vec{B}=0$.

3.4.8 We learned ways to determine whether a work force $\vec{F}$ is the conservative force or not. When the force worked in a closed path is $\oint\vec{F}\cdot d\vec{\ell}=0$, the force is conservative. Find another way to determine using the Stokes' theorem whether a force is a conservative force or not.

Ans. From the Stokes' theorem, this equation is the same as

$$\int(\vec{\nabla}\times\vec{F})\cdot d\vec{A}=0$$

To satisfy the equation for all surfaces, the integrand should be $\vec{\nabla}\times\vec{F}=0$. Thus, when $\vec{\nabla}\times\vec{F}=0$, the force is a conservative force.

3.4.9 Determine whether or not the centripetal force is the conservative. A centripetal force is $\vec{F}=-\frac{k}{r^2}\hat{r}$, where k is a positive constant.

Ans. The curl of the force is expressed in the spherical coordinate system,

$$\vec{\nabla}\times\vec{F}=\frac{1}{h_1h_2h_3}\begin{vmatrix} h_1\hat{r} & h_2\hat{\theta} & h_3\hat{\phi} \\ \frac{\partial}{\partial r} & \frac{\partial}{\partial\theta} & \frac{\partial}{\partial\phi} \\ h_1F_r & h_2F_\theta & h_3F_\phi \end{vmatrix},$$

where $h_1=1$, $h_2=r$ and $h_3=r\sin\theta$. This relation is proved in Chapter 4. Since $F_\theta=0$ and $F_\phi=0$ for the centripetal force, the equation is then

$$\vec{\nabla}\times\vec{F}=\frac{1}{r^2\sin\theta}\begin{vmatrix} \hat{r} & r\hat{\theta} & r\sin\theta\hat{\phi} \\ \frac{\partial}{\partial r} & \frac{\partial}{\partial\theta} & \frac{\partial}{\partial\phi} \\ -\frac{k}{r^2} & 0 & 0 \end{vmatrix}=0$$

Thus, the centripetal force is the conservative.

Since the curl of the gradient of a scalar function is zero, we can define

$$\vec{F}=-\vec{\nabla}U,$$

which is the relation between the force and potential energy.
This is the reason why this equation is only applied for the conservative force.

3.5 Tensor

In this section, we briefly study the concept and examples of a **tensor**. To characterize physical properties of a system, scalar and vector concepts are introduced. As you have learned, scalar quantities have only one component, magnitude, and vector quantities have not only magnitude but also direction which corresponds to three components in a three dimensional space, $3^1 = 3$. A number of components for a tensor of rank n is 3^n in a three dimensional space. Therefore, we can rename the **scalar (vector) as tensor of rank 0 (1)**.
This tensor concept can be applied to higher orders. The **tensor of rank 2 (3)**[7] is called **dyad (triad)** which have magnitude and two (three) directions so that it has $3^2 = 9$ $(3^3 = 27)$ components.

The dyad is a vector times a vector product which is the addition of term-by-term products of each component. If two vectors $\vec{F} = f_1\hat{x} + f_2\hat{y} + f_3\hat{z}$ and $\vec{G} = g_1\hat{x} + g_2\hat{y} + g_3\hat{z}$ are given, the dyad product of them is

$$\begin{aligned}\vec{F}\,\vec{G} &= f_1g_1\hat{x}\hat{x} + f_1g_2\hat{x}\hat{y} + f_1g_3\hat{x}\hat{z} \\ &\quad + f_2g_1\hat{y}\hat{x} + f_2g_2\hat{y}\hat{y} + f_2g_3\hat{y}\hat{z} \\ &\quad + f_3g_1\hat{z}\hat{x} + f_3g_2\hat{z}\hat{y} + f_3g_3\hat{z}\hat{z}\end{aligned}$$

If we define $f_1g_1 = a_{11}$, $f_1g_2 = a_{12}$ and $f_1g_3 = a_{13}$, etc, the dyad of vectors can be expressed as

$$\begin{aligned}\vec{F}\,\vec{G} &= a_{11}\hat{x}\hat{x} + a_{12}\hat{x}\hat{y} + a_{13}\hat{x}\hat{z} \\ &\quad + a_{21}\hat{y}\hat{x} + a_{22}\hat{y}\hat{y} + a_{23}\hat{y}\hat{z} \\ &\quad + a_{31}\hat{z}\hat{x} + a_{32}\hat{z}\hat{y} + a_{33}\hat{z}\hat{z}\end{aligned}$$

Therefore, the dyad can be expressed in the matrix representation and

---

[7] For instance, we need nine components to describe a rotating body. See [Ex. 3.5.2].

has its scalar components in configuration of the matrix. If we express a tensor $\overrightarrow{FG} \equiv \overleftrightarrow{A}$, it is

$$\overleftrightarrow{A} = \begin{pmatrix} a_{11} & a_{12} & a_{13} \\ a_{21} & a_{22} & a_{23} \\ a_{31} & a_{32} & a_{33} \end{pmatrix}$$

The scalar product of the vector $\vec{v} = v_1\hat{x} + v_2\hat{y} + v_3\hat{z}$ with the tensor $\overleftrightarrow{A}$ is then

$$\begin{aligned} \vec{v} \cdot \overleftrightarrow{A} &= (v_1\hat{x} + v_2\hat{y} + v_3\hat{z}) \cdot (a_{11}\hat{x}\hat{x} + a_{12}\hat{x}\hat{y} + a_{13}\hat{x}\hat{z} + a_{21}\hat{y}\hat{x} + a_{22}\hat{y}\hat{y} + a_{23}\hat{y}\hat{z} \\ &\quad + a_{31}\hat{z}\hat{x} + a_{32}\hat{z}\hat{y} + a_{33}\hat{z}\hat{z}) \\ &= v_1(a_{11}\hat{x} + a_{12}\hat{y} + a_{13}\hat{z}) + v_2(a_{21}\hat{x} + a_{22}\hat{y} + a_{23}\hat{z}) + v_3(a_{31}\hat{x} + a_{32}\hat{y} + a_{33}\hat{z}) \\ &= (v_1a_{11} + v_2a_{21} + v_3a_{31})\hat{x} + (v_1a_{12} + v_2a_{22} + v_3a_{32})\hat{y} \\ &\quad + (v_1a_{13} + v_2a_{23} + v_3a_{33})\hat{z} \end{aligned}$$

This can be expressed as

$$(\vec{v} \cdot \overleftrightarrow{A})_i = \sum_j v_j a_{ji}, \tag{3.5.1}$$

where i and j are 1, 2 or 3.

The scalar product of the dyad $\vec{F}\,\vec{G}$ with another vector $\vec{H}$ is

$$\vec{H} \cdot (\vec{F}\,\vec{G}) = (\vec{H} \cdot \vec{F})\vec{G} = A\vec{G} \tag{3.5.2}$$

and its result is another vector with magnitude $A|\vec{G}|$ and its direction is determined by $\vec{G}$. In general, $\vec{H} \cdot (\vec{F}\,\vec{G})$ is not commutative because

$$(\vec{F}\,\vec{G}) \cdot \vec{H} = \vec{F}(\vec{G} \cdot \vec{H}) = \vec{F}B = B\vec{F}, \tag{3.5.3}$$

which has the magnitude $B|\vec{F}|$ and direction is determined by $\vec{F}$.

Example

3.5.1 Prove $(\vec{F}\,\vec{G}) \cdot \vec{H} = \vec{F}(\vec{G} \cdot \vec{H})$ in Eq. (3.5.3), where $\vec{F} = f_1\hat{x} + f_2\hat{y} + f_3\hat{z}$, $\vec{G} = g_1\hat{x} + g_2\hat{y} + g_3\hat{z}$ and $\vec{H} = h_1\hat{x} + h_2\hat{y} + h_3\hat{z}$.

Ans. $(\vec{F}\ \vec{G})\cdot\vec{H} = \begin{pmatrix} f_1g_1 & f_1g_2 & f_1g_3 \\ f_2g_1 & f_2g_2 & f_2g_3 \\ f_3g_1 & f_3g_2 & f_3g_3 \end{pmatrix}\begin{pmatrix} h_1 \\ h_2 \\ h_3 \end{pmatrix} = \begin{pmatrix} f_1g_1h_1 + f_1g_2h_2 + f_1g_3h_3 \\ f_2g_1h_1 + f_2g_2h_2 + f_2g_3h_3 \\ f_3g_1h_1 + f_3g_2h_2 + f_3g_3h_3 \end{pmatrix}$

$$= \begin{pmatrix} f_1\{g_1h_1 + g_2h_2 + g_3h_3\} \\ f_2\{g_1h_1 + g_2h_2 + g_3h_3\} \\ f_3\{g_1h_1 + g_2h_2 + g_3h_3\} \end{pmatrix} = \vec{F}(\vec{G}\cdot\vec{H})$$

3.5.2 Consider a motion of the rigid body, shown in Figure, in a three dimensional space and calculate the moment of inertia about an axis whose direction is arbitrary.

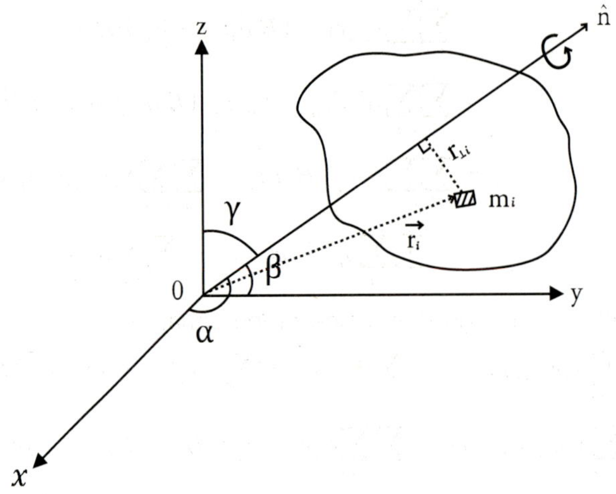

A rigid body rotation about an arbitrary axis with direction $\hat{n}$. The direction cosines of the arbitrary rotation axis are $\cos\alpha$, $\cos\beta$, and $\cos\gamma$, respectively.

Ans. Since the **direction cosines** of a vector are defined as the cosines of angles between the vector and coordinate axes, the direction cosines of an arbitrary axis are $\cos\alpha = \frac{x_i}{r_i}$, $\cos\beta = \frac{y_i}{r_i}$, and $\cos\gamma = \frac{z_i}{r_i}$.

If we define $\hat{n}$ is the direction of the arbitrary rotation axis as shown in Figure, it is written as

$$\hat{n} = \hat{x}\cos\alpha + \hat{y}\cos\beta + \hat{z}\cos\gamma,$$

where $\cos^2\alpha + \cos^2\beta + \cos^2\gamma = 1$.

The moment of inertia about an arbitrary axis is then

$$I = \sum_i m_i r_{\perp i}^2, \qquad (3.5.2.1)$$

where $r_{\perp i}$ is the perpendicular distance from the rotation axis to the particle m_i.

Since $r_{\perp i} = |\vec{r}_i \times \hat{n}|$, it is

$$\vec{r}_{\perp i} = \begin{vmatrix} \hat{x} & \hat{y} & \hat{z} \\ x_i & y_i & z_i \\ \cos\alpha & \cos\beta & \cos\gamma \end{vmatrix}$$

$$= (y_i\cos\gamma - z_i\cos\beta)\hat{x} + (z_i\cos\alpha - x_i\cos\gamma)\hat{y} + (x_i\cos\beta - y_i\cos\alpha)\hat{z}$$

This gives

$$r_{\perp i}^2 = (y_i\cos\gamma - z_i\cos\beta)^2 + (z_i\cos\alpha - x_i\cos\gamma)^2 + (x_i\cos\beta - y_i\cos\alpha)^2 \quad (3.5.2.2)$$

By substituting Eqs. (3.5.2.2) into (3.5.2.1) we have

$$I = \sum_i m_i(y_i^2 + z_i^2)\cos^2\alpha + \sum_i m_i(z_i^2 + x_i^2)\cos^2\beta + \sum_i m_i(x_i^2 + y_i^2)\cos^2\gamma - 2\sum_i m_i y_i z_i \cos\beta\cos\gamma - 2\sum_i m_i z_i x_i \cos\gamma\cos\alpha - 2\sum_i m_i x_i y_i \cos\alpha\cos\beta,$$

which gives

$$I = I_{11}\cos^2\alpha + I_{22}\cos^2\beta + I_{33}\cos^2\gamma + 2I_{23}\cos\beta\cos\gamma + 2I_{31}\cos\gamma\cos\alpha + 2I_{12}\cos\alpha\cos\beta, \quad (3.5.2.3)$$

where $I_{11} = \sum_i m_i(y_i^2 + z_i^2)$, $I_{22} = \sum_i m_i(z_i^2 + x_i^2)$ and $I_{33} = \sum_i m_i(x_i^2 + y_i^2)$ are the moments of inertia about the $x-$, $y-$ and $z-$axis, respectively, and $I_{23} = -\sum_i m_i y_i z_i$, $I_{31} = -\sum_i m_i z_i x_i$ and $I_{12} = -\sum_i m_i x_i y_i$ are the yz, zx and xy products of inertia, respectively. These relations are proved in the following example, [Ex. 3.5.3].

Let's define a matrix

$$\begin{pmatrix} I_{11} & I_{12} & I_{13} \\ I_{21} & I_{22} & I_{23} \\ I_{31} & I_{32} & I_{33} \end{pmatrix} = \overleftrightarrow{I},$$

which is the moment of inertia tensor of rank 2. Since $I_{ij} = I_{ji}$ for $i \neq j$, the matrix is symmetric.

Let's consider $\hat{n} \cdot \overleftrightarrow{I} \cdot \hat{n}$ product. Since it is the product of (1×3), (3×3), and (3×1) matrices, it gives a scalar quantity.

Now let's calculate it:

$$\hat{n} \cdot \overleftrightarrow{I} \cdot \hat{n} = (\cos\alpha \quad \cos\beta \quad \cos\gamma)\begin{pmatrix} I_{11} & I_{12} & I_{13} \\ I_{21} & I_{22} & I_{23} \\ I_{31} & I_{32} & I_{33} \end{pmatrix}\begin{pmatrix} \cos\alpha \\ \cos\beta \\ \cos\gamma \end{pmatrix}$$

$$= I_{11}\cos^2\alpha + I_{22}\cos^2\beta + I_{33}\cos^2\gamma + 2I_{23}\cos\beta\cos\gamma + 2I_{31}\cos\gamma\cos\alpha + 2I_{12}\cos\alpha\cos\beta$$

This is the same result as Eq. (3.5.2.3).

Thus, we can write the moment of inertia about the arbitrary rotation axis $\hat{n}$ as

$$I = \hat{n} \cdot \overleftrightarrow{I} \cdot \hat{n}$$

Then, we can extract each component of the moment of inertia tensor by

$$I_{ij} = \hat{i} \cdot \overleftrightarrow{\mathrm{I}} \cdot \hat{j},$$

where i and j are 1, 2, or 3, which correspond to x, y, or z, respectively.

In general the components of tensor $\overleftrightarrow{\mathrm{T}}$ are obtained by

$$T_{ij} = \hat{i} \cdot \overleftrightarrow{\mathrm{T}} \cdot \hat{j}$$

3.5.3 Prove the relations $I_{11} = \sum_i m_i (y_i^2 + z_i^2)$ and $I_{23} = -\sum_i m_i y_i z_i$ in [Ex. 3.5.2] by calculating a total angular momentum $\vec{L}$ about an arbitrary axis.

$$\vec{L} = \sum_i \vec{L_i} = \sum_i m_i \vec{r_i} \times \vec{v_i} = \sum_i m_i \vec{r_i} \times (\vec{\omega_i} \times \vec{r_i}) \tag{3.5.3.1}$$

Ans. Since angular velocities of any constituents of the rigid body are the same by definition of the rigid body, $\vec{\omega_i} = \vec{\omega}$ is the constant in Eq. (3.5.3.1). The total angular momentum is then expressed as

$$\vec{L} = \sum_i m_i \vec{r_i} \times (\vec{\omega} \times \vec{r_i})$$

The right-hand side of the equation is one of the triple vector product and the equation is

$$\begin{aligned} \vec{L} &= \sum_i m_i [r_i^2 \vec{\omega} - \vec{r_i}(\vec{r_i} \cdot \vec{\omega})] \\ &= [\sum_i m_i r_i^2 \overleftrightarrow{1} - \sum_i m_i \vec{r_i}\vec{r_i}] \cdot \vec{\omega}, \end{aligned} \tag{3.5.3.2}$$

where $\overleftrightarrow{1} = \hat{x}\hat{x} + \hat{y}\hat{y} + \hat{z}\hat{z}$ is the unit tensor.

The equality in Eq. (3.5.3.2) is as follows:

$$\begin{aligned} \overleftrightarrow{1} \cdot \vec{\omega} &= (\hat{x}\hat{x} + \hat{y}\hat{y} + \hat{z}\hat{z}) \cdot \vec{\omega} = \hat{x}(\hat{x} \cdot \vec{\omega}) + \hat{y}(\hat{y} \cdot \vec{\omega}) + \hat{z}(\hat{z} \cdot \vec{\omega}) \\ &= \hat{x}(\omega_x) + \hat{y}(\omega_y) + \hat{z}(\omega_z) = \vec{\omega} \end{aligned}$$

Since the angular momentum is the product of the moment of inertia and angular velocity, $\vec{L} = \overleftrightarrow{\mathrm{I}} \cdot \vec{\omega}$, we get

$$\overleftrightarrow{\mathrm{I}} = \sum_i m_i r_i^2 \overleftrightarrow{1} - \sum_i m_i \vec{r_i}\vec{r_i} \tag{3.5.3.3}$$

from Eq. (3.5.3.2).

Since we can calculate tensor components by $I_{ij} = \hat{i} \cdot \overleftrightarrow{\mathrm{I}} \cdot \hat{j}$ as described in [Ex. 3.5.2], we get

$$I_{11} = \hat{1} \cdot \overleftrightarrow{\mathrm{I}} \cdot \hat{1} = \hat{x} \cdot \overleftrightarrow{\mathrm{I}} \cdot \hat{x} = \hat{x} \cdot [\sum_i m_i r_i^2 \overleftrightarrow{1} - \sum_i m_i \vec{r_i}\vec{r_i}] \cdot \hat{x}$$

$$= \hat{x} \cdot [\sum_i m_i r_i^2 (\hat{x}\hat{x} + \hat{y}\hat{y} + \hat{z}\hat{z}) - \sum_i m_i \overrightarrow{r_i r_i}] \cdot \hat{x}$$

$$= \hat{x} \cdot [\sum_i m_i r_i^2 (\hat{x}\hat{x} \cdot \hat{x} + \hat{y}\hat{y} \cdot \hat{x} + \hat{z}\hat{z} \cdot \hat{x}) - \sum_i m_i \vec{r}_i (\vec{r}_i \cdot \hat{x})]$$

using [Ex. 3.5.1],

where $\hat{x} \cdot \hat{x} = 1$, $\hat{y} \cdot \hat{x} = \hat{z} \cdot \hat{x} = 0$ and $\vec{r}_i \cdot \hat{x} = x_i$.

Thus, we obtain

$$I_{11} = \hat{x} \cdot [\sum_i m_i r_i^2 \hat{x} - \sum_i m_i \vec{r}_i x_i] = \sum_i m_i r_i^2 \hat{x} \cdot \hat{x} - \sum_i m_i (\hat{x} \cdot \vec{r}_i) x_i$$

$$= \sum_i m_i (r_i^2 - x_i^2) = \sum_i m_i (y_i^2 + z_i^2)$$

Similarly, we can calculate yz product of inertia

$$I_{23} = \hat{2} \cdot \overleftrightarrow{I} \cdot \hat{3} = \hat{y} \cdot \overleftrightarrow{I} \cdot \hat{z} = \hat{y} \cdot [\sum_i m_i r_i^2 (\hat{x}\hat{x} + \hat{y}\hat{y} + \hat{z}\hat{z}) - \sum_i m_i \overrightarrow{r_i r_i}] \cdot \hat{z}$$

$$= \hat{y} \cdot [\sum_i m_i r_i^2 \hat{z} - \sum_i m_i \vec{r}_i z_i] = -\sum_i m_i y_i z_i$$

3.5.4 **Maxwell stress tensor** $\overleftrightarrow{T}$ with nine components is expressed in the matrix representation as

$$\overleftrightarrow{T} = \begin{pmatrix} T_{11} & T_{12} & T_{13} \\ T_{21} & T_{22} & T_{23} \\ T_{31} & T_{32} & T_{33} \end{pmatrix}$$

and defined as

$$T_{ij} = \epsilon_0 (E_i E_j - \frac{1}{2}\delta_{ij} E^2) + \frac{1}{\mu_0}(B_i B_j - \frac{1}{2}\delta_{ij} B^2), \tag{3.5.4.1}$$

where $\vec{E}$ and $\vec{B}$ are the electric and magnetic fields, δ_{ij} is the Kronecker delta and $1, 2, 3$ correspond to x, y, z, respectively.

Calculate a diagonal element T_{11}, an off-diagonal element T_{12}, and $\vec{\nabla} \cdot \overleftrightarrow{T}$.

Ans. The diagonal element T_{11} is

$$T_{11} = \epsilon_0 (E_1^2 - \frac{1}{2}E^2) + \frac{1}{\mu_0}(B_1^2 - \frac{1}{2}B^2)$$

$$= \epsilon_0 [E_1^2 - \frac{1}{2}(E_1^2 + E_2^2 + E_3^2)] + \frac{1}{\mu_0}[B_1^2 - \frac{1}{2}(B_1^2 + B_2^2 + B_3^2)]$$

$$= \frac{1}{2}\epsilon_0 (E_1^2 - E_2^2 - E_3^2) + \frac{1}{2\mu_0}(B_1^2 - B_2^2 - B_3^2)$$

The off-diagonal element T_{12} is

$$T_{12} = \epsilon_0 E_1 E_2 + \frac{1}{\mu_0} B_1 B_2$$

From Eq. (3.5.1), we have

$$(\overrightarrow{\nabla}\cdot\overleftrightarrow{T})_i=\sum_j\nabla_jT_{ji}$$

If we insert Eq. (3.5.4.1) into the right-hand side of the equation, we obtain

$$\begin{aligned}(\overrightarrow{\nabla}\cdot\overleftrightarrow{T})_i&=\sum_j\nabla_j[\epsilon_0(E_jE_i-\frac{1}{2}\delta_{ji}E^2)+\frac{1}{\mu_0}(B_jB_i-\frac{1}{2}\delta_{ji}B^2)]\\&=\epsilon_0[\sum_j(\nabla_jE_j)E_i+\sum_j(E_j\nabla_j)E_i-\sum_j\frac{1}{2}\delta_{ji}\nabla_jE^2)]\\&\quad+\frac{1}{\mu_0}[\sum_j(\nabla_jB_j)B_i+\sum_j(B_j\nabla_j)B_i-\sum_j\frac{1}{2}\delta_{ji}\nabla_jB^2)]\\&=\epsilon_0[(\overrightarrow{\nabla}\cdot\overrightarrow{E})E_i+(\overrightarrow{E}\cdot\overrightarrow{\nabla})E_i-\frac{1}{2}\nabla_iE^2]\\&\quad+\frac{1}{\mu_0}[(\overrightarrow{\nabla}\cdot\overrightarrow{B})B_i+(\overrightarrow{B}\cdot\overrightarrow{\nabla})B_i-\frac{1}{2}\nabla_iB^2]\end{aligned}$$

Thus, we obtain

$$\begin{aligned}\overrightarrow{\nabla}\cdot\overleftrightarrow{T}&=\epsilon_0[(\overrightarrow{\nabla}\cdot\overrightarrow{E})\overrightarrow{E}+(\overrightarrow{E}\cdot\overrightarrow{\nabla})\overrightarrow{E}-\frac{1}{2}\overrightarrow{\nabla}E^2]\\&\quad+\frac{1}{\mu_0}[(\overrightarrow{\nabla}\cdot\overrightarrow{B})\overrightarrow{B}+(\overrightarrow{B}\cdot\overrightarrow{\nabla})\overrightarrow{B}-\frac{1}{2}\overrightarrow{\nabla}B^2]\\&=\epsilon_0[(\overrightarrow{\nabla}\cdot\overrightarrow{E})\overrightarrow{E}+(\overrightarrow{E}\cdot\overrightarrow{\nabla})\overrightarrow{E}]+\frac{1}{\mu_0}[(\overrightarrow{\nabla}\cdot\overrightarrow{B})\overrightarrow{B}+(\overrightarrow{B}\cdot\overrightarrow{\nabla})\overrightarrow{B}]\\&\quad-\frac{1}{2}\overrightarrow{\nabla}(\epsilon_0E^2+\frac{1}{\mu_0}B^2)\end{aligned}\qquad(3.5.4.2)$$

Since **Poynting vector** is given by $\overrightarrow{S}=\frac{1}{\mu_0}\overrightarrow{E}\times\overrightarrow{B}$, the electromagnetic force per unit volume which was obtained in [Ex. 3.4.7] can be expressed by

$$\begin{aligned}\overrightarrow{f}&=\epsilon_0[(\overrightarrow{\nabla}\cdot\overrightarrow{E})\overrightarrow{E}+(\overrightarrow{E}\cdot\overrightarrow{\nabla})\overrightarrow{E}]+\frac{1}{\mu_0}[(\overrightarrow{\nabla}\cdot\overrightarrow{B})\overrightarrow{B}+(\overrightarrow{B}\cdot\overrightarrow{\nabla})\overrightarrow{B}]\\&\quad-\frac{1}{2}\overrightarrow{\nabla}(\epsilon_0E^2+\frac{1}{\mu_0}B^2)-\epsilon_0\frac{\partial}{\partial t}(\overrightarrow{E}\times\overrightarrow{B})\\&=\epsilon_0[(\overrightarrow{\nabla}\cdot\overrightarrow{E})\overrightarrow{E}+(\overrightarrow{E}\cdot\overrightarrow{\nabla})\overrightarrow{E}]+\frac{1}{\mu_0}[(\overrightarrow{\nabla}\cdot\overrightarrow{B})\overrightarrow{B}+(\overrightarrow{B}\cdot\overrightarrow{\nabla})\overrightarrow{B}]\\&\quad-\frac{1}{2}\overrightarrow{\nabla}(\epsilon_0E^2+\frac{1}{\mu_0}B^2)-\epsilon_0\mu_0\frac{\partial\overrightarrow{S}}{\partial t}\end{aligned}\qquad(3.5.4.3)$$

From Eqs. (3.5.4.2) and (3.5.4.3), we obtain the relationship between the electromagnetic force per unit volume and the Maxwell stress tensor:

$$\overrightarrow{f}=\overrightarrow{\nabla}\cdot\overleftrightarrow{T}-\epsilon_0\mu_0\frac{\partial\overrightarrow{S}}{\partial t}$$

Chapter 04 Physical Quantities in Coordinate Systems

When considering particular problems, it is convenient to choose a coordinate system adequate to the problem. For instance, it is better to use the spherical coordinate system for a problem of a central force, $\vec{F} = F(r)\hat{r}$. If we want to find a velocity and then an acceleration of a particle of mass m under the central force, we get

$$F(r)\hat{r} = ma_r\hat{r}$$

for the radial component from the second law of Newton's law

$$\vec{F} = m\vec{a}$$

We calculate for velocities and accelerations of a moving particle in the spherical and cylindrical coordinate systems in Section 4.1 and 4.2, respectively. A generalized coordinate system is introduced in Section 4.3, which can be applied to the rectangular, spherical, and cylindrical coordinate systems.

4.1 Spherical Coordinate System

· scale factor

$h_1 = 1$, $h_2 = r$, and $h_3 = r\sin\theta$

· velocity

$$\vec{v} = \dot{r}\hat{r} + r\dot{\theta}\hat{\theta} + r\dot{\phi}\sin\theta\hat{\phi}$$

· acceleration

$$\vec{a} = (\ddot{r} - r\dot{\theta}^2 - r\dot{\phi}^2\sin^2\theta)\hat{r} + (2\dot{r}\dot{\theta} + r\ddot{\theta} - r\dot{\phi}^2\sin\theta\cos\theta)\hat{\theta}$$
$$+ (2\dot{r}\dot{\phi}\sin\theta + 2r\dot{\theta}\dot{\phi}\cos\theta + r\ddot{\phi}\sin\theta)\hat{\phi}$$

· volume element

$$dV = r^2\sin\theta dr d\theta d\phi$$

We first calculate the velocity and acceleration of a particle with a mass m moving in the central force field. The position vector which represents the distance from the origin to the observation point is $\vec{r} = r\hat{r}$. The velocity $\vec{v}$ and acceleration $\vec{a}$ are then

$$\vec{v} = \frac{d\vec{r}}{dt} = \frac{d}{dt}(r\hat{r}) = \dot{r}\hat{r} + r\dot{\hat{r}}$$

and

$$\vec{a} = \frac{d\vec{v}}{dt} = \frac{d}{dt}(\dot{r}\hat{r} + r\dot{\hat{r}}) = \ddot{r}\hat{r} + 2\dot{r}\dot{\hat{r}} + r\ddot{\hat{r}},$$

respectively, where $\dot{r} = \frac{dr}{dt}$, $\dot{\hat{r}} = \frac{d\hat{r}}{dt}$, $\ddot{r} = \frac{d^2r}{dt^2}$, and $\ddot{\hat{r}} = \frac{d^2\hat{r}}{dt^2}$.

Therefore, we have to calculate $\dot{\hat{r}}$ and $\ddot{\hat{r}}$ to find the velocity and acceleration.

For this purpose we need to know the relation between the rectangular and spherical coordinate systems.

Each component x, y and z, of the position vector shown in Fig. 3.1.3, is

$$x = r\sin\theta\cos\phi, \quad y = r\sin\theta\sin\phi, \quad z = r\cos\theta \tag{4.1.1}$$

Since the distance between two points whose coordinates differ by dx, dy, dz is $ds = \sqrt{dx^2 + dy^2 + dz^2}$, it gives

$$ds^2 = dx^2 + dy^2 + dz^2 \tag{4.1.2}$$

Since x, y, and z are functions of r, θ, and ϕ, their derivatives are

$$dx = \frac{\partial x}{\partial r}dr + \frac{\partial x}{\partial \theta}d\theta + \frac{\partial x}{\partial \phi}d\phi,$$

$$dy = \frac{\partial y}{\partial r}dr + \frac{\partial y}{\partial \theta}d\theta + \frac{\partial y}{\partial \phi}d\phi, \text{ and}$$

$$dz = \frac{\partial z}{\partial r}dr + \frac{\partial z}{\partial \theta}d\theta + \frac{\partial z}{\partial \phi}d\phi$$

From Eq. (4.1.1) we obtain

$$\begin{cases} dx = \sin\theta\cos\phi dr + r\cos\theta\cos\phi d\theta - r\sin\theta\sin\phi d\phi \\ dy = \sin\theta\sin\phi dr + r\cos\theta\sin\phi d\theta + r\sin\theta\cos\phi d\phi \\ dz = \cos\theta dr - r\sin\theta d\theta \end{cases} \tag{4.1.3}$$

If we insert these into Eq. (4.1.2), it gives

$$ds^2 = dr^2 + r^2 d\theta^2 + r^2\sin^2\theta d\phi^2 \tag{4.1.4}$$

Since $\hat{r}$, $\hat{\theta}$, and $\hat{\phi}$ are mutually orthogonal and $ds^2 = \vec{ds} \cdot \vec{ds}$, Eq. (4.1.4) means that

$$\vec{ds} = \hat{r}dr + \hat{\theta}rd\theta + \hat{\phi}r\sin\theta d\phi \tag{4.1.5}$$

The value of each component on the right-hand side of the equation is called **scale factor**. We find that the scale factors in the spherical coordinate system are 1, r and $r\sin\theta$.

For convenience, we denote them as

$$h_1 = 1, \quad h_2 = r, \quad \text{and} \quad h_3 = r\sin\theta$$

On the other hand, Eq. (4.1.2) in the rectangular coordinate system gives

$$\vec{ds} = \hat{x}dx + \hat{y}dy + \hat{z}dz \tag{4.1.6}$$

By inserting Eqs. (4.1.3) into (4.1.6), we get

$$\begin{aligned} \vec{ds} &= (\hat{x}\sin\theta\cos\phi + \hat{y}\sin\theta\sin\phi + \hat{z}\cos\theta)dr \\ &\quad + r(\hat{x}\cos\theta\cos\phi + \hat{y}\cos\theta\sin\phi - \hat{z}\sin\theta)d\theta \\ &\quad + r\sin\theta(-\hat{x}\sin\phi + \hat{y}\cos\phi)d\phi \end{aligned} \tag{4.1.7}$$

By comparing this with Eq. (4.1.5), we obtain the following relations between the spherical and rectangular coordinate systems:

$$\begin{cases} \hat{r} = \hat{x}\sin\theta\cos\phi + \hat{y}\sin\theta\sin\phi + \hat{z}\cos\theta \\ \hat{\theta} = \hat{x}\cos\theta\cos\phi + \hat{y}\cos\theta\sin\phi - \hat{z}\sin\theta \\ \hat{\phi} = -\hat{x}\sin\phi + \hat{y}\cos\phi \end{cases} \tag{4.1.8}$$

Since $\hat{x}$, $\hat{y}$, and $\hat{z}$ are not changed with the time, we have

$$\dot{\hat{x}} = 0, \ \dot{\hat{y}} = 0, \text{ and } \dot{\hat{z}} = 0$$

If we calculate the time derivative of $\hat{r}$ in Eq. (4.1.8), we get

$$\begin{aligned} \dot{\hat{r}} &= \frac{\partial \hat{r}}{\partial t} = \frac{\partial \hat{r}}{\partial \theta}\frac{\partial \theta}{\partial t} + \frac{\partial \hat{r}}{\partial \phi}\frac{\partial \phi}{\partial t} \\ &= (\hat{x}\cos\theta\cos\phi + \hat{y}\cos\theta\sin\phi - \hat{z}\sin\theta)\dot{\theta} + (-\hat{x}\sin\theta\sin\phi + \hat{y}\sin\theta\cos\phi)\dot{\phi} \\ &= \hat{\theta}\dot{\theta} + \sin\theta\hat{\phi}\dot{\phi} \end{aligned}$$

In a similar way, we can get for $\dot{\hat{\theta}}$, $\dot{\hat{\phi}}$ and then have

$$\begin{cases} \dot{\hat{r}} = \dot{\theta}\hat{\theta} + \dot{\phi}\sin\theta\hat{\phi} \\ \dot{\hat{\theta}} = -\dot{\theta}\hat{r} + \dot{\phi}\cos\theta\hat{\phi} \\ \dot{\hat{\phi}} = -\dot{\phi}(\hat{r}\sin\theta + \hat{\theta}\cos\theta) \end{cases} \tag{4.1.9}$$

Then, the velocity of a particle is

$$\vec{v} = \dot{r}\hat{r} + r\dot{\hat{r}} = \dot{r}\hat{r} + r(\dot{\theta}\hat{\theta} + \dot{\phi}\sin\theta\hat{\phi}) = \dot{r}\hat{r} + r\dot{\theta}\hat{\theta} + r\dot{\phi}\sin\theta\hat{\phi} \tag{4.1.10}$$

The acceleration is given by differentiating Eq. (4.1.10) with respect to time,

$$\begin{aligned} \vec{a} &= \dot{\vec{v}} \\ &= \ddot{r}\hat{r} + \dot{r}\dot{\hat{r}} + \dot{r}\dot{\theta}\hat{\theta} + r\ddot{\theta}\hat{\theta} + r\dot{\theta}\dot{\hat{\theta}} + \dot{r}\dot{\phi}\sin\theta\hat{\phi} \\ &\quad + r\ddot{\phi}\sin\theta\hat{\phi} + r\dot{\phi}\dot{\theta}\cos\theta\hat{\phi} + r\dot{\phi}\sin\theta\dot{\hat{\phi}} \end{aligned} \tag{4.1.11}$$

By substituting Eqs. (4.1.9) into (4.1.11), we obtain the acceleration

$$\begin{aligned} \vec{a} &= (\ddot{r} - r\dot{\theta}^2 - r\dot{\phi}^2\sin^2\theta)\hat{r} + (2\dot{r}\dot{\theta} + r\ddot{\theta} - r\dot{\phi}^2\sin\theta\cos\theta)\hat{\theta} \\ &\quad + (2\dot{r}\dot{\phi}\sin\theta + 2r\dot{\theta}\dot{\phi}\cos\theta + r\ddot{\phi}\sin\theta)\hat{\phi} \end{aligned}$$

Therefore, the radial component of the acceleration is given by

$$a_r = \ddot{r} - r\dot{\theta}^2 - r\dot{\phi}^2\sin^2\theta$$

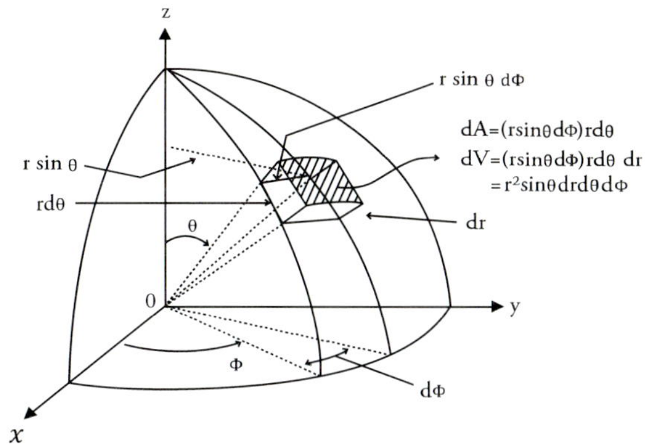

Fig. 4.1.1 A volume element dV in the spherical coordinate system.

The volume element as shown in Fig. 4.1.1 is

$$dV = r^2 \sin\theta dr d\theta d\phi = h_1 h_2 h_3 dr d\theta d\phi, \tag{4.1.12}$$

where h_1, h_2 and h_3 are the scale factors in the spherical coordinate system. The **solid angle** Ω subtended by a surface A is defined as $\Omega = \frac{A}{r^2}$ (measured in **steradians**). As shown in Fig. 4.1.2 an elemental surface area is $dA = rd\theta \times r\sin\theta d\phi$.

If we project the surface onto a sphere of radius r, it gives the differential solid angle

$$d\Omega = \frac{dA}{r^2} = \frac{r^2 \sin\theta d\theta d\phi}{r^2} = \sin\theta d\theta d\phi$$

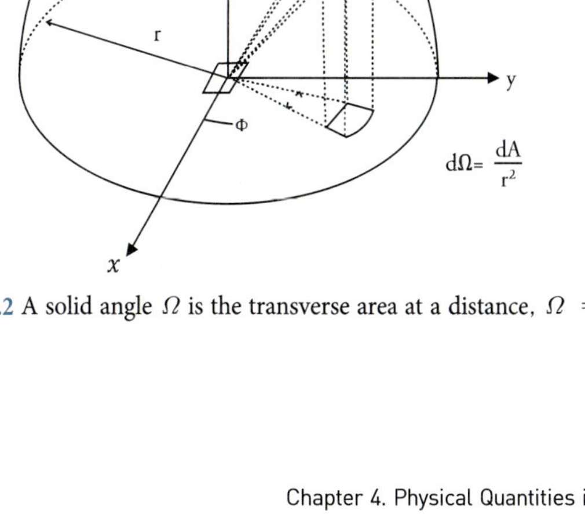

Fig 4.1.2 A solid angle Ω is the transverse area at a distance, $\Omega = \frac{Area}{r^2}$.

If we integrate the differential solid angle, we obtain the solid angle subtended by a whole sphere is

$$\iint d\Omega = \int_0^{\pi} \sin\theta d\theta \int_0^{2\pi} d\phi = -2\pi\cos\theta|_{\theta=0}^{\theta=\pi} = 4\pi$$

Example

4.1.1 Calculate the volume V of a sphere with radius a.

Ans. $V = \int dV = \int_{r=0}^{a}\int_{\theta=0}^{\pi}\int_{\phi=0}^{2\pi} r^2 \sin\theta dr d\theta d\phi$

$= 2\pi \int_0^{\pi} \sin\theta d\theta \int_0^a r^2 dr = 2\pi[-\cos\theta]_0^{\pi}\frac{1}{3}[r^3]_0^a = \frac{4}{3}\pi a^3$.

4.1.2 Find a motion in a plane of a particle with mass m moving under the central force field $\vec{F} = F(r)\hat{r}$.

Ans. Let's start with Newton's second law $\vec{F} = m\vec{a}$ to obtain the equation of the motion and then we have

$$F(r)\hat{r} = m\ddot{\vec{r}} = m(\ddot{r}\hat{r} + 2\dot{r}\dot{\hat{r}} + r\ddot{\hat{r}}) \quad (4.1.2.1)$$

We need to calculate $\dot{\hat{r}}$ and $\ddot{\hat{r}}$ to solve this equation. In this case, it is easy to represent its motion in the plane polar coordinates, $x = r\cos\theta$ and $y = r\sin\theta$, which give

$$dx = \cos\theta dr - r\sin\theta d\theta \text{ and } dy = \sin\theta dr + r\cos\theta d\theta \quad (4.1.2.2)$$

This gives

$$ds^2 = dx^2 + dy^2 = dr^2 + r^2 d\theta^2,$$

which means

$$\vec{ds} = \hat{r}dr + \hat{\theta}rd\theta \quad (4.1.2.3)$$

If we insert Eq. (4.1.2.2) into $\vec{ds} = \hat{x}dx + \hat{y}dy$, we get

$$\vec{ds} = (\hat{x}\cos\theta + \hat{y}\sin\theta)dr + r(-\hat{x}\sin\theta + \hat{y}\cos\theta)d\theta \quad (4.1.2.4)$$

By comparing Eqs. (4.1.2.3) and (4.1.2.4), we obtain

$$\begin{cases} \hat{r} = \hat{x}\cos\theta + \hat{y}\sin\theta \\ \hat{\theta} = -\hat{x}\sin\theta + \hat{y}\cos\theta \end{cases} \quad (4.1.2.5)$$

By differentiating the equations with respect to time, we get

$$\begin{cases} \dot{\hat{r}} = (-\hat{x}\sin\theta + \hat{y}\cos\theta)\dot{\theta} = \dot{\theta}\hat{\theta} \\ \dot{\hat{\theta}} = -(\hat{x}\cos\theta + \hat{y}\sin\theta)\dot{\theta} = -\dot{\theta}\hat{r} \end{cases} \tag{4.1.2.6}$$

and

$$\ddot{\hat{r}} = \ddot{\theta}\hat{\theta} + \dot{\theta}\dot{\hat{\theta}} = \ddot{\theta}\hat{\theta} - \dot{\theta}^2\hat{r} \tag{4.1.2.7}$$

As expected, we can see that Eq. (4.1.2.6) is the same as Eq. (4.1.9) for the spherical coordinates with $\theta = \frac{\pi}{2}$ and $\phi \to \theta$.

We then obtain the following equation by substituting Eqs. (4.1.2.6) and (4.1.2.7) into Eq. (4.1.2.1):

$$\begin{aligned} F(r)\hat{r} &= m(\ddot{r}\hat{r} + 2\dot{r}\dot{\hat{r}} + r\ddot{\hat{r}}) \\ &= m\ddot{r}\hat{r} + 2m\dot{r}\dot{\theta}\hat{\theta} + mr(\ddot{\theta}\hat{\theta} - \dot{\theta}^2\hat{r}) \\ &= m(\ddot{r} - r\dot{\theta}^2)\hat{r} + m(2\dot{r}\dot{\theta} + r\ddot{\theta})\hat{\theta} \end{aligned}$$

Since each component of the vector of the left- and right-hand sides of the equation should be same, for the radial part it gives

$$m(\ddot{r} - r\dot{\theta}^2) = F(r),$$

which is the equation in [Ex. 1.5.3],
while for the angular part it gives

$$m(2\dot{r}\dot{\theta} + r\ddot{\theta}) = 0$$

The equation for the angular part is the same as $\frac{1}{r}\frac{d}{dt}(mr^2\dot{\theta}) = 0$. Therefore, $mr^2\dot{\theta}$ is the constant with respect to time and we can define this constant as L because it has an unit of the angular momentum: $mr^2\dot{\theta} = L$

If we define $r = \frac{1}{u}$ and ℓ as the angular momentum per unit mass, we get

$$\frac{1}{u^2}\dot{\theta} = r^2\dot{\theta} = \frac{L}{m} = \ell$$

This is the equation appeared in [Ex. 1.5.3].

4.2 Cylindrical Coordinate System

· scale factor

$h_1 = 1,\ h_2 = \rho$ and $h_3 = 1$

· velocity

$\vec{v} = \dot{\rho}\hat{\rho} + \rho\dot{\phi}\hat{\phi} + \dot{z}\hat{z}$

· acceleration

$\vec{a} = (\ddot{\rho} - \rho\dot{\phi}^2)\hat{\rho} + (2\dot{\rho}\dot{\phi} + \rho\ddot{\phi})\hat{\phi} + \ddot{z}\hat{z}$

· volume element

$dV = \rho d\rho d\phi dz$

Let's calculate the velocity and acceleration in the cylindrical coordinate system in a similar way as we did in the previous section for the spherical coordinate system.

Since $x = \rho\cos\phi$ and $y = \rho\sin\phi$ in the cylindrical coordinate system, shown in Fig. 3.1.4, their derivatives are

$$dx = \cos\phi d\rho - \rho\sin\phi d\phi \text{ and } dy = \sin\phi d\rho + \rho\cos\phi d\phi, \tag{4.2.1}$$

which give

$$ds^2 = dx^2 + dy^2 + dz^2 = d\rho^2 + \rho^2 d\phi^2 + dz^2 \tag{4.2.2}$$

Thus, we have

$$\vec{ds} = \hat{\rho}d\rho + \hat{\phi}\rho d\phi + \hat{z}dz \tag{4.2.3}$$

Then, the scale factors in the cylindrical coordinate system are 1, ρ and 1, respectively. We denote the scale factors in the cylindrical coordinate system as

$$h_1 = 1,\ h_2 = \rho \text{ and } h_3 = 1$$

If we insert Eq. (4.2.1) into $\vec{ds} = \hat{x}dx + \hat{y}dy + \hat{z}dz$, we have

$$\vec{ds} = (\hat{x}\cos\phi + \hat{y}\sin\phi)d\rho + \rho(-\hat{x}\sin\phi + \hat{y}\cos\phi)d\phi + \hat{z}dz \tag{4.2.4}$$

By comparing Eqs. (4.2.3) and (4.2.4), we get

$$\hat{\rho} = \hat{x}\cos\phi + \hat{y}\sin\phi \text{ and } \hat{\phi} = -\hat{x}\sin\phi + \hat{y}\cos\phi \tag{4.2.5}$$

The position vector in the cylindrical coordinates can be expressed as

$$\begin{aligned}\vec{r} &= x\hat{x}+y\hat{y}+z\hat{z} = \hat{x}\rho\cos\phi+\hat{y}\rho\sin\phi+\hat{z}z \\ &= \rho(\hat{x}\cos\phi+\hat{y}\sin\phi)+\hat{z}z = \rho\hat{\rho}+z\hat{z}\end{aligned} \tag{4.2.6}$$

using Eq. (4.2.5).

The velocity is then given by

$$\vec{v} = \dot{\vec{r}} = \dot{\rho}\hat{\rho}+\rho\dot{\hat{\rho}}+\dot{z}\hat{z}+z\dot{\hat{z}} = \dot{\rho}\hat{\rho}+\rho\dot{\hat{\rho}}+\dot{z}\hat{z} \tag{4.2.7}$$

because $\dot{\hat{z}} = 0$.

If we differentiate Eq. (4.2.5) with respect to time, we have

$$\dot{\hat{\rho}} = (-\hat{x}\sin\phi+\hat{y}\cos\phi)\dot{\phi} = \dot{\phi}\hat{\phi} \quad \text{and} \quad \dot{\hat{\phi}} = -(\hat{x}\cos\phi+\hat{y}\sin\phi)\dot{\phi} = -\dot{\phi}\hat{\rho} \tag{4.2.8}$$

By inserting Eqs. (4.2.8) into (4.2.7) for $\dot{\hat{\rho}}$, we have

$$\vec{v} = \dot{\vec{r}} = \dot{\rho}\hat{\rho}+\rho\dot{\phi}\hat{\phi}+\dot{z}\hat{z} \tag{4.2.9}$$

By differentiating Eq. (4.2.9) with respect to time, it gives the acceleration

$$\begin{aligned}\vec{a} &= \ddot{\vec{r}} = \ddot{\rho}\hat{\rho}+\dot{\rho}\dot{\hat{\rho}}+\dot{\rho}\dot{\phi}\hat{\phi}+\rho\ddot{\phi}\hat{\phi}+\rho\dot{\phi}\dot{\hat{\phi}}+\ddot{z}\hat{z} \\ &= (\ddot{\rho}-\rho\dot{\phi}^2)\hat{\rho}+(2\dot{\rho}\dot{\phi}+\rho\ddot{\phi})\hat{\phi}+\ddot{z}\hat{z}\end{aligned}$$

Since volume element is given by $dV = h_1h_2h_3 d\rho d\phi dz$, we have

$$dV = \rho d\rho d\phi dz$$

in the cylindrical coordinate system.

Example

4.2.1 Calculate the volume of a cylinder with radius a and height h.

Ans. $$\begin{aligned}V &= \int dV = \int_{\rho=0}^{a}\int_{\phi=0}^{2\pi}\int_{z=0}^{h} \rho d\rho d\phi dz \\ &= 2\pi h\int_0^a \rho d\rho = 2\pi h\frac{1}{2}[\rho^2]_0^a = \pi a^2 h\end{aligned}$$

$$\cdot\ \overrightarrow{\nabla}\Psi = \frac{1}{h_1}\frac{\partial \Psi}{\partial q_1}\hat{q_1} + \frac{1}{h_2}\frac{\partial \Psi}{\partial q_2}\hat{q_2} + \frac{1}{h_3}\frac{\partial \Psi}{\partial q_3}\hat{q_3}$$

$$\cdot\ \overrightarrow{\nabla}\cdot\vec{f} = \frac{1}{h_1h_2h_3}\left[\frac{\partial}{\partial q_1}(h_2h_3f_1) + \frac{\partial}{\partial q_2}(h_3h_1f_2) + \frac{\partial}{\partial q_3}(h_1h_2f_3)\right]$$

$$\cdot\ \overrightarrow{\nabla}\times\vec{f} = \frac{1}{h_1h_2h_3}\begin{vmatrix} h_1\hat{q_1} & h_2\hat{q_2} & h_3\hat{q_3} \\ \dfrac{\partial}{\partial q_1} & \dfrac{\partial}{\partial q_2} & \dfrac{\partial}{\partial q_3} \\ h_1f_1 & h_2f_2 & h_3f_3 \end{vmatrix}$$

$$\cdot\ \nabla^2\Psi = \frac{1}{h_1h_2h_3}\left[\frac{\partial}{\partial q_1}\left(h_2h_3\frac{1}{h_1}\frac{\partial \Psi}{\partial q_1}\right) + \frac{\partial}{\partial q_2}\left(h_3h_1\frac{1}{h_2}\frac{\partial \Psi}{\partial q_2}\right) + \frac{\partial}{\partial q_3}\left(h_1h_2\frac{1}{h_3}\frac{\partial \Psi}{\partial q_3}\right)\right]$$

Let's try to express the rectangular, spherical and cylindrical coordinate systems in one general coordinate called **curvilinear coordinate system**, let's say, whose coordinates are q_1, q_2 and q_3, which do not need to be orthogonal to each other. But, we only consider for the case of q_1, q_2, and q_3 to be orthogonal to each other because all coordinate systems we are interested in are orthogonal.

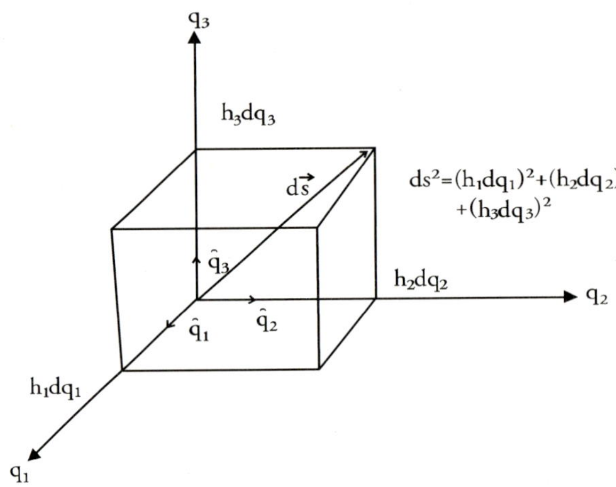

Fig. 4.3.1 $\vec{ds} = h_1dq_1\hat{q_1} + h_2dq_2\hat{q_2} + h_3dq_3\hat{q_3}$ in q_1, q_2, and q_3 of the curvilinear coordinate system.

Since a small change of a position vector in this new coordinate system is $\vec{ds} = h_1 dq_1 \hat{q_1} + h_2 dq_2 \hat{q_2} + h_3 dq_3 \hat{q_3}$ with scale factors h_i $(i = 1, 2, 3)$, shown in Fig. 4.3.1, the distance square from the corner to the opposite corner across the cube is $ds^2 = h_1^2 dq_1^2 + h_2^2 dq_2^2 + h_3^2 dq_3^2$.

First, we consider how the gradient of a function is expressed in the curvilinear coordinate system.

(a) q_1 component of the gradient of a scalar function Ψ is expressed as

$$(\overrightarrow{\nabla \Psi})_{q_1} = \hat{q_1} \cdot (\overrightarrow{\nabla \Psi})$$

$$= \lim_{dq_1 \to 0} \frac{\Psi(A) - \Psi(0)}{h_1 dq_1} = \frac{1}{h_1} \lim_{dq_1 \to 0} \frac{\Psi(A) - \Psi(0)}{dq_1} = \frac{1}{h_1} \frac{\partial \Psi}{\partial q_1},$$

where A is the distance from the origin in q_1 direction.

Similarly, we have for q_2 and q_3 components

$$(\overrightarrow{\nabla \Psi})_{q_2} = \frac{1}{h_2} \frac{\partial \Psi}{\partial q_2} \text{ and } (\overrightarrow{\nabla \Psi})_{q_3} = \frac{1}{h_3} \frac{\partial \Psi}{\partial q_3}$$

Then, we obtain the gradient of a scalar function Ψ in the curvilinear coordinate to be

$$\overrightarrow{\nabla \Psi} = (\overrightarrow{\nabla \Psi})_{q_1} \hat{q_1} + (\overrightarrow{\nabla \Psi})_{q_2} \hat{q_2} + (\overrightarrow{\nabla \Psi})_{q_3} \hat{q_3}$$

$$= \frac{1}{h_1} \frac{\partial \Psi}{\partial q_1} \hat{q_1} + \frac{1}{h_2} \frac{\partial \Psi}{\partial q_2} \hat{q_2} + \frac{1}{h_3} \frac{\partial \Psi}{\partial q_3} \hat{q_3} \tag{4.3.1}$$

Let's apply Eq. (4.3.1) into the rectangular, spherical and cylindrical coordinate systems.

(i) For the rectangular coordinates, we have $h_1 = h_2 = h_3 = 1$, $q_1 = x$, $q_2 = y$ and $q_3 = z$. This gives

$$\overrightarrow{\nabla \Psi} = \frac{\partial \Psi}{\partial x} \hat{x} + \frac{\partial \Psi}{\partial y} \hat{y} + \frac{\partial \Psi}{\partial z} \hat{z}$$

(ii) For the spherical coordinates, we have $h_1 = 1$, $h_2 = r$, $h_3 = r\sin\theta$, $q_1 = r$, $q_2 = \theta$ and $q_3 = \phi$. This gives

$$\overrightarrow{\nabla \Psi} = \frac{\partial \Psi}{\partial r} \hat{r} + \frac{1}{r} \frac{\partial \Psi}{\partial \theta} \hat{\theta} + \frac{1}{r\sin\theta} \frac{\partial \Psi}{\partial \phi} \hat{\phi}$$

(iii) For the cylindrical coordinates, we have $h_1 = 1$, $h_2 = \rho$, $h_3 = 1$, $q_1 = \rho$, $q_2 = \phi$ and $q_3 = z$, This gives

$$\vec{\nabla}\Psi = \frac{\partial \Psi}{\partial \rho}\hat{\rho} + \frac{1}{\rho}\frac{\partial \Psi}{\partial \phi}\hat{\phi} + \frac{\partial \Psi}{\partial z}\hat{z}$$

Secondly, we consider how the divergence of a vector function is expressed in the curvilinear coordinate system.

(b) The divergence of a vector function $\vec{f}$ is expressed as

$$\vec{\nabla}\cdot\vec{f} = (\vec{\nabla}\cdot\vec{f})_{q_1} + (\vec{\nabla}\cdot\vec{f})_{q_2} + (\vec{\nabla}\cdot\vec{f})_{q_3}$$

Since $\vec{f} = f_1\hat{q_1} + f_2\hat{q_2} + f_3\hat{q_3} = h_2h_3f_1\frac{\hat{q_1}}{h_2h_3} + h_3h_1f_2\frac{\hat{q_2}}{h_3h_1} + h_1h_2f_3\frac{\hat{q_3}}{h_1h_2}$, the first term of the right-hand side of the equation in the divergence of a vector function $\vec{f}$ can be expressed as

$$\begin{aligned}(\vec{\nabla}\cdot\vec{f})_{q_1} &= \vec{\nabla}\cdot\left(h_2h_3f_1\frac{\hat{q_1}}{h_2h_3}\right)\\ &= \frac{\hat{q_1}}{h_2h_3}\cdot\vec{\nabla}(h_2h_3f_1) + h_2h_3f_1\vec{\nabla}\cdot\left(\frac{\hat{q_1}}{h_2h_3}\right)\end{aligned} \tag{4.3.2}$$

Let's consider $\vec{\nabla}q_1$ to calculate the second term on the right-hand side of Eq. (4.3.2). By substituting q_1 instead of Ψ in Eq. (4.3.1) we obtain

$$\vec{\nabla}q_1 = \frac{1}{h_1}\frac{\partial q_1}{\partial q_1}\hat{q_1} + \frac{1}{h_2}\frac{\partial q_1}{\partial q_2}\hat{q_2} + \frac{1}{h_3}\frac{\partial q_1}{\partial q_3}\hat{q_3} = \frac{1}{h_1}\hat{q_1}$$

Similarly, we get

$$\vec{\nabla}q_2 = \frac{1}{h_2}\hat{q_2} \text{ and } \vec{\nabla}q_3 = \frac{1}{h_3}\hat{q_3}$$

Then, it gives

$$\vec{\nabla}q_1 \times \vec{\nabla}q_2 = \frac{1}{h_1}\hat{q_1} \times \frac{1}{h_2}\hat{q_2} = \frac{1}{h_1h_2}\hat{q_1}\times\hat{q_2} = \frac{1}{h_1h_2}\hat{q_3} \tag{4.3.3}$$

If we take the divergence of Eq. (4.3.3), it gives

$$\vec{\nabla}\cdot(\vec{\nabla}q_1 \times \vec{\nabla}q_2) = \frac{1}{h_1h_2}\vec{\nabla}\cdot\hat{q_3} \tag{4.3.4}$$

Since the divergence of the curl is zero, the left-hand side of the equation is zero. Thus, Eq. (4.3.4) is

$$\frac{1}{h_1 h_2}\vec{\nabla}\cdot\hat{q_3}=0 \tag{4.3.5}$$

Similarly, we obtain

$$\frac{1}{h_3 h_1}\vec{\nabla}\cdot\hat{q_2}=0 \text{ and } \frac{1}{h_2 h_3}\vec{\nabla}\cdot\hat{q_1}=0 \tag{4.3.6}$$

Therefore, the second term on the right-hand side of Eq. (4.3.2) is zero. Eq. (4.3.2) is then

$$\begin{aligned}(\vec{\nabla}\cdot\vec{f})_{q_1} &= \frac{\hat{q_1}}{h_2 h_3}\cdot\vec{\nabla}(h_2 h_3 f_1) \\ &= \frac{1}{h_2 h_3}[\vec{\nabla}(h_2 h_3 f_1)]_{q_1} \\ &= \frac{1}{h_2 h_3}[\frac{1}{h_1}\frac{\partial}{\partial q_1}(h_2 h_3 f_1)] \quad \text{(using Eq. (4.3.1))} \\ &= \frac{1}{h_1 h_2 h_3}\frac{\partial}{\partial q_1}(h_2 h_3 f_1)\end{aligned}$$

Similarly, we can obtain q_2 and q_3 components of the divergence of a vector function f to be

$$(\vec{\nabla}\cdot\vec{f})_{q_2} = \frac{1}{h_1 h_2 h_3}\frac{\partial}{\partial q_2}(h_3 h_1 f_2)$$

and

$$(\vec{\nabla}\cdot\vec{f})_{q_3} = \frac{1}{h_1 h_2 h_3}\frac{\partial}{\partial q_3}(h_1 h_2 f_3),$$

respectively.

Since $\vec{\nabla}\cdot\vec{f}=(\vec{\nabla}\cdot\vec{f})_{q_1}+(\vec{\nabla}\cdot\vec{f})_{q_2}+(\vec{\nabla}\cdot\vec{f})_{q_3}$, we obtain the divergence of a vector function f in the curvilinear coordinate as

$$\begin{aligned}\vec{\nabla}\cdot\vec{f} &= \frac{1}{h_1 h_2 h_3}\frac{\partial}{\partial q_1}(h_2 h_3 f_1) \\ &+ \frac{1}{h_1 h_2 h_3}\frac{\partial}{\partial q_2}(h_3 h_1 f_2)+\frac{1}{h_1 h_2 h_3}\frac{\partial}{\partial q_3}(h_1 h_2 f_3)\end{aligned} \tag{4.3.7}$$

Let's apply Eq. (4.3.7) into the rectangular, spherical and cylindrical coordinate systems.

(i) For the rectangular coordinates, we have $h_1 = h_2 = h_3 = 1$, and $q_1 = x$, $q_2 = y$, $q_3 = z$ and $f_1 = f_x$, $f_2 = f_y$ and $f_3 = f_z$.

This gives

$$\vec{\nabla} \cdot \vec{f} = \frac{\partial}{\partial x} f_x + \frac{\partial}{\partial y} f_y + \frac{\partial}{\partial z} f_z$$

(ii) For the spherical coordinates, we have $h_1 = 1$, $h_2 = r$, $h_3 = r\sin\theta$, and $q_1 = r$, $q_2 = \theta$, $q_3 = \phi$ and $f_1 = f_r$, $f_2 = f_\theta$ and $f_3 = f_\phi$.

This gives

$$\begin{aligned}\vec{\nabla} \cdot \vec{f} &= \frac{1}{r^2\sin\theta}\frac{\partial}{\partial r}(r^2\sin\theta f_r) + \frac{1}{r^2\sin\theta}\frac{\partial}{\partial \theta}(r\sin\theta f_\theta) + \frac{1}{r^2\sin\theta}\frac{\partial}{\partial \phi}(r f_\phi) \\ &= \frac{1}{r^2\sin\theta}[\sin\theta\frac{\partial}{\partial r}(r^2 f_r) + r\frac{\partial}{\partial \theta}(\sin\theta f_\theta) + r\frac{\partial}{\partial \phi} f_\phi]\end{aligned}$$

(iii) For the cylindrical coordinates, we have $h_1 = 1$, $h_2 = \rho$, $h_3 = 1$, and

$q_1 = \rho$, $q_2 = \phi$, $q_3 = z$, $f_1 = f_\rho$, $f_2 = f_\phi$ and $f_3 = f_z$.

This gives

$$\begin{aligned}\vec{\nabla} \cdot \vec{f} &= \frac{1}{\rho}[\frac{\partial}{\partial \rho}(\rho f_\rho) + \frac{\partial}{\partial \phi} f_\phi + \frac{\partial}{\partial z}(\rho f_z)] \\ &= \frac{1}{\rho}[\frac{\partial}{\partial \rho}(\rho f_\rho) + \frac{\partial}{\partial \phi} f_\phi + \rho\frac{\partial}{\partial z} f_z]\end{aligned}$$

Now, we consider how the curl of a vector function is expressed in the curvilinear coordinate system.

(c) The curl of a vector function $\vec{f}$ is $\vec{\nabla}\times\vec{f}$. The q_1 component of $\vec{\nabla}\times\vec{f}$ is expressed as

$$(\vec{\nabla}\times\vec{f})_{q_1} = \vec{\nabla}_{q_2} f_3 - \vec{\nabla}_{q_3} f_2,$$

where the first and second terms on the right-hand side of the equation can be expressed by $(\vec{\nabla} f_3)_{q_2}$ and $(\vec{\nabla} f_2)_{q_3}$ using Eq. (4.3.1).

Therefore, the q_1 component of $\vec{\nabla}\times\vec{f}$ is then

$$(\vec{\nabla}\times\vec{f})_{q_1}=(\vec{\nabla}f_3)_{q_2}-(\vec{\nabla}f_2)_{q_3}=\frac{1}{h_2}\frac{\partial f_3}{\partial q_2}-\frac{1}{h_3}\frac{\partial f_2}{\partial q_3}$$

$$=\frac{1}{h_2h_3}[\frac{\partial(h_3f_3)}{\partial q_2}-\frac{\partial(h_2f_2)}{\partial q_3}]$$

Similarly, we obtain the q_2 and q_3 components of $\vec{\nabla}\times\vec{f}$ are

$$\frac{1}{h_3h_1}[\frac{\partial(h_1f_1)}{\partial q_3}-\frac{\partial(h_3f_3)}{\partial q_1}] \text{ and } \frac{1}{h_1h_2}[\frac{\partial(h_2f_2)}{\partial q_1}-\frac{\partial(h_1f_1)}{\partial q_2}],$$

respectively.

Thus, we obtain the curl of a vector function $\vec{f}$ is

$$\vec{\nabla}\times\vec{f}=\frac{\hat{q_1}}{h_2h_3}[\frac{\partial(h_3f_3)}{\partial q_2}-\frac{\partial(h_2f_2)}{\partial q_3}]+\frac{\hat{q_2}}{h_3h_1}[\frac{\partial(h_1f_1)}{\partial q_3}-\frac{\partial(h_3f_3)}{\partial q_1}]$$

$$+\frac{\hat{q_3}}{h_1h_2}[\frac{\partial(h_2f_2)}{\partial q_1}-\frac{\partial(h_1f_1)}{\partial q_2}] \qquad (4.3.8)$$

This equation can be expressed using the determinant:

$$\vec{\nabla}\times\vec{f}=\frac{1}{h_1h_2h_3}\begin{vmatrix} h_1\hat{q_1} & h_2\hat{q_2} & h_3\hat{q_3} \\ \frac{\partial}{\partial q_1} & \frac{\partial}{\partial q_2} & \frac{\partial}{\partial q_3} \\ h_1f_1 & h_2f_2 & h_3f_3 \end{vmatrix} \qquad (4.3.9)$$

Let's apply Eq. (4.3.9) into the rectangular, spherical and cylindrical coordinate systems.

(i) For the rectangular coordinates, we have $h_1=h_2=h_3=1$, and $q_1=x$, $q_2=y$, $q_3=z$, $f_1=f_x$, $f_2=f_y$ and $f_3=f_z$.

This gives

$$\vec{\nabla}\times\vec{f}=\begin{vmatrix} \hat{x} & \hat{y} & \hat{z} \\ \frac{\partial}{\partial x} & \frac{\partial}{\partial y} & \frac{\partial}{\partial z} \\ f_x & f_y & f_z \end{vmatrix}$$

$$=(\frac{\partial f_z}{\partial y}-\frac{\partial f_y}{\partial z})\hat{x}+(\frac{\partial f_x}{\partial z}-\frac{\partial f_z}{\partial x})\hat{y}+(\frac{\partial f_y}{\partial x}-\frac{\partial f_x}{\partial y})\hat{z}$$

(ii) For the spherical coordinates, we have $h_1 = 1$, $h_2 = r$, $h_3 = r\sin\theta$, and $q_1 = r$, $q_2 = \theta$, $q_3 = \phi$, $f_1 = f_r$, $f_2 = f_\theta$ and $f_3 = f_\phi$.

This gives

$$\begin{aligned}
\vec{\nabla}\times\vec{f} &= \frac{1}{r^2\sin\theta}\begin{vmatrix} \hat{r} & r\hat{\theta} & r\sin\theta\hat{\phi} \\ \frac{\partial}{\partial r} & \frac{\partial}{\partial \theta} & \frac{\partial}{\partial \phi} \\ f_r & rf_\theta & r\sin\theta f_\phi \end{vmatrix} \\
&= \frac{1}{r^2\sin\theta}[(\frac{\partial(r\sin\theta f_\phi)}{\partial\theta} - \frac{\partial(rf_\theta)}{\partial\phi})\hat{r} \\
&\quad + (\frac{\partial f_r}{\partial\phi} - \frac{\partial(r\sin\theta f_\phi)}{\partial r})r\hat{\theta} \\
&\quad + (\frac{\partial(rf_\theta)}{\partial r} - \frac{\partial f_r}{\partial\theta})r\sin\theta\hat{\phi}] \\
&= \frac{1}{r\sin\theta}[\frac{\partial(f_\phi\sin\theta)}{\partial\theta} - \frac{\partial f_\theta}{\partial\phi}]\hat{r} \\
&\quad + \frac{1}{r\sin\theta}[\frac{\partial f_r}{\partial\phi} - \sin\theta\frac{\partial(rf_\phi)}{\partial r}]\hat{\theta} \\
&\quad + \frac{1}{r}[\frac{\partial(rf_\theta)}{\partial r} - \frac{\partial f_r}{\partial\theta}]\hat{\phi}
\end{aligned}$$

(iii) For the cylindrical coordinates, we have $h_1 = 1$, $h_2 = \rho$, $h_3 = 1$, and $q_1 = \rho$, $q_2 = \phi$, $q_3 = z$, $f_1 = f_\rho$, $f_2 = f_\phi$ and $f_3 = f_z$.

This gives

$$\begin{aligned}
\vec{\nabla}\times\vec{f} &= \frac{1}{\rho}\begin{vmatrix} \hat{\rho} & \rho\hat{\phi} & \hat{z} \\ \frac{\partial}{\partial\rho} & \frac{\partial}{\partial\phi} & \frac{\partial}{\partial z} \\ f_\rho & \rho f_\phi & f_z \end{vmatrix} \\
&= \frac{1}{\rho}[(\frac{\partial f_z}{\partial\phi} - \frac{\partial(\rho f_\phi)}{\partial z})\hat{\rho} + (\frac{\partial f_\rho}{\partial z} - \frac{\partial f_z}{\partial\rho})\rho\hat{\phi} + (\frac{\partial(\rho f_\phi)}{\partial\rho} - \frac{\partial f_\rho}{\partial\phi})\hat{z}] \\
&= (\frac{1}{\rho}\frac{\partial f_z}{\partial\phi} - \frac{\partial f_\phi}{\partial z})\hat{\rho} + (\frac{\partial f_\rho}{\partial z} - \frac{\partial f_z}{\partial\rho})\hat{\phi} + \frac{1}{\rho}(\frac{\partial(\rho f_\phi)}{\partial\rho} - \frac{\partial f_\rho}{\partial\phi})\hat{z}
\end{aligned}$$

The Laplacian in the curvilinear coordinate system can be easily obtained by using the previously obtained results for the gradient and divergence.

(d) For the Laplacian in the curvilinear coordinate system, we have

$$\nabla^2\Psi = \vec{\nabla}\cdot\vec{\nabla}\Psi$$
$$= \frac{1}{h_1h_2h_3}[\frac{\partial}{\partial q_1}(h_2h_3(\vec{\nabla}\Psi)_{q_1}) + \frac{\partial}{\partial q_2}(h_3h_1(\vec{\nabla}\Psi)_{q_2})$$
$$+ \frac{\partial}{\partial q_3}(h_1h_2(\vec{\nabla}\Psi)_{q_3})], \qquad (4.3.10)$$

by substituting $\vec{\nabla}\Psi$ instead of $\vec{f}$ in Eq. (4.3.7), where $(\vec{\nabla}\Psi)_{q_1}$ is the q_1 component of the gradient of a function Ψ, therefore, $(\vec{\nabla}\Psi)_{q_1} = \frac{1}{h_1}\frac{\partial\Psi}{\partial q_1}$ from Eq. (4.3.1). Similarly, we have $(\vec{\nabla}\Psi)_{q_2} = \frac{1}{h_2}\frac{\partial\Psi}{\partial q_2}$ and $(\vec{\nabla}\Psi)_{q_3} = \frac{1}{h_3}\frac{\partial\Psi}{\partial q_3}$. Thus, Eq. (4.3.10) is given by

$$\nabla^2\Psi = \frac{1}{h_1h_2h_3}[\frac{\partial}{\partial q_1}(h_2h_3\frac{1}{h_1}\frac{\partial\Psi}{\partial q_1})$$
$$+ \frac{\partial}{\partial q_2}(h_3h_1\frac{1}{h_2}\frac{\partial\Psi}{\partial q_2}) + \frac{\partial}{\partial q_3}(h_1h_2\frac{1}{h_3}\frac{\partial\Psi}{\partial q_3})] \qquad (4.3.11)$$

Let's apply Eq. (4.3.11) into the rectangular, spherical and cylindrical coordinate systems.

(i) For the rectangular coordinates, we have $h_1 = h_2 = h_3 = 1$ and $q_1 = x$, $q_2 = y$ and $q_3 = z$.

This gives

$$\nabla^2\Psi = \frac{\partial}{\partial x}(\frac{\partial\Psi}{\partial x}) + \frac{\partial}{\partial y}(\frac{\partial\Psi}{\partial y}) + \frac{\partial}{\partial z}(\frac{\partial\Psi}{\partial z})$$
$$= \frac{\partial^2\Psi}{\partial x^2} + \frac{\partial^2\Psi}{\partial y^2} + \frac{\partial^2\Psi}{\partial z^2}$$

(ii) For the spherical coordinates, we have $h_1 = 1$, $h_2 = r$, $h_3 = r\sin\theta$, and $q_1 = r$, $q_2 = \theta$ and $q_3 = \phi$.

This gives

$$\nabla^2\Psi = \frac{1}{r^2\sin\theta}[\frac{\partial}{\partial r}(r^2\sin\theta\frac{\partial\Psi}{\partial r}) + \frac{\partial}{\partial\theta}(\sin\theta\frac{\partial\Psi}{\partial\theta}) + \frac{\partial}{\partial\phi}(\frac{1}{\sin\theta}\frac{\partial\Psi}{\partial\phi})]$$
$$= \frac{1}{r^2}\frac{\partial}{\partial r}(r^2\frac{\partial\Psi}{\partial r}) + \frac{1}{r^2\sin\theta}\frac{\partial}{\partial\theta}(\sin\theta\frac{\partial\Psi}{\partial\theta}) + \frac{1}{r^2\sin^2\theta}\frac{\partial^2\Psi}{\partial\phi^2} \qquad (4.3.12)$$

(iii) For the cylindrical coordinates, we have $h_1 = 1$, $h_2 = \rho$, $h_3 = 1$, and $q_1 = \rho$, $q_2 = \phi$ and $q_3 = z$.

This gives

$$\begin{aligned}\nabla^2\Psi &= \frac{1}{\rho}[\frac{\partial}{\partial\rho}(\rho\frac{\partial\Psi}{\partial\rho}) + \frac{\partial}{\partial\phi}(\frac{1}{\rho}\frac{\partial\Psi}{\partial\phi}) + \frac{\partial}{\partial z}(\rho\frac{\partial\Psi}{\partial z})] \\ &= \frac{1}{\rho}\frac{\partial}{\partial\rho}(\rho\frac{\partial\Psi}{\partial\rho}) + \frac{1}{\rho^2}\frac{\partial^2\Psi}{\partial\phi^2} + \frac{\partial^2\Psi}{\partial z^2}\end{aligned}$$

Table 4.1 Summary of coordinate systems.

Coordinate systems	Coordinates			Scale factors		
	q_1	q_2	q_3	h_1	h_2	h_3
rectangular	x	y	z	1	1	1
spherical	r	θ	ϕ	1	r	$r\sin\theta$
cylindrical	ρ	ϕ	z	1	ρ	1

Example

4.3.1 Represent the following eigenvalue equation,

$$[\frac{p_{op}^2}{2m} + U(r)]\Psi(r,\theta,\phi) = E\Psi(r,\theta,\phi)$$

where $\vec{p}_{op} = \frac{\hbar}{i}\vec{\nabla}$ is the momentum operator, in the spherical coordinate system.

Ans. By inserting $\vec{p}_{op} = \frac{\hbar}{i}\vec{\nabla}$, the equation is expressed as

$$-\frac{\hbar^2}{2m}\nabla^2\Psi + U\Psi = E\Psi \qquad (4.3.1.1)$$

If we substituting Eqs. (4.3.12) into (4.3.1.1), we get

$$-\frac{\hbar^2}{2m}[\frac{1}{r^2}\frac{\partial}{\partial r}(r^2\frac{\partial}{\partial r}) + \frac{1}{r^2\sin\theta}\frac{\partial}{\partial\theta}(\sin\theta\frac{\partial}{\partial\theta}) + \frac{1}{r^2\sin^2\theta}\frac{\partial^2}{\partial\phi^2}]\Psi + U\Psi = E\Psi$$

4.3.2 Represent the angular momentum $\vec{L}$ in the spherical coordinate system.

Ans. Since $\vec{L}_{op} = \vec{r} \times \vec{p}_{op}$ and $\vec{p}_{op} = \frac{\hbar}{i}\vec{\nabla}$ in Quantum Mechanics, the angular momentum is

$$\vec{L}_{op} = \vec{r} \times \vec{p}_{op} = -i\hbar\vec{r} \times \vec{\nabla}, \qquad (4.3.2.1)$$

where $\vec{r} = r\hat{r}$, so that

$$\vec{r} \times \vec{\nabla} = \frac{1}{r^2\sin\theta}\begin{vmatrix} \hat{r} & r\hat{\theta} & r\sin\theta\hat{\phi} \\ r & 0 & 0 \\ \frac{\partial}{\partial r} & \frac{\partial}{\partial\theta} & \frac{\partial}{\partial\phi} \end{vmatrix}$$

$$= \frac{1}{\sin\theta}(-\hat{\theta}\frac{\partial}{\partial\phi} + \hat{\phi}\sin\theta\frac{\partial}{\partial\theta}) \qquad (4.3.2.2)$$

By substituting Eqs. (4.3.2.2) into (4.3.2.1), we obtain the angular momentum is

$$\vec{L}_{op} = -i\hbar\frac{1}{\sin\theta}(-\hat{\theta}\frac{\partial}{\partial\phi} + \hat{\phi}\sin\theta\frac{\partial}{\partial\theta}) \qquad (4.3.2.3)$$

This equation appears very often as the angular momentum operator in problems of Quantum Mechanics.

Then, we have

$$L_{op}^2 = \vec{L}_{op} \cdot \vec{L}_{op}$$

$$= -\hbar^2\frac{1}{\sin^2\theta}(-\hat{\theta}\frac{\partial}{\partial\phi} + \hat{\phi}\sin\theta\frac{\partial}{\partial\theta}) \cdot (-\hat{\theta}\frac{\partial}{\partial\phi} + \hat{\phi}\sin\theta\frac{\partial}{\partial\theta})$$

$$= -\hbar^2\frac{1}{\sin^2\theta}[\frac{\partial^2}{\partial\phi^2} + \sin\theta\frac{\partial}{\partial\theta}(\sin\theta\frac{\partial}{\partial\theta})]$$

$$= -\hbar^2[\frac{1}{\sin\theta}\frac{\partial}{\partial\theta}(\sin\theta\frac{\partial}{\partial\theta}) + \frac{1}{\sin^2\theta}\frac{\partial^2}{\partial\phi^2}]$$

Therefore, we obtain

$$L_{op}^2 = -\hbar^2[\frac{1}{\sin\theta}\frac{\partial}{\partial\theta}(\sin\theta\frac{\partial}{\partial\theta}) + \frac{1}{\sin^2\theta}\frac{\partial^2}{\partial\phi^2}]$$

4.3.3 Find z component of the angular momentum operator $\vec{L}_{op}$ in the spherical coordinate system.

Ans. To find z component of the angular momentum operator we have to calculate

$$(\vec{L}_{op})_z = \hat{z} \cdot \vec{L}_{op} \qquad (4.3.3.1)$$

If we multiply the first equation of Eq. (4.1.8) by $\cos\theta$, we get

$$\hat{r}\cos\theta = \hat{x}\sin\theta\cos\theta\cos\phi + \hat{y}\sin\theta\cos\theta\sin\phi + \hat{z}\cos^2\theta \qquad (4.3.3.2)$$

If we multiply the second equation of Eq. (4.1.8) by $\sin\theta$, we get

$$\hat{\theta}\sin\theta = \hat{x}\cos\theta\sin\theta\cos\phi + \hat{y}\cos\theta\sin\theta\sin\phi - \hat{z}\sin^2\theta \tag{4.3.3.3}$$

If we subtract Eqs. (4.3.3.3) from Eq. (4.3.3.2), it gives

$$\hat{r}\cos\theta - \hat{\theta}\sin\theta = \hat{z} \tag{4.3.3.4}$$

By substituting Eqs. (4.3.2.3) and (4.3.3.4) into (4.3.3.1), we get

$$\begin{aligned}(\vec{L}_{op})_z &= (\hat{r}\cos\theta - \hat{\theta}\sin\theta)\cdot\frac{-i\hbar}{\sin\theta}\left(-\hat{\theta}\frac{\partial}{\partial\phi} + \hat{\phi}\sin\theta\frac{\partial}{\partial\theta}\right)\\ &= -i\hbar\frac{\partial}{\partial\phi}\end{aligned}$$

Then, we have the following relationship for $\Phi(\phi) = e^{im\phi}$

$$\begin{aligned}L_z\Phi &= -i\hbar\frac{\partial}{\partial\phi}e^{im\phi} = -i\hbar\frac{d}{d\phi}e^{im\phi}\\ &= m\hbar e^{im\phi} = m\hbar\Phi\end{aligned}$$

4.3.4 By following a similar manner as the previous example, find L_x and L_y.

Ans. From Eq. (4.1.8), we get

$$\hat{x} = \hat{r}\sin\theta\cos\phi + \hat{\theta}\cos\theta\cos\phi - \hat{\phi}\sin\phi$$

and

$$\hat{y} = \hat{r}\sin\theta\sin\phi + \hat{\theta}\cos\theta\sin\phi + \hat{\phi}\cos\phi$$

Then, x and y components of the angular momentum operator are

$$\begin{aligned}L_x &= \hat{x}\cdot\vec{L}\\ &= (\hat{r}\sin\theta\cos\phi + \hat{\theta}\cos\theta\cos\phi - \hat{\phi}\sin\phi)\cdot\frac{-i\hbar}{\sin\theta}\left(-\hat{\theta}\frac{\partial}{\partial\phi} + \hat{\phi}\sin\theta\frac{\partial}{\partial\theta}\right)\\ &= i\hbar\left(\sin\phi\frac{\partial}{\partial\theta} + \cot\theta\cos\phi\frac{\partial}{\partial\phi}\right)\end{aligned}$$

and

$$\begin{aligned}L_y &= \hat{y}\cdot\vec{L}\\ &= (\hat{r}\sin\theta\sin\phi + \hat{\theta}\cos\theta\sin\phi + \hat{\phi}\cos\phi)\cdot\frac{-i\hbar}{\sin\theta}\left(-\hat{\theta}\frac{\partial}{\partial\phi} + \hat{\phi}\sin\theta\frac{\partial}{\partial\theta}\right)\\ &= i\hbar\left(-\cos\phi\frac{\partial}{\partial\theta} + \cot\theta\sin\phi\frac{\partial}{\partial\phi}\right),\end{aligned}$$

respectively.

Chapter 05 Infinite Series

It is very convenient to approximate a function with an infinite series in some approximation limit. This method provides a way to solve complicate integrals or differentiations in physics problems which can be approximated in the existing limit.

In Section 5.1 we discuss a method to test for convergence of an infinite series. Section 5.2 covers Taylor series which is a series expansion of a function about a point. For a multi-variable function, its expansion is described in Section 5.3.

5.1 Infinite Series

- For geometric series,

  $\sum_{n=1}^{\infty} ar^{n-1}$ converges if $|r| < 1$, otherwise it diverges

- For power series,

  $\sum_{n=1}^{\infty} \frac{1}{n^k}$ converges if $|k| > 1$, otherwise it diverges

- Ratio test

  $\sum a_n$ converges, diverges, and needs to use another test

  if $\lim_{n\to\infty} \left|\frac{a_{n+1}}{a_n}\right|$ is less, larger than, and equal to 1, respectively.

- L'Hopital's rule

  $\lim_{x\to a} \frac{f(x)}{g(x)} = \lim_{x\to a} \frac{f'(x)}{g'(x)}$ for $\lim_{x\to a} \frac{f(x)}{g(x)} = \frac{0}{0}$ or $\frac{\infty}{\infty}$

When a sequence of values in an infinite series is an addition of a common constant, let's say it is d, from the previous term, it is called an **arithmetic series**. If we define a partial sum as S_n, we have

$$S_n = a + (a+d) + (a+2d) + \cdots\cdots + [a+(n-1)d] = \frac{n}{2}(a_1 + a_n),$$

where a_1 is the first term and a_n is the nth term of the series.
Then, the infinite sum is $S = \lim_{n\to\infty} S_n$ and we get

$$S = \sum_{n=1}^{\infty} [a + (n-1)d]$$

For an infinite series

$$S = a + ar + ar^2 + ar^3 + \cdots\cdots + ar^{n-1} + \cdots\cdots = \sum_{n=1}^{\infty} ar^{n-1}, \qquad (5.1.1)$$

it is called a **geometric series** when the ratio $\frac{u_{n+1}}{u_n} = r$ (in here $u_n = ar^{n-1}$) is independent of n.
If we multiply the partial sum S_n of Eq. (5.1.1) by r, it gives

$$rS_n = ar + ar^2 + ar^3 + \cdots\cdots + ar^n \qquad (5.1.2)$$

By subtracting this equation from the partial sum, we get

$$S_n - rS_n = a - ar^n = a(1 - r^n)$$

This gives

$$S_n = \frac{a(1-r^n)}{1-r}$$

We then obtain

$$S = \lim_{n\to\infty} S_n = \lim_{n\to\infty} \frac{a(1-r^n)}{1-r}$$

$$= \frac{a}{1-r} \text{ for } |r| < 1$$

Therefore, the geometric series converges and diverges for $|r| < 1$ and $|r| \geq 1$, respectively.

On the other hand, it is called a **power series** when an infinite series is expressed as

$$S = a_0 + a_1x + a_2x^2 + a_3x^3 + \cdots\cdots + a_nx^n + \cdots\cdots = \sum_{n=0}^{\infty} a_nx^n,$$

where a_0, a_1, a_2, a_3, $\cdots$, a_n, $\cdots$ are constants.

For any infinite series, the series is called **convergent** when the partial sum S_n approaches a definite value as $n \to \infty$. The series is called **divergent** when the partial sum S_n does not tend to be a definite value as $n \to \infty$. When an infinite series satisfies $\lim_{n\to\infty}\left|\frac{a_{n+1}}{a_n}\right| < 1$, **d'Alembert ratio test** (or **ratio test**), the series $\sum a_n$ with positive terms is convergent. In the case of $\lim_{n\to\infty}\left|\frac{a_{n+1}}{a_n}\right| = 1$, the series is called **indeterminate** or the test is inconclusive and we have to use another method to decide its convergence. Otherwise, an infinite series is divergent.

We can think the ratio test as a generalization of the geometric series test because the ratio test for the geometric series is $\lim_{n\to\infty}\left|\frac{a_{n+1}}{a_n}\right| = \frac{ar^{n+1}}{ar^n} = r$ and the series converges if and only if $|r| < 1$.

Let's consider differentiable functions $f(x)$ and $g(x)$. When we have an indeterminate form such as $\frac{0}{0}$ or $\frac{\infty}{\infty}$ for $\lim_{x\to a}\frac{f(x)}{g(x)}$, we have

$$\lim_{x\to a}\frac{f(x)}{g(x)} = \lim_{x\to a}\frac{f'(x)}{g'(x)} \tag{5.1.3}$$

if the limit on the right-hand side of the equation exists. This is called **L'Hopital's rule**.

Example

5.1.1 Consider an infinite series,

$$(1-x)^{-1} = \frac{1}{1-x} = 1 + x + x^2 + \cdots\cdots + x^{n-1} + \cdots\cdots,$$

determine whether it is convergent.

Ans. This corresponds to $a = 1$ and $r = x$ in the geometric series. Thus, this series is convergent for $|x| < 1$.

5.1.2 Determine if the following series,

$$1 + \frac{1}{2!} + \frac{1}{3!} + \frac{1}{4!} + \cdots\cdots + \frac{1}{n!},$$

is convergent or divergent. In here $n!$ is called n **factorial** and $n! = n \cdot (n-1) \cdot (n-2) \cdots\cdots 2 \cdot 1$.

Ans. Since $a_{n+1} = \frac{1}{(n+1)!}$ and $a_n = \frac{1}{n!}$, we have

$$\lim_{n\to\infty} \left|\frac{a_{n+1}}{a_n}\right| = \lim_{n\to\infty} \frac{1/(n+1)!}{1/n!} = \lim_{n\to\infty} \frac{1}{n+1} = 0$$

for the ratio test.

Thus, the series is convergent because $\lim_{n\to\infty} \left|\frac{a_{n+1}}{a_n}\right| < 1$.

5.1.3 Determine $\sum \frac{1}{n}$ is convergent or divergent.

Ans. The ratio test is

$$\lim_{n\to\infty} \left|\frac{a_{n+1}}{a_n}\right| = \lim_{n\to\infty} \frac{1/(n+1)}{1/n} = \lim_{n\to\infty} \frac{n}{n+1} = 1$$

Thus, the series is inconclusive and it may converge or diverge.

5.1.4 Determine $\sum \frac{1}{n^2}$ is convergent or divergent.

Ans. The ratio test is

$$\lim_{n\to\infty} \left|\frac{a_{n+1}}{a_n}\right| = \lim_{n\to\infty} \frac{1/(n+1)^2}{1/n^2} = \lim_{n\to\infty} \frac{n^2}{(n+1)^2}$$

$$= \lim_{n\to\infty} \frac{n^2}{n^2 + 2n + 1} = 1$$

Thus, the series is inconclusive and it may converge or diverge.

5.1.5 Calculate $\lim_{x\to 0}\frac{\sin x}{x}$.

Ans. Since $\lim_{x\to 0}\frac{\sin x}{x}$ has an indeterminate form of $\frac{0}{0}$, we can apply L'Hopital's rule and then we have

$$\lim_{x\to 0}\frac{(\sin x)'}{(x)'}=\lim_{x\to 0}\frac{\cos x}{1}=1$$

5.1.6 Calculate $\lim_{x\to\infty}\frac{ln x}{\sqrt{x}}$.

Ans. Since $\lim_{x\to\infty}\frac{ln x}{\sqrt{x}}$ has an indeterminate form of $\frac{\infty}{\infty}$, we can apply L'Hopital's rule and then we have

$$\lim_{x\to\infty}\frac{\{ln(x)\}'}{\{\sqrt{x}\}'}=\lim_{x\to\infty}\frac{\frac{1}{x}}{\frac{1}{2}x^{-\frac{1}{2}}}=2\lim_{x\to\infty}x^{-\frac{1}{2}}=0$$

5.2 Taylor Series

- $f(x+a)=\sum_{n=0}^{\infty}\frac{(x-a)^n}{n!}f^{(n)}(a)$, Taylor series
- $f(x)=\sum_{n=0}^{\infty}\frac{x^n}{n!}f^{(n)}(0)$, Maclaurin series
- $e^x=\sum_{n=0}^{\infty}\frac{x^n}{n!}$ for all x
- $\cos x=\sum_{n=0}^{\infty}(-1)^n\frac{1}{(2n)!}x^{2n}$ for all x
- $\sin x=\sum_{n=0}^{\infty}(-1)^n\frac{x^{2n+1}}{(2n+1)!}$ for all x

$$\cdot\ (1+x)^n = 1+nx+\frac{n(n-1)}{2!}x^2+\frac{n(n-1)(n-2)}{3!}x^3+\cdots\cdots$$

$$= \sum_{k=0}^{n}\binom{n}{k}x^k, \text{ Binomial expansion}$$

$$\frac{1}{1+x} = \sum_{n=0}^{\infty}(-1)^n x^n$$

$$\frac{1}{1-x} = \sum_{n=0}^{\infty}x^n \text{ for } |x|<1$$

$$\cdot\ \ln(1+x) = \sum_{n=0}^{\infty}(-1)^n\frac{x^{n+1}}{n+1} \text{ for } |x|<1$$

When a continuous function is represented as an infinite sum of terms which are obtained from derivatives of a function at a certain point, this infinite sum is called **Taylor series** (or **Taylor expansion**). Taylor series is very useful in solving physics problems when the first couple of terms give an adequate approximation.

When $\frac{d^n f(x)}{dx^n} \equiv f^{(n)}(x)$ denotes the nth derivative of a function $f(x)$, we have

$$\int_a^x f^{(n)}(x_1)dx_1 = \int_a^x \frac{d}{dx_1}f^{(n-1)}(x_1)dx_1$$

$$= f^{(n-1)}(x_1)|_{x_1=a}^{x_1=x} = f^{(n-1)}(x) - f^{(n-1)}(a)$$

If we integrate this again, we get

$$\int_a^x dx_2 \int_a^{x_2} f^{(n)}(x_1)dx_1 = \int_a^x dx_2[f^{(n-1)}(x_2) - f^{(n-1)}(a)]$$

$$= \int_a^x f^{(n-1)}(x_2)dx_2 \ - \ f^{(n-1)}(a)\int_a^x dx_2$$

$$= f^{(n-2)}(x_2)|_{x_2=a}^{x_2=x} - f^{(n-1)}(a)[x_2]_{x_2=a}^{x_2=x}$$

$$= f^{(n-2)}(x) - f^{(n-2)}(a) - (x-a)f^{(n-1)}(a)$$

If we integrate this again, we get

$$\int_a^x dx_3 \int_a^{x_3} dx_2 \int_a^{x_2} f^{(n)}(x_1)dx_1$$
$$= \int_a^x dx_3 [f^{(n-2)}(x_3) - f^{(n-2)}(a) - (x_3 - a)f^{(n-1)}(a)] \tag{5.2.1}$$

The first and second terms on the right-hand side of the equation give

$$f^{(n-3)}(x_3)|_{x_3=a}^{x_3=x} = f^{(n-3)}(x) - f^{(n-3)}(a)$$

and

$$f^{(n-2)}(a)[x_3]_{x_3=a}^{x_3=x} = (x-a)f^{(n-2)}(a),$$

respectively.

Let's calculate the third term on the right-hand side of the equation:

$$f^{(n-1)}(a)\int_a^x (x_3 - a)dx_3 = f^{(n-1)}(a)[\frac{1}{2}(x^2 - a^2) - a(x-a)]$$
$$= \frac{1}{2}(x^2 - 2ax + a^2)f^{(n-1)}(a)$$
$$= \frac{1}{2!}(x-a)^2 f^{(n-1)}(a)$$

By substituting all these terms into Eq. (5.2.1), we obtain

$$\int_a^x dx_3 \int_a^{x_3} dx_2 \int_a^{x_2} f^{(n)}(x_1)dx_1 =$$
$$f^{(n-3)}(x) - f^{(n-3)}(a) - (x-a)f^{(n-2)}(a) - \frac{1}{2!}(x-a)^2 f^{(n-1)}(a), \tag{5.2.2}$$

$$\vdots$$
$$\vdots$$

Therefore, we know that if we integrate the equation n times, we get

$$\int_a^x dx_n \cdots \int_a^{x_2} dx_1 f^{(n)}(x_1)$$
$$= f(x) - f(a) - (x-a)f^{(1)}(a) - \frac{1}{2!}(x-a)^2 f^{(2)}(a)$$
$$-\cdots\cdots - \frac{1}{(n-1)!}(x-a)^{n-1} f^{(n-1)}(a) - \cdots\cdots \tag{5.2.3}$$

Let the left-hand side of the equation to be R_n:

$$\int_a^x dx_n \cdots \int_a^{x_2} dx_1 f^{(n)}(x_1) \equiv R_n$$

Since $\lim_{n\to\infty} R_n = 0$, Eq. (5.2.3) gives

$$f(x) = f(a) + (x-a)f^{(1)}(a) + \frac{1}{2!}(x-a)^2 f^{(2)}(a) + \cdots\cdots + \frac{1}{(n-1)!}(x-a)^{n-1} f^{(n-1)}(a) + \cdots\cdots, \tag{5.2.4}$$

where $f^{(1)}(a) = \frac{df(x)}{dx}|_{x=a} = f'(a)$ and $f^{(2)}(a) = \frac{d^2 f(x)}{dx^2}|_{x=a} = f''(a)$ if $f(x)$ is
well-behaved function in the vicinity of $x = a$.

Thus, Eq. (5.2.4) is expressed as

$$f(x) = f(a) + (x-a)f'(a) + \frac{(x-a)^2}{2!}f''(a) + \cdots\cdots + \frac{(x-a)^n}{n!}f^{(n)}(a) + \cdots\cdots$$
$$= \sum_{n=0}^{\infty} \frac{(x-a)^n}{n!} f^{(n)}(a), \tag{5.2.5}$$

where $f^{(n)}(a) = \frac{d^n f}{dx^n}|_{x=a}$ means that a is substituted for x after n differentiations with respect to the function $f(x)$.

The linear approximation, $f(a) + (x-a)f'(a)$, in Eq. (5.2.5) fits $f(x)$ well with a tangent line through $x = a$ to a graph. In general we can add quadratic, third order and higher order terms to approximate $f(x)$ better near $x = a$.

Eq. (5.2.5) is Taylor series about a point $x = a$ while **Maclaurin series** is an expansion at $a = 0$.

Thus, Maclaurin series is expressed as

$$f(x) = f(0) + xf'(0) + \frac{x^2}{2!}f''(0) + \frac{x^3}{3!}f'''(0) + \cdots\cdots + \frac{x^n}{n!}f^{(n)}(0) + \cdots\cdots$$
$$= \sum_{n=0}^{\infty} \frac{x^n}{n!} f^{(n)}(0) \tag{5.2.6}$$

Let's consider an expansion of powers of a binomial. For $(a+b)^n$, where $a > b$ and n is a positive integer, we have

$$(a+b)^n = a^n[1 + n(\frac{b}{a}) + \frac{1}{2}n(n-1)(\frac{b}{a})^2 + \cdots\cdots]$$

$$= a^n + na^{n-1}b + \frac{n(n-1)}{2}a^{n-2}b^2 + \cdots\cdots + nab^{n-1} + b^n$$

$$= \sum_{k=0}^{n}\binom{n}{k}a^{n-k}b^k, \tag{5.2.7}$$

where $\binom{n}{k} = \frac{n!}{k!(n-k)!} = \frac{n(n-1)\cdots\cdots(n-k+1)}{k!}$

This series is called the **binomial theorem** (or **binomial formula**).

(a) If we insert $a = 1$ and $b = x$ into Eq. (5.2.7), we obtain

$$(1+x)^n = \sum_{k=0}^{n}\binom{n}{k}x^k \tag{5.2.8}$$

Then, d'Alembert ratio test gives

$$\lim_{k\to\infty}\left|\frac{a_{k+1}}{a_k}\right|$$

$$= \lim_{k\to\infty}\left|\frac{n(n-1)\cdots\cdots(n-k+1)(n-k)x^{k+1}}{(k+1)!}\frac{k!}{n(n-1)\cdots\cdots(n-k+1)x^k}\right|$$

$$= |x|\lim_{k\to\infty}\left|\frac{n-k}{k+1}\right| = |x|$$

Therefore, we get $\lim_{k\to\infty}\left|\frac{a_{k+1}}{a_k}\right| < 1$ for $|x| < 1$.

Thus, the binomial series converges for $|x| < 1$.

(b) Let's express a function $f(x) = (1+x)^n$ in the Maclaurin series. Its derivatives are

$$f'(x)|_{x=0} = n(1+x)^{n-1}|_{x=0} = n$$

$$f''(x)|_{x=0} = n(n-1)(1+x)^{n-2}|_{x=0} = n(n-1)$$

$$f'''(x)|_{x=0} = n(n-1)(n-2)(1+x)^{n-3}|_{x=0} = n(n-1)(n-2)$$

$$\vdots$$

$$\vdots$$

By inserting these into Eq. (5.2.6), we get

$$(1+x)^n = 1+nx+\frac{n(n-1)}{2!}x^2+\frac{n(n-1)(n-2)}{3!}x^3+\cdots\cdots$$
$$+\frac{n(n-1)(n-2)\cdots\cdots(n-k+1)}{k!}x^k+\cdots\cdots+x^n,$$

where $\dfrac{n(n-1)(n-2)\cdots\cdots(n-k+1)}{k!}x^k = \dfrac{n!}{(n-k)!k!}x^k = \dbinom{n}{k}x^k$

Thus, we obtain

$$(1+x)^n = 1+nx+\frac{n(n-1)}{2!}x^2+\frac{n(n-1)(n-2)}{3!}x^3+\cdots\cdots+\binom{n}{k}x^k+\cdots\cdots+x^n$$
$$=\sum_{k=0}^{n}\binom{n}{k}x^k,$$

which is the same as Eq. (5.2.8).

Example

5.2.1 Express a function e^x in the Maclaurin series.

Ans. Its derivatives are

$$f'(x)|_{x=0} = e^x|_{x=0} = 1$$
$$f''(x)|_{x=0} = e^x|_{x=0} = 1$$
$$f'''(x)|_{x=0} = e^x|_{x=0} = 1$$
$$\vdots$$
$$\vdots$$

By inserting these into Eq. (5.2.6), we get

$$e^x = 1+x+\frac{x^2}{2!}+\cdots\cdots+\frac{x^n}{n!}+\cdots\cdots \quad = \sum_{n=0}^{\infty}\frac{x^n}{n!}$$

5.2.2 Calculate $\displaystyle\lim_{x\to 0}\frac{e^x-1-x}{x^2}$.

Ans. Let's calculate this by using two different ways:

(i) using the Maclaurin series for e^x,

$$\lim_{x\to 0}\frac{e^x-1-x}{x^2} = \lim_{x\to 0}\frac{1}{x^2}(1+x+\frac{x^2}{2!}+\frac{x^3}{3!}+\cdots\cdots-1-x)$$

$$= \lim_{x\to 0}\frac{1}{x^2}(\frac{x^2}{2!}+\frac{x^3}{3!}+\cdots\cdots)$$

$$= \lim_{x\to 0}(\frac{1}{2!}+\frac{x}{3!}+\cdots\cdots) = \frac{1}{2}$$

(ii) using L'Hopital's rule,

Since $\lim_{x\to 0}\frac{e^x-1-x}{x^2}$ is in the form of $\frac{0}{0}$, we can apply L'Hopital's rule and get

$$\lim_{x\to 0}\frac{e^x-1-x}{x^2} = \lim_{x\to 0}\frac{e^x-1}{2x}$$

The right-hand side of the equation is in the form of $\frac{0}{0}$.

By applying L'Hopital's rule again and we obtain

$$\lim_{x\to 0}\frac{e^x-1}{2x} = \lim_{x\to 0}\frac{e^x}{2} = \frac{1}{2}$$

Thus, the two methods give the same answer.

5.2.3 Calculate $\int e^{x^2}dx$.

Ans. Since $e^x = 1+x+\frac{x^2}{2!}+\cdots\cdots+\frac{x^n}{n!}+\cdots\cdots = \sum_{n=0}^{\infty}\frac{x^n}{n!}$, we have

$$e^{x^2} = 1+x^2+\frac{x^4}{2!}+\cdots\cdots+\frac{x^{2n}}{n!}+\cdots\cdots = \sum_{n=0}^{\infty}\frac{x^{2n}}{n!}$$

The integral is then

$$\int e^{x^2}dx = \int(1+x^2+\frac{x^4}{2!}+\frac{x^6}{3!}+\cdots\cdots+\frac{x^{2n}}{n!}+\cdots\cdots)dx$$

$$= x+\frac{x^3}{3\cdot 1!}+\frac{x^5}{5\cdot 2!}+\frac{x^7}{7\cdot 3!}+\cdots\cdots+C,$$

where C is the integration constant.

Thus, we obtain

$$\int e^{x^2}dx = \sum_{n=0}^{\infty}\frac{x^{2n+1}}{(2n+1)n!}+C$$

5.2.4 Find the binomial series for $\dfrac{1}{1+x}$ and $\dfrac{1}{1-x}$ with $|x|<1$.

Ans. We have

$$\begin{aligned}\frac{1}{1+x} &= (1+x)^{-1} \\ &= 1-x+\frac{1}{2!}(-1)(-2)x^2+\frac{1}{3!}(-1)(-2)(-3)x^3+\cdots\cdots \\ &= 1-x+x^2-x^3+\cdots\cdots \\ &= \sum_{n=0}^{\infty}(-1)^n x^n\end{aligned}$$

Similarly, we obtain

$$\begin{aligned}\frac{1}{1-x} &= (1-x)^{-1} \\ &= 1+x+\frac{1}{2!}(-1)(-2)(-x)^2+\frac{1}{3!}(-1)(-2)(-3)(-x)^3+\cdots\cdots \\ &= \sum_{n=0}^{\infty}x^n\end{aligned}$$

5.2.5 Find $ln(1+x)$ in an infinite series.

Ans. (i) We can express it by

$$ln(1+x)=\int\frac{1}{1+x}dx$$

We know that $\dfrac{1}{1+x} = \sum_{n=0}^{\infty}(-1)^n x^n$ and the equation is then

$$\begin{aligned}ln(1+x) &= \sum_{n=0}^{\infty}(-1)^n\int x^n dx \\ &= \sum_{n=0}^{\infty}(-1)^n\frac{x^{n+1}}{n+1} \quad \left(\text{or } \sum_{n=1}^{\infty}(-1)^{n-1}\frac{x^n}{n}\right)\end{aligned}$$

(ii) We can use the Maclaurin series for $f(x)=ln(1+x)$.
It's derivatives are

$$f(x)|_{x=0}=ln1=0$$

$$f'(x)|_{x=0}=\frac{1}{1+x}|_{x=0}=1$$

$$f''(x)|_{x=0}=\frac{-1}{(1+x)^2}|_{x=0}=-1$$

$$f'''(x)|_{x=0} = \frac{2}{(1+x)^3}|_{x=0} = 2$$

$\vdots$

$\vdots$

By inserting these into Eq. (5.2.6), we get

$$ln(1+x) = x - \frac{x^2}{2} + \frac{x^3}{3} - \cdots\cdots = \sum_{n=0}^{\infty}(-1)^n \frac{x^{n+1}}{n+1}$$

Thus, (i) and (ii) give the same answer.

Similarly, we obtain

$$\ln(1-x) = -\int \frac{1}{1-x}dx = -x - \frac{x^2}{2} - \frac{x^3}{3} - \cdots\cdots$$

$$= -\sum_{n=0}^{\infty}\frac{x^{n+1}}{n+1} \quad (\text{or} \quad -\sum_{n=1}^{\infty}\frac{x^n}{n})$$

5.2.6 Calculate $\frac{1}{2}ln\frac{1+x}{1-x}$.

Ans. If we use the results obtained in [Ex. 5.2.5], we have

$$\frac{1}{2}ln\frac{1+x}{1-x} = \frac{1}{2}[\ln(1+x) - \ln(1-x)]$$

$$= \frac{1}{2}2(x + \frac{x^3}{3} + \frac{x^5}{5} + \cdots\cdots)$$

$$= \sum_{n=0}^{\infty}\frac{x^{2n+1}}{2n+1}$$

5.2.7 Einstein equation for energy is $E = \frac{mc^2}{\sqrt{1-\frac{v^2}{c^2}}} - mc^2$. Show that this is the classical kinetic energy equation $E = \frac{1}{2}mv^2$ for $v << c$.

Ans. If we use the binomial theorem, we get

$$E = \frac{mc^2}{\sqrt{1-\frac{v^2}{c^2}}} - mc^2 = mc^2[(1-\frac{v^2}{c^2})^{-\frac{1}{2}} - 1]$$

$$= mc^2[1 - \frac{1}{2}(-\frac{v^2}{c^2}) + (-\frac{1}{2})(-\frac{3}{2})\frac{1}{2!}(-\frac{v^2}{c^2})^2 + \cdots\cdots - 1]$$

$$= mc^2\frac{1}{2}\frac{v^2}{c^2} = \frac{1}{2}mv^2,$$

neglecting higher order terms greater than $(\frac{v}{c})^2$.

5.2.8 Find the Maclaurin series for $f(x) = \cos x$.

Ans. Its derivatives are

$$f(x)|_{x=0} = \cos 0 = 1$$
$$f'(x)|_{x=0} = -\sin 0 = 0$$
$$f''(x)|_{x=0} = -\cos 0 = -1$$
$$f'''(x)|_{x=0} = \sin 0 = 0$$
$$f''''(x)|_{x=0} = \cos 0 = 1$$
$$\vdots$$
$$\vdots$$

By inserting these into Eq. (5.2.6), we get

$$\begin{aligned}\cos x &= 1 + \frac{x^2}{2!}(-1) + \frac{x^4}{4!}(1) + \cdots\cdots \\ &= 1 - \frac{x^2}{2!} + \frac{x^4}{4!} - \cdots\cdots = \sum_{n=0}^{\infty}(-1)^n \frac{x^{2n}}{(2n)!}\end{aligned}$$

5.2.9 Find the Maclaurin expansion for $f(x) = \sin x$.

Ans. It's derivatives are

$$f(x)|_{x=0} = \sin 0 = 0$$
$$f'(x)|_{x=0} = \cos 0 = 1$$
$$f''(x)|_{x=0} = -\sin 0 = 0$$
$$f'''(x)|_{x=0} = -\cos 0 = -1$$
$$f''''(x)|_{x=0} = \sin 0 = 0$$
$$\vdots$$
$$\vdots$$

By inserting these into Eq. (5.2.6), we get

$$\sin x = x - \frac{x^3}{3!} + \frac{x^5}{5!} - \cdots\cdots = \sum_{n=0}^{\infty}(-1)^n \frac{x^{2n+1}}{(2n+1)!}$$

and some of the terms in the Maclaurin series are shown in Figure.

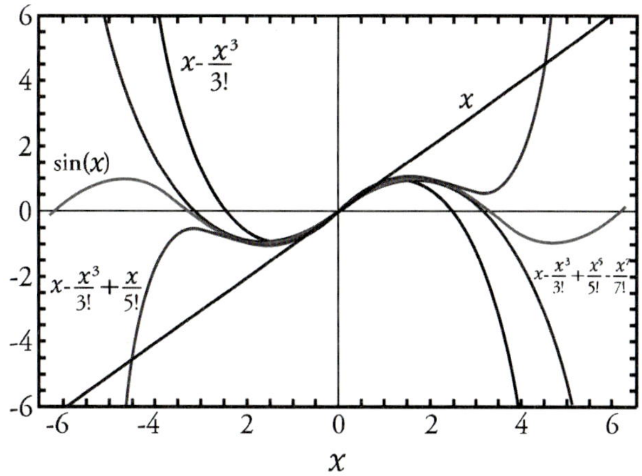

A Comparison between $\sin x$ and its Maclaurin series up to the third term.

5.2.10 Calculate $\lim_{x\to 0}\frac{\sin x}{x}$ in [Ex. 5.1.5] using the result of [Ex. 5.2.9].

Ans. We know that $\sin x = \sum_{n=0}^{\infty}(-1)^n \frac{x^{2n+1}}{(2n+1)!}$ and the problem is then expressed as

$$\lim_{x\to 0}\frac{\sin x}{x} = \lim_{x\to 0}\frac{1}{x}(x - \frac{x^3}{3!} + \frac{x^5}{5!} + \cdots\cdots)$$

$$= \lim_{x\to 0}(1 - \frac{x^2}{3!} + \frac{x^4}{5!} + \cdots\cdots) = 1$$

Thus, we obtain the same answer as that in [Ex. 5.1.5].

5.2.11 Consider the **simple pendulum problem**.

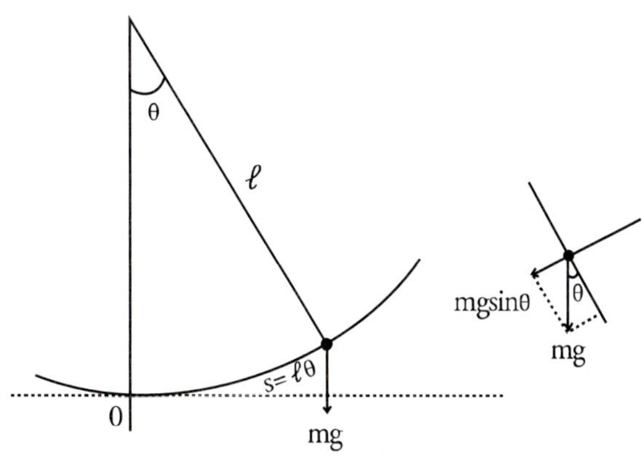

A pendulum with mass m and length ℓ is suspended from a frictionless pivot. We do not consider the string mass and air resistance.

Ans. From Newton's second law $\vec{F} = m\vec{a}$, we have

$$-mg\sin\theta = m\frac{d^2s}{dt^2}, \tag{5.2.11.1}$$

where the minus sign on the left-hand side of the equation indicates the force is the restoring force.

For small oscillation, we have $s = \ell\theta$ and $\sin\theta = \theta$.

Then, the equation is

$$m\ell\frac{d^2\theta}{dt^2} + mg\theta = 0,$$

which gives

$$\frac{d^2\theta}{dt^2} + \frac{g}{\ell}\theta = 0,$$

where g and ℓ are constants.

If we define $\frac{g}{\ell} = \omega_0^2$, it gives

$$\frac{d^2\theta}{dt^2} + \omega_0^2\theta = 0$$

This is the second order ordinary homogeneous differential equation corresponding to $a = 1$, $b = 0$ and $c = \omega_0^2$ of Eq. (2.3.2).

Then, we have

$$D_{1,2} = \frac{\pm\sqrt{-4\omega_0^2}}{2} = \pm i\omega_0$$

The general solution is then

$$\theta(t) = A_1 e^{+i\omega_0 t} + A_2 e^{-i\omega_0 t} \text{ or } \theta(t) = \theta_0 \sin(\omega_0 t + \phi)$$

For $\theta = 0$ at $t = 0$, $\theta(0) = \theta_0 \sin\phi = 0$ gives $\phi = 0$.

Thus, the solution is

$$\theta(t) = \theta_0 \sin\omega_0 t$$

For period T, the solution satisfies $\theta(t + T) = \theta(t)$.

This gives

$$\theta_0 \sin\omega_0 (t + T) = \theta_0 \sin\omega_0 t,$$

which is

$$\sin(\omega_0 t + \omega_0 T) = \sin\omega_0 t$$

To satisfy this equation, the following relation is required.

$$\omega_0 T = 2\pi$$

Thus, the period is given by

$$T = \frac{2\pi}{\omega_0} = 2\pi\sqrt{\frac{\ell}{g}}$$

5.3 Taylor Series for More than One Variable

When f is a function of two variables, x and y, and $f(x,y)$ is a well-defined function about $x = x_0$ and $y = y_0$, the expansion is expressed as

$$\begin{aligned}
f(x,y) &= f(x_0,y_0) + (x - x_0)\frac{\partial f}{\partial x}\Big|_{x = x_0, y = y_0} + (y - y_0)\frac{\partial f}{\partial y}\Big|_{x = x_0, y = y_0} \\
&+ \frac{1}{2!}\left\{(x - x_0)^2\frac{\partial^2 f}{\partial x^2}\Big|_{x = x_0, y = y_0} + 2(x - x_0)(y - y_0)\frac{\partial^2 f}{\partial x \partial y}\Big|_{x = x_0, y = y_0}\right. \\
&\qquad \left. + (y - y_0)^2\frac{\partial^2 f}{\partial y^2}\Big|_{x = x_0, y = y_0}\right\} \\
&+ \frac{1}{3!}\left\{(x - x_0)^3\frac{\partial^3 f}{\partial x^3}\Big|_{x = x_0, y = y_0} + 3(x - x_0)^2(y - y_0)\frac{\partial^3 f}{\partial x^2 \partial y}\Big|_{x = x_0, y = y_0}\right. \\
&\left. + 3(x - x_0)(y - y_0)^2\frac{\partial^3 f}{\partial x \partial y^2}\Big|_{x = x_0, y = y_0} + (y - y_0)^3\frac{\partial^3 f}{\partial y^3}\Big|_{x = x_0, y = y_0}\right\} \\
&+ \cdots\cdots\cdots
\end{aligned} \tag{5.3.1}$$

For convenience, let's define

$$\frac{\partial f}{\partial x}\Big|_{x = x_0, y = y_0} = f_x(x_0, y_0), \quad \frac{\partial f}{\partial y}\Big|_{x = x_0, y = y_0} = f_y(x_0, y_0),$$

$$\frac{\partial^2 f}{\partial x^2}\Big|_{x = x_0, y = y_0} = f_{xx}(x_0, y_0), \quad \frac{\partial^2 f}{\partial y^2}\Big|_{x = x_0, y = y_0} = f_{yy}(x_0, y_0)$$

$$\frac{\partial^2 f}{\partial x \partial y}\Big|_{x = x_0, y = y_0} = f_{xy}(x_0, y_0), \text{ and } \frac{\partial^2 f}{\partial y \partial x}\Big|_{x = x_0, y = y_0} = f_{yx}(x_0, y_0),$$

$$\vdots$$

Eq. (5.3.1) is then expressed as

$$f(x,y) = f(x_0, y_0) + (x - x_0)f_x(x_0, y_0) + (y - y_0)f_y(x_0, y_0)$$

$$+\frac{1}{2!}\{(x-x_0)^2 f_{xx}(x_0,y_0)+2(x-x_0)(y-y_0)f_{xy}(x_0,y_0)$$

$$+(y-y_0)^2 f_{yy}(x_0,y_0)\}+\cdots\cdots \quad (5.3.2)$$

If we define $(x,y)=\vec{x}$ and $(x_0,y_0)=\vec{x_0}$, we can express the equation in more compact form:

$$f(\vec{x})=f(\vec{x_0})+[(\vec{x}-\vec{x_0})\cdot\vec{\nabla}f(\vec{x_0})]$$

$$+\frac{1}{2!}[(\vec{x}-\vec{x_0})^T\circ H(\vec{x_0})\circ(\vec{x}-\vec{x_0})]+\cdots\cdots, \quad (5.3.3)$$

where $H(\vec{x_0})$ is called **Hessian square matrix** defined by

$$H(\vec{x_0})=H(x_0,y_0)=\begin{pmatrix} f_{xx}(x_0,y_0) & f_{xy}(x_0,y_0) \\ f_{yx}(x_0,y_0) & f_{yy}(x_0,y_0) \end{pmatrix}$$

When $\frac{\partial f(x,y)}{\partial x}|_{x_0,y_0}=0$ and $\frac{\partial f(x,y)}{\partial y}|_{x_0,y_0}=0$, we say that a function $f(x,y)$ has an **equilibrium position** at $x=x_0$ and $y=y_0$.

For instance, consider a conservative system in one degree of freedom. When a force acting on a system vanishes, it is said that the system is in equilibrium. Since the force is the first derivative of the potential energy, $F(x)=-\frac{dU(x)}{dx}$,
an equilibrium position x_0 satisfies $F(x)|_{x=x_0}=-\frac{dU(x)}{dx}|_{x=x_0}=0$. When a small disturbance from the equilibrium position is given, it is said that the equilibrium position is **stable** (**unstable**) if a system returns (moves away from equilibrium) to its original equilibrium. If a system is not in the stable or unstable equilibrium, it is said **neutral** equilibrium.

When we express the potential $U(x)$ in Taylor series about the equilibrium position x_0, we get

$$U(x)=U(x_0)+x\frac{dU}{dx}|_{x_0}+\frac{1}{2!}x^2\frac{d^2U}{dx^2}|_{x_0}+\cdots\cdots$$

$$=\frac{1}{2!}x^2\frac{d^2U}{dx^2}|_{x_0}$$

because we can choose $U(x_0)=0$ without affecting results and the linear term is zero at the equilibrium position.

Then, we have

$$F(x)=-\frac{dU(x)}{dx}=-[\frac{d^2U}{dx^2}|_{x_0}]\frac{d}{dx}(\frac{1}{2!}x^2)=-[\frac{d^2U}{dx^2}|_{x_0}]x$$

Therefore, the force is the restoring (anti-restoring) and the equilibrium is the stable (unstable) if $\frac{d^2U}{dx^2}|_{x_0}$ is positive (negative).

Depending on a sign of the second derivative of the potential, the equilibrium is classified as

$$\frac{d^2U}{dx^2}|_{x_0}=\begin{cases} >0 & \text{stable (local minimum)} \\ <0 & \text{unstable (local maximum)} \\ =0 & \text{neutral} \end{cases}$$

For a system with two degrees of freedom, it is possible that a system is in stable in one direction but is not stable in the other direction. In this case, it is said the equilibrium position is **saddle**.

Therefore, a function can have a local minimum, local maximum or saddle point at the equilibrium position $(x_0,\ y_0)$, as shown in Fig. 5.3.1, depending on the second derivatives of a function.

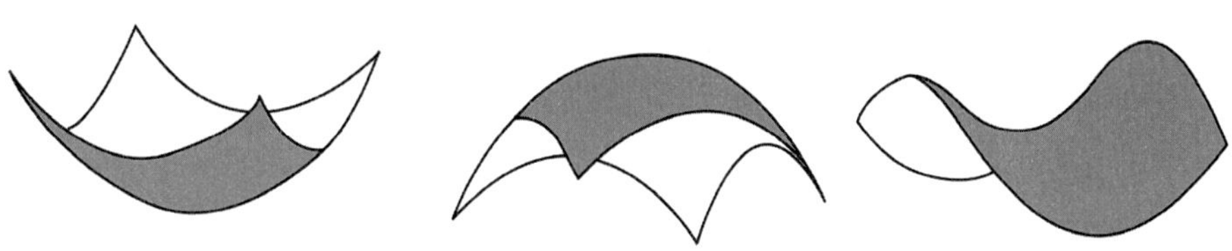

(a) Minimum (b) Maximum (c) Saddle point

Fig 5.3.1 For local extremum: (a) minimum, (b) maximum, and (c) saddle points for equilibrium.

Let's assume $(x_0,\ y_0)$ is the equilibrium position and choose $f(x_0,\ y_0)=0$. Then, the only relevant term determining stationary types is the second derivatives in Eq. (5.3.3). If we define the determinant of the Hessian matrix, $D=f_{xx}(x_0,y_0)f_{yy}(x_0,y_0)-f_{xy}(x_0,y_0)f_{yx}(x_0,y_0)$, a system is in stable equilibrium for $D>0$ and

$$\text{a function has} \begin{cases} \bullet \text{ minimum point if } D>0 \text{ and } f_{xx}(x_0,y_0)>0 \\ \bullet \text{ maximum point if } D>0 \text{ and } f_{xx}(x_0,y_0)<0 \\ \bullet \text{ saddle point if } D<0 \end{cases}$$

Example

5.3.1 Expand a function $f(x,y) = y\sin x + \cos y$ up to the third term on the right-hand side of Eq. (5.3.3) about the position $\overrightarrow{x_0} = (0,0)$ in the neighbourhood of $\overrightarrow{x_0}$.

Ans. (i) The function of up to the third term is then

$$f(\vec{x}) = f(\overrightarrow{x_0}) + [(\vec{x} - \overrightarrow{x_0}) \cdot \vec{\nabla} f(\overrightarrow{x_0})] + \frac{1}{2!}[(\vec{x} - \overrightarrow{x_0})^T \circ H(\overrightarrow{x_0}) \circ (\vec{x} - \overrightarrow{x_0})] \tag{5.3.1.1}$$

The first and second terms on the right-hand side of the equation are

$$f(\overrightarrow{x_0}) = f(0,0) = 0\sin 0 + \cos 0 = 1$$

and

$$(\vec{x} - \overrightarrow{x_0}) \cdot \vec{\nabla} f(\overrightarrow{x_0}) = (x-0) \cdot \frac{\partial f}{\partial x}|_{x=0,y=0} + (y-0) \cdot \frac{\partial f}{\partial y}|_{x=0,y=0}$$
$$= x[y\cos x]_{x=0,y=0} + y[\sin x - \sin y]_{x=0,y=0} = 0,$$

respectively.

Let's calculate the second derivatives of the function to find the Hessian matrix:

$$f_{xx}(0,0) = \frac{\partial^2 f}{\partial x^2}|_{x=0,y=0} = -y\sin x|_{x=0,y=0} = 0,$$

$$f_{xy}(0,0) = \frac{\partial^2 f}{\partial x \partial y}|_{x=0,y=0} = \cos x|_{x=0,y=0} = 1 = f_{yx}(0,0)$$

$$f_{yy}(0,0) = \frac{\partial^2 f}{\partial y^2}|_{x=0,y=0} = -\cos y|_{x=0,y=0} = -1$$

Then, the third term on the right-hand side of the equation is

$$\frac{1}{2!}[(\vec{x} - \overrightarrow{x_0})^T \circ H(\overrightarrow{x_0}) \circ (\vec{x} - \overrightarrow{x_0})] = \frac{1}{2}(x \quad y)\begin{pmatrix} 0 & 1 \\ 1 & -1 \end{pmatrix}\begin{pmatrix} x \\ y \end{pmatrix}$$

$$= \frac{1}{2}[xy + y(x-y)] = \frac{1}{2}(2xy - y^2)$$

By substituting all these into Eq. (5.3.1.1), we obtain

$$f(\vec{x}) = f(x,y) = 1 + \frac{1}{2}(2xy - y^2) = 1 + xy - \frac{1}{2}y^2 \tag{5.3.1.2}$$

(ii) If we apply Taylor series for $\sin x$ and $\cos y$ in the function $f(x,y)$, we have

$$f(x,y) = y\sin x + \cos y = y(x - \frac{1}{3!}x^3 + \cdots\cdots) + (1 - \frac{1}{2!}y^2 + \cdots\cdots)$$

To compare this with Eq. (5.3.1.2), let's keep only same order terms on the right-hand side of the equation and we get

$$f(x,y) = 1 + xy - \frac{1}{2}y^2$$

Thus, we obtain the same answer as Eq. (5.3.1.2).

Chapter 06 Matrices

A matrix is an array of numbers or any expressions represented in rows and columns. When a matrix has m rows and n columns, it is called m-by-n (or $m \times n$) matrix. Each item in each row and column of a matrix is called its element. Therefore a_{mn} is an element at the mth row and n^{th} column of a matrix A. Such a matrix is called a rectangular matrix.
The matrix representation of physical quantities such as vectors, operators, linear transformations, and coordinate changes is a very convenient method to solve problems. This chapter covers concepts and calculations of matrices which are very useful in studying Classical Mechanics and Quantum Mechanics.

6.1 Matrix Representation

· For a given matrix A
  - transpose matrix A^T : defined by swapping the rows and columns
  - minor $|M_{ij}|$: determinant of the matrix left after i^{th} row and j^{th} column are deleted
  - cofactor C_{ij} : $(-1)^{i+j}|M_{ij}|$

· Cofactor matrix C: every element corresponding its cofactor

· Adjoint matrix C^T : transpose of the cofactor matrix C

· Inverse matrix A^{-1} : satisfies $A^{-1}A = AA^{-1} = I$ and is obtained by

$$A^{-1} = \frac{1}{|A|}C^T$$

All positions in the rectangular coordinate system are expressed by

$$\overrightarrow{X} = a_1\hat{x}+a_2\hat{y}+a_3\hat{z} = \sum_{k=1}^{3} a_k\widehat{x_k}, \tag{6.1.1}$$

assuming $\hat{x}=\widehat{x_1}$, $\hat{y}=\widehat{x_2}$, and $\hat{z}=\widehat{x_3}$.

In this equation, $-\infty < a_k < \infty$ $(k=1,2,3)$ are the coordinates of the vector $\overrightarrow{X}$ in the **basis** $\widehat{x_k}$.

The basis is expressed in the matrix representation as follows:

$$\widehat{x_1} = (1\ 0\ 0),\ \widehat{x_2} = (0\ 1\ 0),\ \text{and}\ \widehat{x_3} = (0\ 0\ 1)$$

Then, vector $\overrightarrow{X}$ in Eq.(6.1.1) is written as $\overrightarrow{X}=(a_1\ a_2\ a_3)$ in the basis $\widehat{x_k}$.

A size of a matrix is called the **order** which is in general specified as $m\times n$ for number of rows and number of columns are m and n, respectively.

Then, $1\times n$ is called **row matrix** and $n\times 1$ is called **column matrix**.

For instance, $(a_1\ a_2\ \cdots\cdots\ a_n)$ is the $1\times n$ row matrix and $\begin{pmatrix} a_1 \\ a_2 \\ \vdots \\ \vdots \\ a_n \end{pmatrix}$ is the $n\times 1$ column matrix.

When number of rows and number of columns of a matrix are m and n, respectively, an order of a matrix is $m\times n$ and it is called a **rectangular matrix**. For $m=n$, it is called a **square matrix**.

For $m\times n$ matrices

$$A=\begin{pmatrix} a_{11} & a_{12} & \cdots\cdots & a_{1n} \\ a_{21} & a_{22} & \cdots\cdots & a_{2n} \\ \vdots & \vdots & \ddots & \vdots \\ a_{m1} & a_{m2} & \cdots\cdots & a_{mn} \end{pmatrix} \text{ and } B=\begin{pmatrix} b_{11} & b_{12} & \cdots\cdots & b_{1n} \\ b_{21} & b_{22} & \cdots\cdots & b_{2n} \\ \vdots & \vdots & \ddots & \vdots \\ b_{m1} & b_{m2} & \cdots\cdots & b_{mn} \end{pmatrix},$$

two matrices A and B can be added or subtracted element-by-element when thcy have the same numbers of rows and columns.

Therefore, for a_{ij} and b_{ij} are the elements at i^{th} row and j^{th} column of the matrices A and B respectively, we get $(A\pm B)_{ij}=a_{ij}\pm b_{ij}$.

For matrix multiplication, the product of $A \circ B$ is in general not the same as the product of $B \circ A$.

If a_{ij} represents an element at i^{th} row and j^{th} column of a matrix A, we can write the 3×3 matrix $A=\begin{pmatrix} a_{11} & a_{12} & a_{13} \\ a_{21} & a_{22} & a_{23} \\ a_{31} & a_{32} & a_{33} \end{pmatrix}$.

Its product with the basis $\hat{x_k}$ is expressed by

$$A \circ \hat{x_k} = \sum_{i=1}^{3} a_{ik}\hat{x_i},$$

which gives each column of the matrix A.

For the 3×1 column matrix $X=\begin{pmatrix} x_1 \\ x_2 \\ x_3 \end{pmatrix}$, it is expressed by

$$\begin{pmatrix} x_1 \\ x_2 \\ x_3 \end{pmatrix} = x_1\begin{pmatrix} 1 \\ 0 \\ 0 \end{pmatrix} + x_2\begin{pmatrix} 0 \\ 1 \\ 0 \end{pmatrix} + x_3\begin{pmatrix} 0 \\ 0 \\ 1 \end{pmatrix} = \sum_{k=1}^{3} x_k\hat{x_k}$$

Then, we have

$$A \circ X = A \circ \sum_{k=1}^{3} x_k\hat{x_k} = \sum_{k=1}^{3} x_k A \circ \hat{x_k}$$
$$= \sum_{i=1}^{3}\left(\sum_{k=1}^{3} a_{ik}x_k\right)\hat{x_i},$$

which gives

$$(A \circ X)_i = \sum_{k=1}^{3} a_{ik}x_k$$

Thus, **a number of columns of the first matrix should be the same as a number of rows of the second matrix in the matrix multiplication.**

The elements of the product are

$$(A \circ X)_1 = a_{11}x_1 + a_{12}x_2 + a_{13}x_3,\ (A \circ X)_2 = a_{21}x_1 + a_{22}x_2 + a_{23}x_3$$

and

$$(A \circ X)_3 = a_{31}x_1 + a_{32}x_2 + a_{33}x_3$$

Therefore, the product of two matrices gives

$$A \circ X = \begin{pmatrix} a_{11} & a_{12} & a_{13} \\ a_{21} & a_{22} & a_{23} \\ a_{31} & a_{32} & a_{33} \end{pmatrix} \begin{pmatrix} x_1 \\ x_2 \\ x_3 \end{pmatrix}$$

$$= \begin{pmatrix} a_{11}x_1 + a_{12}x_2 + a_{13}x_3 \\ a_{21}x_1 + a_{22}x_2 + a_{23}x_3 \\ a_{31}x_1 + a_{32}x_2 + a_{33}x_3 \end{pmatrix}$$

We know that the product of the 3×3 matrix A with 3×1 matrix X results in 3×1 matrix.

In general, we have the matrix multiplication for the $m \times n$ matrix A and $n \times \ell$ matrix B,

$$A \circ B = \begin{pmatrix} a_{11} & a_{12} & \cdots\cdots & a_{1n} \\ a_{21} & a_{22} & \cdots\cdots & a_{2n} \\ \vdots & \vdots & \ddots & \vdots \\ a_{m1} & a_{m2} & \cdots\cdots & a_{mn} \end{pmatrix} \circ \begin{pmatrix} b_{11} & b_{12} & \cdots\cdots & b_{1\ell} \\ b_{21} & b_{22} & \cdots\cdots & b_{2\ell} \\ \vdots & \vdots & \ddots & \vdots \\ b_{n1} & b_{n2} & \cdots\cdots & b_{n\ell} \end{pmatrix}$$

$$= \begin{pmatrix} \sum_{k=1}^{n} a_{1k}b_{k1} & \sum_{k=1}^{n} a_{1k}b_{k2} & \cdots\cdots & \sum_{k=1}^{n} a_{1k}b_{k\ell} \\ \sum_{k=1}^{n} a_{2k}b_{k1} & \sum_{k=1}^{n} a_{2k}b_{k2} & \cdots\cdots & \sum_{k=1}^{n} a_{2k}b_{k\ell} \\ \vdots & \vdots & \ddots & \vdots \\ \sum_{k=1}^{n} a_{mk}b_{k1} & \sum_{k=1}^{n} a_{mk}b_{k2} & \cdots\cdots & \sum_{k=1}^{n} a_{mk}b_{k\ell} \end{pmatrix},$$

It gives

$$(A \circ B)_{ij} = \sum_{k=1}^{n} a_{ik}b_{kj}, \tag{6.1.2}$$

where $i = 1, 2, \cdots\cdots, m$ and $j = 1, 2, \cdots\cdots, \ell$. We obtain the product of the $m \times n$ matrix and $n \times \ell$ matrix results in the $m \times \ell$ matrix.

For the square matrix A, the sum of diagonal elements is called **trace** (Tr) of the matrix. Therefore, we have

$$TrA = a_{11} + a_{22} + \cdots\cdots + a_{nn} = \sum_{i=1}^{n} a_{ii}$$

When $A^{-1}A = AA^{-1} = I$ for the square matrix A, A^{-1} is called an inverse matrix to A and I is an **unit** (or **identity**) matrix. When all diagonal elements are 1 and all off-diagonal elements are zero, it is the unit matrix. For instance, the 3×3 unit matrix is

$$I=\begin{pmatrix}1&0&0\\0&1&0\\0&0&1\end{pmatrix}$$

When all elements except the diagonal elements of a matrix are zero, it is called a **diagonal** matrix. For instance, the unit matrix is the diagonal matrix.

A **transpose** matrix to a matrix A is written as A^T (or A^t) and is defined by swapping the rows and columns.
If a matrix is not square, its transpose alters numbers of rows and columns. When the 3×3 matrix A is given by

$$A=\begin{pmatrix}a_{11}&a_{12}&a_{13}\\a_{21}&a_{22}&a_{23}\\a_{31}&a_{32}&a_{33}\end{pmatrix},$$

the **transpose** matrix is

$$A^T=\begin{pmatrix}a_{11}&a_{21}&a_{31}\\a_{12}&a_{22}&a_{32}\\a_{13}&a_{23}&a_{33}\end{pmatrix}$$

When $A=A^T$, the square matrix A is called **symmetric**.

The **minor** $|M_{ij}|$ of the matrix A is defined as the determinant of the matrix left after i^{th} row and j^{th} column are deleted. For simplicity, consider the 3×3 matrix A. Then, we get

$$|M_{21}|=\begin{vmatrix}a_{12}&a_{13}\\a_{32}&a_{33}\end{vmatrix}=a_{12}a_{33}-a_{13}a_{32}$$

Since the determinant of the matrix A, denoted as $|A|$ or $\det A$, is defined as

$$|A|=a_{11}\begin{vmatrix}a_{22}&a_{23}\\a_{32}&a_{33}\end{vmatrix}-a_{12}\begin{vmatrix}a_{21}&a_{23}\\a_{31}&a_{33}\end{vmatrix}+a_{13}\begin{vmatrix}a_{21}&a_{22}\\a_{31}&a_{32}\end{vmatrix},$$

we can express it with the minors

$$|A| = a_{11}|M_{11}| - a_{12}|M_{12}| + a_{13}|M_{13}|$$

On the other hand, a **cofactor** C_{ij} is defined as $(-1)^{i+j}|M_{ij}|$.

For instance,

$$C_{21} = (-1)^{2+1}\ |M_{21}| = -\begin{vmatrix} a_{12} & a_{13} \\ a_{32} & a_{33} \end{vmatrix}$$

Therefore, the determinant of the matrix A can be expressed in terms of the cofactors,

$$|A| = a_{11}C_{11} + a_{12}C_{12} + a_{13}C_{13}$$

We can generalize this equation for a matrix A of order n,

$$|A| = \sum_{i,j=1}^{i,j=n} a_{ij}C_{ij}$$

When we replace every element in a matrix with its corresponding cofactor, the resulting matrix is called a **cofactor matrix** C.

The cofactor matrix C of the 3×3 matrix A is then

$$C = \begin{pmatrix} C_{11} & C_{12} & C_{13} \\ C_{21} & C_{22} & C_{23} \\ C_{31} & C_{32} & C_{33} \end{pmatrix} = \begin{pmatrix} |M_{11}| & -|M_{12}| & |M_{13}| \\ -|M_{21}| & |M_{22}| & -|M_{23}| \\ |M_{31}| & -|M_{32}| & |M_{33}| \end{pmatrix}$$

$$= \begin{pmatrix} a_{22}a_{33} - a_{23}a_{32} & a_{23}a_{31} - a_{21}a_{33} & a_{21}a_{32} - a_{22}a_{31} \\ a_{13}a_{32} - a_{12}a_{33} & a_{11}a_{33} - a_{13}a_{31} & a_{12}a_{31} - a_{11}a_{32} \\ a_{12}a_{23} - a_{13}a_{22} & a_{13}a_{21} - a_{11}a_{23} & a_{11}a_{22} - a_{12}a_{21} \end{pmatrix}$$

The **adjoint** matrix is defined as the transpose of the cofactor matrix C. The adjoint matrix is then

$$C^T = \begin{pmatrix} C_{11} & C_{21} & C_{31} \\ C_{12} & C_{22} & C_{32} \\ C_{13} & C_{23} & C_{33} \end{pmatrix}$$

Let's consider the set of two simultaneous equations which have two unknown variables x and y:

$$\begin{cases} a_1 x + b_1 y = p_1 \\ a_2 x + b_2 y = p_2 \end{cases}$$

We want to find the unknown variables by solving the equations. These two equations can be expressed in the matrix representation as

$$\begin{pmatrix} a_1 & b_1 \\ a_2 & b_2 \end{pmatrix}\begin{pmatrix} x \\ y \end{pmatrix} = \begin{pmatrix} p_1 \\ p_2 \end{pmatrix} \tag{6.1.3}$$

This is in the form of $A \circ X = P$, where each of them is

$$A = \begin{pmatrix} a_1 & b_1 \\ a_2 & b_2 \end{pmatrix}, \quad X = \begin{pmatrix} x \\ y \end{pmatrix}, \quad \text{and} \quad P = \begin{pmatrix} p_1 \\ p_2 \end{pmatrix}$$

To solve the equations, let's consider **Cramer's rule** which calculates values of the unknown variables one at a time. Then we will consider **inverse matrix method**.

(a) Cramer's Rule

From the second equation of the two simultaneous equations,

$$y = \frac{p_2 - a_2 x}{b_2} \tag{6.1.4}$$

By substituting this into the first equation of the two simultaneous equations, we have

$$x = \frac{b_2 p_1 - b_1 p_2}{a_1 b_2 - a_2 b_1} = \frac{1}{|A|}(b_2 p_1 - b_1 p_2), \tag{6.1.5}$$

where $|A|$ is the determinant of the matrix A.

If we substitute this into Eq. (6.1.4), we get

$$y = \frac{a_1 p_2 - p_1 a_2}{a_1 b_2 - a_2 b_1} = \frac{1}{|A|}(a_1 p_2 - p_1 a_2) \tag{6.1.6}$$

Cramer's rule is defined as

$$x_i = \frac{|A_i|}{|A|},$$

where $|A_i|$ is the determinant of the matrix A with column i replaced by

$$P=\begin{pmatrix} p_1 \\ p_2 \end{pmatrix}.$$

Then, we get

$$x_1 = x = \frac{|A_1|}{|A|} = \frac{1}{|A|}\begin{vmatrix} p_1 & b_1 \\ p_2 & b_2 \end{vmatrix} = \frac{1}{|A|}(b_2p_1 - b_1p_2)$$

and

$$x_2 = y = \frac{|A_2|}{|A|} = \frac{1}{|A|}\begin{vmatrix} a_1 & p_1 \\ a_2 & p_2 \end{vmatrix} = \frac{1}{|A|}(a_1p_2 - p_1a_2)$$

Therefore, Cramer's rule gives the same answers as Eqs. (6.1.5) and (6.1.6).

(b) **Inverse Matrix Method**

Since the simultaneous equations can be expressed in the matrix representation by $AX=P$[8], it gives

$$A^{-1}AX=A^{-1}P \Rightarrow X=A^{-1}P$$

Therefore, the unknown variables x and y can be obtained from the matrix multiplication $A^{-1}P$ so that we need to find the inverse matrix A^{-1}. Let's find out a way to calculate its elements with assumption of the inverse matrix $A^{-1}=\begin{pmatrix} f_1 & g_1 \\ f_2 & g_2 \end{pmatrix}$ in this problem.

By the definition of the inverse matrix A^{-1} we know $AA^{-1}=I$, which gives

$$\begin{pmatrix} a_1 & b_1 \\ a_2 & b_2 \end{pmatrix}\begin{pmatrix} f_1 & g_1 \\ f_2 & g_2 \end{pmatrix}=\begin{pmatrix} 1 & 0 \\ 0 & 1 \end{pmatrix}$$

It gives the following simultaneous equations:

$$\begin{cases} a_1f_1+b_1f_2=1 \\ a_1g_1+b_1g_2=0 \\ a_2f_1+b_2f_2=0 \\ a_2g_1+b_2g_2=1 \end{cases} \tag{6.1.7}$$

Let's first find f_1 and f_2.

---

[8] Throughout this book, " ∘ " is omitted in the matrix multiplication for simplicity.

From the first equation we have

$$f_1 = \frac{1 - b_1 f_2}{a_1} \tag{6.1.8}$$

By substituting this into the third equation in Eq. (6.1.7), we have

$$f_2 = -\frac{a_2}{a_1 b_2 - b_1 a_2} \tag{6.1.9}$$

If we insert this into Eq. (6.1.8), we get

$$f_1 = \frac{b_2}{a_1 b_2 - b_1 a_2} \tag{6.1.10}$$

Similarly, we can obtain g_1 and g_2 from the second and fourth equations in Eq. (6.1.7):

$$g_1 = \frac{-b_1}{a_1 b_2 - b_1 a_2} \tag{6.1.11}$$

and

$$g_2 = \frac{a_1}{a_1 b_2 - b_1 a_2} \tag{6.1.12}$$

From Eqs. (6.1.9–12), we obtain the inverse matrix

$$A^{-1} = \frac{1}{a_1 b_2 - b_1 a_2} \begin{pmatrix} b_2 & -b_1 \\ -a_2 & a_1 \end{pmatrix},$$

where $a_1 b_2 - b_1 a_2$ is the determinant of A.

On the other hand, the adjoint matrix of A is

$$C^T = \begin{pmatrix} C_{11} & C_{21} \\ C_{12} & C_{22} \end{pmatrix} = \begin{pmatrix} (-1)^{1+1}|M_{11}| & (-1)^{2+1}|M_{21}| \\ (-1)^{1+2}|M_{12}| & (-1)^{2+2}|M_{22}| \end{pmatrix}$$
$$= \begin{pmatrix} b_2 & -b_1 \\ -a_2 & a_1 \end{pmatrix}$$

Thus, the inverse matrix is given by

$$A^{-1} = \frac{1}{|A|} C^T$$

Let's find the unknown variables x and y by using this inverse matrix formula.

Since $X = A^{-1}P$, we have

$$\begin{pmatrix} x \\ y \end{pmatrix} = \frac{1}{|A|}\begin{pmatrix} b_2 & -b_1 \\ -a_2 & a_1 \end{pmatrix}\begin{pmatrix} p_1 \\ p_2 \end{pmatrix}$$

This gives

$$x = \frac{1}{|A|}(b_2 p_1 - b_1 p_2)$$

and

$$y = \frac{1}{|A|}(a_1 p_2 - a_2 p_1)$$

Therefore, we obtain the same answers as those obtained using Cramer's rule.

Example

6.1.1 Find the determinant of the matrix A given by

$$A = \begin{pmatrix} 0 & 10 & -1 & 1 \\ 2 & -2 & -4 & 0 \\ 4 & 2 & 0 & 4 \\ 3 & 2 & 0 & 3 \end{pmatrix}$$

Ans. From the expression of the determinant of the matrix using the cofactors, we have

$$\begin{aligned}
|A| &= a_{11}C_{11} + a_{12}C_{12} + a_{13}C_{13} + a_{14}C_{14} \\
&= a_{11}|M_{11}| - a_{12}|M_{12}| + a_{13}|M_{13}| - a_{14}|M_{14}| \\
&= 0(-1)^{1+1}\begin{vmatrix} -2 & -4 & 0 \\ 2 & 0 & 4 \\ 2 & 0 & 3 \end{vmatrix} + 10(-1)^{1+2}\begin{vmatrix} 2 & -4 & 0 \\ 4 & 0 & 4 \\ 3 & 0 & 3 \end{vmatrix} \\
&\quad + (-1)(-1)^{1+3}\begin{vmatrix} 2 & -2 & 0 \\ 4 & 2 & 4 \\ 3 & 2 & 3 \end{vmatrix} + 1(-1)^{1+4}\begin{vmatrix} 2 & -2 & -4 \\ 4 & 2 & 0 \\ 3 & 2 & 0 \end{vmatrix} \\
&= -10\left\{2\begin{vmatrix} 0 & 4 \\ 0 & 3 \end{vmatrix} - 4(-1)\begin{vmatrix} 4 & 4 \\ 3 & 3 \end{vmatrix} + 0\begin{vmatrix} 4 & 0 \\ 3 & 0 \end{vmatrix}\right\} \\
&\quad + (-1)\left\{2\begin{vmatrix} 2 & 4 \\ 2 & 3 \end{vmatrix} - 2(-1)\begin{vmatrix} 4 & 4 \\ 3 & 3 \end{vmatrix} + 0\begin{vmatrix} 4 & 2 \\ 3 & 2 \end{vmatrix}\right\} \\
&\quad + (-1)\left\{2\begin{vmatrix} 2 & 0 \\ 2 & 0 \end{vmatrix} - 2(-1)\begin{vmatrix} 4 & 0 \\ 3 & 0 \end{vmatrix} - 4\begin{vmatrix} 4 & 2 \\ 3 & 2 \end{vmatrix}\right\} \\
&= 12
\end{aligned}$$

6.1.2 Solve the following simultaneous equations,

$$\begin{cases} 2x - z = 2 \\ 6x + 5y + 3z = 7, \\ 2x - y = 4 \end{cases}$$

using Cramer's rule and the inverse matrix method, show that two methods give the same answers.

Ans. (i) Let's solve the equations using Cramer's rule, $x_i = \dfrac{|A_i|}{|A|}$.

These simultaneous equations are expressed in the matrix representation, $AX = P$:

$$\begin{pmatrix} 2 & 0 & -1 \\ 6 & 5 & 3 \\ 2 & -1 & 0 \end{pmatrix}\begin{pmatrix} x \\ y \\ z \end{pmatrix} = \begin{pmatrix} 2 \\ 7 \\ 4 \end{pmatrix}$$

The determinant of A is

$$\begin{vmatrix} 2 & 0 & -1 \\ 6 & 5 & 3 \\ 2 & -1 & 0 \end{vmatrix} = 2\begin{vmatrix} 5 & 3 \\ -1 & 0 \end{vmatrix} - 0\begin{vmatrix} 6 & 3 \\ 2 & 0 \end{vmatrix} - 1\begin{vmatrix} 6 & 5 \\ 2 & -1 \end{vmatrix} = 22$$

From Cramer's rule, we have

$$\begin{aligned} x &= \frac{1}{|A|}\begin{vmatrix} 2 & 0 & -1 \\ 7 & 5 & 3 \\ 4 & -1 & 0 \end{vmatrix} \\ &= \frac{1}{22}\left\{2\begin{vmatrix} 5 & 3 \\ -1 & 0 \end{vmatrix} - 0\begin{vmatrix} 7 & 3 \\ 4 & 0 \end{vmatrix} - 1\begin{vmatrix} 7 & 5 \\ 4 & -1 \end{vmatrix}\right\} \\ &= \frac{33}{22} = \frac{3}{2} \\ y &= \frac{1}{|A|}\begin{vmatrix} 2 & 2 & -1 \\ 6 & 7 & 3 \\ 2 & 4 & 0 \end{vmatrix} = \frac{-22}{22} = -1 \\ z &= \frac{1}{|A|}\begin{vmatrix} 2 & 0 & 2 \\ 6 & 5 & 7 \\ 2 & -1 & 4 \end{vmatrix} = \frac{22}{22} = 1 \end{aligned}$$

Thus, we obtain that $x = \frac{3}{2}$, $y = -1$, and $z = 1$.

(ii) Let's solve the equations by using the inverse matrix method.

Since $A = \begin{pmatrix} 2 & 0 & -1 \\ 6 & 5 & 3 \\ 2 & -1 & 0 \end{pmatrix}$, the cofactor matrix is

$$C = \begin{pmatrix} C_{11} & C_{12} & C_{13} \\ C_{21} & C_{22} & C_{23} \\ C_{31} & C_{32} & C_{33} \end{pmatrix} = \begin{pmatrix} (-1)^{1+1}|M_{11}| & (-1)^{1+2}|M_{12}| & (-1)^{1+3}|M_{13}| \\ (-1)^{2+1}|M_{21}| & (-1)^{2+2}|M_{22}| & (-1)^{2+3}|M_{23}| \\ (-1)^{3+1}|M_{31}| & (-1)^{3+2}|M_{32}| & (-1)^{3+3}|M_{33}| \end{pmatrix}$$

$$= \begin{pmatrix} \begin{vmatrix} 5 & 3 \\ -1 & 0 \end{vmatrix} & -\begin{vmatrix} 6 & 3 \\ 2 & 0 \end{vmatrix} & \begin{vmatrix} 6 & 5 \\ 2 & -1 \end{vmatrix} \\ -\begin{vmatrix} 0 & -1 \\ -1 & 0 \end{vmatrix} & \begin{vmatrix} 2 & -1 \\ 2 & 0 \end{vmatrix} & -\begin{vmatrix} 2 & 0 \\ 2 & -1 \end{vmatrix} \\ \begin{vmatrix} 0 & -1 \\ 5 & 3 \end{vmatrix} & -\begin{vmatrix} 2 & -1 \\ 6 & 3 \end{vmatrix} & \begin{vmatrix} 2 & 0 \\ 6 & 5 \end{vmatrix} \end{pmatrix} = \begin{pmatrix} 3 & 6 & -16 \\ 1 & 2 & 2 \\ 5 & -12 & 10 \end{pmatrix}$$

Since the adjoint matrix C^T is the transpose of the cofactor matrix C, we get

$$C^T = \begin{pmatrix} 3 & 1 & 5 \\ 6 & 2 & -12 \\ -16 & 2 & 10 \end{pmatrix}$$

Since the inverse matrix is given by $A^{-1} = \frac{1}{|A|} C^T$, we get

$$A^{-1} = \frac{1}{|A|} \begin{pmatrix} 3 & 1 & 5 \\ 6 & 2 & -12 \\ -16 & 2 & 10 \end{pmatrix}$$

Since $X = A^{-1}P$, we have

$$\begin{pmatrix} x \\ y \\ z \end{pmatrix} = \frac{1}{|A|} \begin{pmatrix} 3 & 1 & 5 \\ 6 & 2 & -12 \\ -16 & 2 & 10 \end{pmatrix} \begin{pmatrix} 2 \\ 7 \\ 4 \end{pmatrix},$$

which gives

$$x = \frac{1}{|A|}(6+7+20) = \frac{33}{22} = \frac{3}{2}$$

$$y = \frac{1}{|A|}(12+14-48) = \frac{-22}{22} = -1$$

$$z = \frac{1}{|A|}(-32+14+40) = \frac{22}{22} = 1$$

Thus, we obtain the same answers as those by Cramer's rule.

6.2 Operator

· Hermitian conjugate $A^\dagger$: consists of transposed and complex conjugated elements of a complex square operator A in the matrix representation

- A is Hermitian if $A^\dagger = A$
- A is orthogonal if $A^{-1} = A^T$
- A is unitary if $A^{-1} = A^\dagger$

An operator is very useful and an important concept in Quantum Mechanics.

When an operator corresponding to any physical observable acts on any physical state, it results in another physical state along with some measurable physical quantity. An expression of an operator in terms of a matrix is very powerful method to find out other physical state and observable quantity.

For a complex square operator A, the operator $A^\dagger$ is called **Hermitian conjugate** (or conjugate transpose) to the operator A in the matrix representation:

$$(A)_{ij}^\dagger = (A^*)_{ji},$$

where $(A)_{ij}$ is the element at i^{th} row and j^{th} column of the matrix A. Therefore, the Hermitian conjugate is the transposed elements which are complex conjugates for elements of the matrix.

(a) If $A = A^\dagger$ for the operator A in the matrix representation, the operator A is called **Hermitian** (or **self-adjoint**). Therefore,

$$(A)_{ij} = (A)_{ij}^\dagger = (A^*)_{ji},$$

so that the diagonal elements of the Hermitian operator should be real. The Hermitian has an important concept[9] in Quantum Mechanics. Examples of this operator are energy and momentum operators which provide measurable quantities, energy and momentum, respectively.

(b) When the operator A is called **invertible** if the operator in the matrix representation has the inverse matrix. If its inverse is equal to its transpose, $A^{-1} = A^T$, the operator is called **orthogonal**.

(c) If $U^\dagger U = UU^\dagger = I$ for an operator U, the operator is called the **unitary**.

---

[9] See Eqs. (6.4.4) and (6.4.5).

Therefore, the unitary operator in the matrix representation satisfies

$$U^{-1} = U^{\dagger}$$

Since a unitary transformation preserves norms[10], it is important in Quantum Mechanics.

Example

6.2.1 This is an example of the **coordinate transformation**.

A vector $\vec{A}$ is given by $\vec{A} = A_x\hat{x} + A_y\hat{y} + A_z\hat{z}$ in the rectangular coordinate system. When the coordinates are changed by rotation, $\vec{A}$ can be expressed in a new coordinate system (let's say it is a primed coordinate system):

$$\vec{A} = A_{x'}\hat{x}' + A_{y'}\hat{y}' + A_{z'}\hat{z}'$$

Each component of the $\vec{A}$ in the primed coordinate system is obtained by the scalar product:

$$\vec{A} \cdot \hat{x}' = A_{x'}\hat{x}' \cdot \hat{x}' + A_{y'}\hat{y}' \cdot \hat{x}' + A_{z'}\hat{z}' \cdot \hat{x}' = A_{x'}$$

because $\hat{x}' \cdot \hat{x}' = 1$ and $\hat{y}' \cdot \hat{x}' = \hat{z}' \cdot \hat{x}' = 0$.

Similarly, we get

$$\vec{A} \cdot \hat{y}' = A_{y'} \text{ and } \vec{A} \cdot \hat{z}' = A_{z'}$$

Since $\vec{A} = A_x\hat{x} + A_y\hat{y} + A_z\hat{z}$, we also have

$$A_{x'} = \vec{A} \cdot \hat{x}' = A_x\hat{x} \cdot \hat{x}' + A_y\hat{y} \cdot \hat{x}' + A_z\hat{z} \cdot \hat{x}'$$

$$A_{y'} = \vec{A} \cdot \hat{y}' = A_x\hat{x} \cdot \hat{y}' + A_y\hat{y} \cdot \hat{y}' + A_z\hat{z} \cdot \hat{y}'$$

$$A_{z'} = \vec{A} \cdot \hat{z}' = A_x\hat{x} \cdot \hat{z}' + A_y\hat{y} \cdot \hat{z}' + A_z\hat{z} \cdot \hat{z}'$$

These three equations can be expressed in the matrix representation as

$$\begin{pmatrix} A_{x'} \\ A_{y'} \\ A_{z'} \end{pmatrix} = \begin{pmatrix} \hat{x} \cdot \hat{x}' & \hat{y} \cdot \hat{x}' & \hat{z} \cdot \hat{x}' \\ \hat{x} \cdot \hat{y}' & \hat{y} \cdot \hat{y}' & \hat{z} \cdot \hat{y}' \\ \hat{x} \cdot \hat{z}' & \hat{y} \cdot \hat{z}' & \hat{z} \cdot \hat{z}' \end{pmatrix} \begin{pmatrix} A_x \\ A_y \\ A_z \end{pmatrix},$$

where the 3×3 matrix in the right-hand side of the equation is called a **transformation (or rotation** in this example) **matrix** (R) from one to another coordinates.

[10] See (2) and (3) in Sec. 6.4.

Consider the coordinate system rotated counterclockwise by an angle θ about the $z-$axis, shown in Figure.

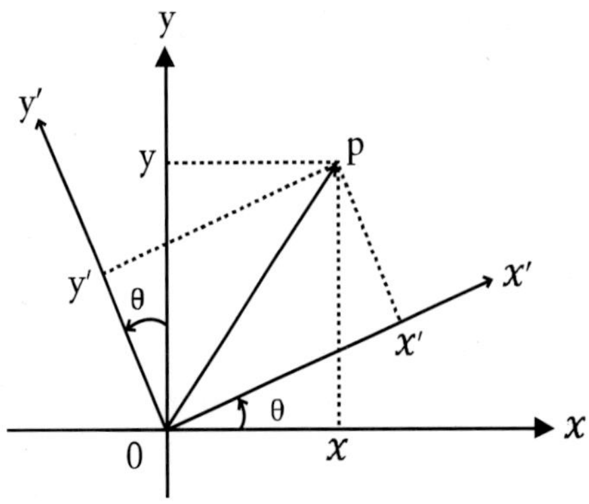

A new coordinate system $Ox'y'z'$ is obtained by rotating $Oxyz$ coordinate system through θ about the $z-$axis. The $Ox'y'z'$ has the same origin as the $Oxyz$.

Ans. Since the $z-$axis is not rotated, we have $\hat{z} = \hat{z}'$.

Thus, we get

$$\hat{z} \cdot \hat{z}' = \hat{z} \cdot \hat{z} = \cos 0 = 1$$

and

$$\hat{x} \cdot \hat{z}' = \hat{y} \cdot \hat{z}' = \cos\frac{\pi}{2} = 0$$

For other elements, we get

$$\hat{x} \cdot \hat{x}' = \cos\theta, \quad \hat{y} \cdot \hat{x}' = \cos\left(\frac{\pi}{2} - \theta\right) = \sin\theta,$$

and

$$\hat{z} \cdot \hat{x}' = \hat{z}' \cdot \hat{x}' = \cos\frac{\pi}{2} = 0$$

On the other hand, we get

$$\hat{x} \cdot \hat{y}' = \cos\left(\frac{\pi}{2} + \theta\right) = -\sin\theta, \quad \hat{y} \cdot \hat{y}' = \cos\theta,$$

and

$$\hat{z} \cdot \hat{y}' = \hat{z}' \cdot \hat{y}' = \cos\frac{\pi}{2} = 0$$

If we substitute these into the rotation matrix, we obtain the rotation matrix

$$R_z(\theta) = \begin{pmatrix} \cos\theta & \sin\theta & 0 \\ -\sin\theta & \cos\theta & 0 \\ 0 & 0 & 1 \end{pmatrix}$$

for the counterclockwise rotation by an angle θ about the $z-$axis.

If we substitute $-\theta$ instead of θ in the rotation matrix, we can obtain the rotation matrix for the clockwise rotation about the $z-$axis,

$$R_z(-\theta)=\begin{pmatrix}\cos\theta & -\sin\theta & 0\\ \sin\theta & \cos\theta & 0\\ 0 & 0 & 1\end{pmatrix}$$

6.2.2 When xyz coordinates are rotated clockwise by θ about the $z-$axis, new coordinates are expressed in the matrix representation using the rotation matrix obtained in the previous example:

$$\begin{pmatrix}x'\\ y'\\ z'\end{pmatrix}=R_z(-\theta)\begin{pmatrix}x\\ y\\ z\end{pmatrix}=\begin{pmatrix}\cos\theta & -\sin\theta & 0\\ \sin\theta & \cos\theta & 0\\ 0 & 0 & 1\end{pmatrix}\begin{pmatrix}x\\ y\\ z\end{pmatrix}$$

Since the coordinate in the $z-$axis is not changed, let's consider only x and y components. The cofactor and adjoint matrices of the rotation matrix are then

$$C=\begin{pmatrix}(-1)^{1+1}|M_{11}| & (-1)^{1+2}|M_{12}|\\ (-1)^{2+1}|M_{21}| & (-1)^{2+2}|M_{22}|\end{pmatrix}=\begin{pmatrix}\cos\theta & -\sin\theta\\ \sin\theta & \cos\theta\end{pmatrix}$$

and

$$C^T=\begin{pmatrix}\cos\theta & \sin\theta\\ -\sin\theta & \cos\theta\end{pmatrix},$$

respectively,

Since the inverse matrix of matrix A is given by $A^{-1}=\frac{1}{|A|}C^T$,

we obtain the inverse matrix of the rotation matrix is

$$\begin{aligned}R_z^{-1}(\theta) &= \frac{1}{\cos^2\theta+\sin^2\theta}\begin{pmatrix}\cos\theta & \sin\theta\\ -\sin\theta & \cos\theta\end{pmatrix}\\ &=\begin{pmatrix}\cos\theta & \sin\theta\\ -\sin\theta & \cos\theta\end{pmatrix}\end{aligned}$$

If we take complex conjugate for each element in the rotation matrix and transpose it, we get its Hermitian conjugate and transpose matrices are

$$(R^*)^T=R^\dagger=\begin{pmatrix}\cos\theta & \sin\theta\\ -\sin\theta & \cos\theta\end{pmatrix}$$

and

$$R^T=\begin{pmatrix}\cos\theta & \sin\theta\\ -\sin\theta & \cos\theta\end{pmatrix},$$

respectively.

Since $R^{-1}=R^T$ and $R^{-1}=R^\dagger$, the rotation matrix is the orthogonal and unitary matrix.

Since the orthogonal matrix is real, the orthogonal group forms a subgroup of the unitary. The rotation matrix has the determinant +1 and is the orthogonal, so that it is called the **SO(2) group**. The "S" denotes "Special" having the determinant of the matrix $+1$, "O" denotes for the "Orthogonal" and "2" denotes for the 2×2 matrix. Since the rotation matrix is also the unitary, it is also **SU(2) group**, in here "U" stands for the "Unitary".

Similarly, we can obtain the rotation matrices of the clockwise rotations by θ about the $x-$ and $y-$axis are

$$R_x(\theta)=\begin{pmatrix}1 & 0 & 0\\ 0 & \cos\theta & -\sin\theta\\ 0 & \sin\theta & \cos\theta\end{pmatrix}$$

and

$$R_y(\theta)=\begin{pmatrix}\cos\theta & 0 & \sin\theta\\ 0 & 1 & 0\\ -\sin\theta & 0 & \cos\theta\end{pmatrix},$$

respectively,

6.2.3 If xyz coordinates are rotated clockwise by θ_1 about the $z-$axis and rotation θ_2 is followed, we can express this operation by

$$R_z[-(\theta_1+\theta_2)]=R_z(-\theta_2)R_z(-\theta_1)^{[11]},$$

which gives

$$\begin{pmatrix}\cos(\theta_1+\theta_2) & -\sin(\theta_1+\theta_2)\\ \sin(\theta_1+\theta_2) & \cos(\theta_1+\theta_2)\end{pmatrix}=\begin{pmatrix}\cos\theta_2 & -\sin\theta_2\\ \sin\theta_2 & \cos\theta_2\end{pmatrix}\begin{pmatrix}\cos\theta_1 & -\sin\theta_1\\ \sin\theta_1 & \cos\theta_1\end{pmatrix}$$

$$=\begin{pmatrix}\cos\theta_1\cos\theta_2-\sin\theta_1\sin\theta_2 & -\sin\theta_1\cos\theta_2-\cos\theta_1\sin\theta_2\\ \cos\theta_1\sin\theta_2+\sin\theta_1\cos\theta_2 & -\sin\theta_1\sin\theta_2+\cos\theta_1\cos\theta_2\end{pmatrix}$$

By comparing elements in the matrices of the left- and right-hand sides of the equation, we obtain

$$\cos(\theta_1+\theta_2)=\cos\theta_1\cos\theta_2-\sin\theta_1\sin\theta_2$$
$$\sin(\theta_1+\theta_2)=\sin\theta_1\cos\theta_2+\cos\theta_1\sin\theta_2,$$

Thus, the matrix multiplication for the rotational operations gives the same formulas as those we obtained for the trigonometric functions in Section 1.1.

6.2.4 Since we know that $R(\theta_1+\theta_2)=R(\theta_2+\theta_1)$ and $R(\theta_1+\theta_2)=R(\theta_2)R(\theta_1)$

---

[11] The order of operations in the right-hand side of the equation is important. The operator for the first operation is positioned at the far right.

from the previous example, we get $R(\theta_1)R(\theta_2) = R(\theta_2)R(\theta_1)$.

The order of rotations does not effect the result, which is called **commutative** or **Abelian** in Group theory.

6.3 Eigenvalue Equation

- Solving an eigenvalue equation $AX = \lambda X$ means to diagonalize the matrix A
- Eigenvalues and eigenvectors corresponding to the matrix A are obtained by solving the characteristic equation,
  $\det(A - \lambda I) = |A - \lambda I| = 0$, where λ is the diagonal matrix.
- Eigenvector X diagonalizes the matrix A by the matrix multiplication $X^{-1}AX$

When $X^{-1}AX = \lambda$ and a square matrix λ is the diagonal matrix, the square matrix A is called **diagonalizable**.

Then, the diagonal elements of the matrix λ are called **eigenvalues** and corresponding column vectors of X are called **eigenvectors** for those eigenvalues.

For instance, if column vectors corresponding to their eigenvalues, λ_1 and λ_2, are $\begin{pmatrix} x_{11} \\ x_{21} \end{pmatrix}$ and $\begin{pmatrix} x_{12} \\ x_{22} \end{pmatrix}$, respectively, we get

$$\lambda = \begin{pmatrix} \lambda_1 & 0 \\ 0 & \lambda_2 \end{pmatrix} \text{ and } X = \begin{pmatrix} x_{11} & x_{12} \\ x_{21} & x_{22} \end{pmatrix}$$

Since the equation $X^{-1}AX = \lambda$ can be written as $AX = X\lambda$, it gives an **eigenvalue equation** in the form of

$$AX = \lambda X$$

The equation can be expressed as

$$(A-\lambda)X=0$$

The **invertible matrix theorem** in linear algebra says that if the matrix $(A-\lambda)$ is **invertible**, the equation $(A-\lambda)X=0$ has only the trivial solution $X=0$. Therefore, the matrix $(A-\lambda)$ should not be invertible to have a non-trivial solution X. This happens if and only if its determinant is zero. Thus, we have for the non-trivial solution

$$\det(A-\lambda)=|A-\lambda|=0,$$

which is called the **characteristic equation.**

Thus, solving the eigenvalue equation $AX=\lambda X$ means to diagonalize the matrix A.

Example

6.3.1 Consider the following simultaneous equations (or eigenvalue equation) which come from a coupled harmonic oscillator's problem in Classical Mechanics:

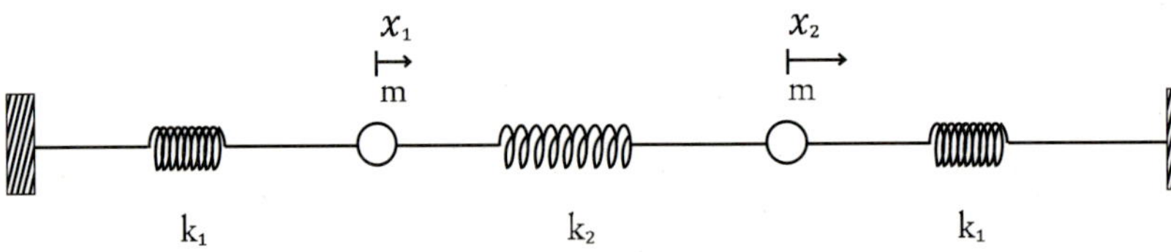

Two identical masses m are attached with springs whose spring constants are different.

We have the following relation, which is derived in [Ex. 9.1.13]:

$$\begin{pmatrix} k_1+k_2 & -k_2 \\ -k_2 & k_1+k_2 \end{pmatrix}\begin{pmatrix} a_1 \\ a_2 \end{pmatrix}=\lambda\begin{pmatrix} a_1 \\ a_2 \end{pmatrix}$$

By solving the characteristic equation, let's find eigenvalues and eigenvectors corresponding to those eigenvalues.

Ans. The original equation can be expressed as

$$\begin{pmatrix} k_1+k_2-\lambda & -k_2 \\ -k_2 & k_1+k_2-\lambda \end{pmatrix}\begin{pmatrix} a_1 \\ a_2 \end{pmatrix}=0$$

It's characteristic equation is then

$$\begin{vmatrix} k_1+k_2-\lambda & -k_2 \\ -k_2 & k_1+k_2-\lambda \end{vmatrix} = 0,$$

which gives

$$(k_1-\lambda)(k_1+2k_2-\lambda)=0$$

Therefore, we have two eigenvalues, $\lambda_1 = k_1$ and $\lambda_2 = k_1 + 2k_2$.

Thus, we obtain

$$\lambda = \begin{pmatrix} k_1 & 0 \\ 0 & k_1+2k_2 \end{pmatrix}$$

Let's find an eigenvector corresponding to each eigenvalue.

(i) If we insert $\lambda_1 = k_1$ into the original equation, we have

$$\begin{pmatrix} k_1+k_2-k_1 & -k_2 \\ -k_2 & k_1+k_2-k_1 \end{pmatrix}\begin{pmatrix} a_1 \\ a_2 \end{pmatrix} = 0,$$

which gives

$$\begin{pmatrix} k_2 & -k_2 \\ -k_2 & k_2 \end{pmatrix}\begin{pmatrix} a_1 \\ a_2 \end{pmatrix} = 0$$

From this, we get

$$k_2a_1 - k_2a_2 = 0 \Rightarrow a_1 = a_2$$

This gives the eigenvector $\begin{pmatrix} 1 \\ 1 \end{pmatrix}$.

Since a normalized eigenvector $C\begin{pmatrix} 1 \\ 1 \end{pmatrix}$ with a normalization constant C satisfies $|C|^2(1\ \ 1)\begin{pmatrix} 1 \\ 1 \end{pmatrix} = 1$, we get $C = \frac{1}{\sqrt{2}}$.

Thus, we obtain the normalized eigenvector $\frac{1}{\sqrt{2}}\begin{pmatrix} 1 \\ 1 \end{pmatrix}$.

(ii) For $\lambda_2 = k_1 + 2k_2$, we have

$$\begin{pmatrix} k_1+k_2-k_1-2k_2 & -k_2 \\ -k_2 & k_1+k_2-k_1-2k_2 \end{pmatrix}\begin{pmatrix} a_1 \\ a_2 \end{pmatrix} = 0,$$

which gives

$$\begin{pmatrix} -k_2 & -k_2 \\ -k_2 & -k_2 \end{pmatrix}\begin{pmatrix} a_1 \\ a_2 \end{pmatrix} = 0$$

From this, we get

$$-k_2a_1 - k_2a_2 = 0 \Rightarrow a_1 = -a_2$$

This gives the eigenvector $\begin{pmatrix} 1 \\ -1 \end{pmatrix}$.

Thus, the normalized eigenvector is $\frac{1}{\sqrt{2}}\begin{pmatrix} 1 \\ -1 \end{pmatrix}$.

Since the column vectors in X are the eigenvectors corresponding to the eigenvalues, we obtain

$$X=\begin{pmatrix}1 & 1\\ 1 & -1\end{pmatrix},$$

which gives its determinant $|X|$ is -2.

Let's check the relation $X^{-1}AX=\lambda$, where

$$A=\begin{pmatrix}k_1+k_2 & -k_2\\ -k_2 & k_1+k_2\end{pmatrix}$$

First, let's find the inverse matrix of X by using $X^{-1}=\frac{1}{|X|}C^T$:

$$X^{-1} = -\frac{1}{2}\begin{pmatrix}|M_{11}| & -|M_{12}|\\ -|M_{21}| & |M_{22}|\end{pmatrix} = -\frac{1}{2}\begin{pmatrix}-1 & -1\\ -1 & 1\end{pmatrix}$$

Then, we get

$$\begin{aligned}X^{-1}AX &= -\frac{1}{2}\begin{pmatrix}-1 & -1\\ -1 & 1\end{pmatrix}\begin{pmatrix}k_1+k_2 & -k_2\\ -k_2 & k_1+k_2\end{pmatrix}\begin{pmatrix}1 & 1\\ 1 & -1\end{pmatrix}\\ &= -\frac{1}{2}\begin{pmatrix}-1 & -1\\ -1 & 1\end{pmatrix}\begin{pmatrix}k_1 & k_1+2k_2\\ k_1 & -k_1-2k_2\end{pmatrix}\\ &= \begin{pmatrix}k_1 & 0\\ 0 & k_1+2k_2\end{pmatrix}\end{aligned}$$

Therefore, we can see that the matrix X diagonalizes the matrix A and gives the diagonalized matrix

$$\lambda=\begin{pmatrix}\lambda_1 & 0\\ 0 & \lambda_2\end{pmatrix},$$

whose each element corresponds to the eigenvalue.

6.3.2 Consider a problem which can be appeared in Quantum Mechanics, in which the Hamiltonian is given by

$$H=\begin{pmatrix}m_1\cos^2\alpha+m_2\sin^2\alpha & (m_1-m_2)\cos\alpha\sin\alpha\\ (m_1-m_2)\cos\alpha\sin\alpha & m_1\sin^2\alpha+m_2\cos^2\alpha\end{pmatrix}$$

Find eigenvalues and their corresponding eigenvectors of the Hamiltonian.

Ans. Let's find eigenvalues from the characteristic equation,

$$\begin{vmatrix}m_1\cos^2\alpha+m_2\sin^2\alpha-\lambda & (m_1-m_2)\cos\alpha\sin\alpha\\ (m_1-m_2)\cos\alpha\sin\alpha & m_1\sin^2\alpha+m_2\cos^2\alpha-\lambda\end{vmatrix}=0$$

It gives

$$\lambda^2-(m_1+m_2)\lambda+m_1m_2=0,$$

which gives two eigenvalues, $\lambda_1=m_1$ and $\lambda_2=m_2$.

Thus, we obtain

$$\lambda = \begin{pmatrix} m_1 & 0 \\ 0 & m_2 \end{pmatrix}$$

Now let's find an eigenvector corresponding to each eigenvalue.

(i) For $\lambda_1 = m_1$, we have

$$\begin{pmatrix} m_1\cos^2\alpha + m_2\sin^2\alpha - m_1 & (m_1 - m_2)\cos\alpha\sin\alpha \\ (m_1 - m_2)\cos\alpha\sin\alpha & m_1\sin^2\alpha + m_2\cos^2\alpha - m_1 \end{pmatrix}\begin{pmatrix} x_1 \\ x_2 \end{pmatrix} = 0.$$

This gives two simultaneous equations:

$$\begin{cases} x_1(m_2 - m_1)\sin^2\alpha = - x_2(m_1 - m_2)\cos\alpha\sin\alpha \\ x_1(m_1 - m_2)\cos\alpha\sin\alpha = - x_2(m_2 - m_1)\cos^2\alpha \end{cases} \qquad (6.3.2.1)$$

From the second equation of Eq. (6.3.2.1), we have

$$-(m_1 - m_2)\cos\alpha\sin\alpha = \frac{x_2(m_2 - m_1)\cos^2\alpha}{x_1}$$

If we insert this into the first equation of Eq. (6.3.2.1), it gives

$x_1 = \cos\alpha$ and $x_2 = \sin\alpha$.

Thus, the eigenvector corresponding to the eigenvalue $\lambda_1 = m_1$ is

$$\begin{pmatrix} \cos\alpha \\ \sin\alpha \end{pmatrix}$$

(ii) For $\lambda_2 = m_2$, similarly, we can obtain the corresponding eigenvector

$$\begin{pmatrix} \sin\alpha \\ -\cos\alpha \end{pmatrix}$$

Thus, the eigenvector matrix of the Hamiltonian corresponding to the eigenvalues, m_1 and m_2, is

$$X = \begin{pmatrix} \cos\alpha & \sin\alpha \\ \sin\alpha & -\cos\alpha \end{pmatrix}$$

6.4 Notation in Quantum Mechanics

· Dirac notation

- $< X|$ and $|X>$ are "bra" and "ket" vectors, respectively.

- The bra and ket in the matrix representation correspond to the row and column matrices, respectively.

· $< A|B> = \int A^*(x)B(x)dx$ (Inner product)

· $|A>< B|$ (Outer product)

· $\sum_n |\phi_n >< \phi_n| = 1$ (Completeness relation)

The Hermitian and unitary operators (or matrices) are important physical concepts in Quantum Mechanics and it is meaningful to look at their properties. Let's start with frequently used notations in Quantum Mechanics.

In the eigenvalue equation $AX = \lambda X$, an operator A and eigenvector X can be expressed in terms of the matrices. If A is an $n \times n$ square matrix, X is consisted of eigenvectors corresponding to eigenvalues in columns. The column matrix is written as $|X>$. On the other hand, the row matrix is a transpose of the column matrix and can be written as $< X|$.

Since "$<\ \ >$" is called "braket", $< X|$ and $|X>$ are called "**bra**" and "**ket**" vectors, respectively. We call this representation **Dirac notation.**

Since the matrix multiplication of the row vector A times column vector B results in a number, it is possible to write the inner product of A and B (which is expressed as $\int A^*(x)B(x)dx$ in one-dimensional complex vector space) by the bra and ket, $< A|B>$.

Therefore, the inner product of A and B is

$$< A|B> = \int A^*(x)B(x)dx = [\int B^*(x)A(x)dx]^* = < B|A >^*$$

Thus, we obtain

$$<A|B>=<B|A>^*$$

in the complex vector space.

For $A=\begin{pmatrix} a_1 \\ a_2 \\ \vdots \\ \vdots \\ a_n \end{pmatrix}$ and $B=\begin{pmatrix} b_1 \\ b_2 \\ \vdots \\ \vdots \\ b_n \end{pmatrix}$,

we get

$$<A|B>=(a_1^*,\ a_2^*,\cdots\cdots,a_n^*)\begin{pmatrix} b_1 \\ b_2 \\ \vdots \\ \vdots \\ b_n \end{pmatrix}=a_1^*b_1+a_2^*b_2+\cdots\cdots+a_n^*b_n$$

Keep in mind that the bra vector is the row matrix having complex conjugating elements.

When $<A|B>=0$, we call that two vectors, A and B, are orthogonal. We have

$$<A|A>=a_1^*a_1+a_2^*a_2+\cdots\cdots+a_n^*a_n=\sum_n |a_n|^2$$

Then, $\sqrt{<A|A>}=\|A\|$ is called the **norm** (or **magnitude**) of the vector A. For instance, the norm of the vector $A=a_x\hat{x}+a_y\hat{y}+a_z\hat{z}$ is

$$\|A\|=\sqrt{a_x^2+a_y^2+a_z^2}$$

When the **orthonormal** basis satisfying $<\phi_m|\phi_n>=\delta_{mn}$ is defined, an arbitrary vector $|\Psi>$ can be written by using these basis states $|\phi_n>$:

$$\begin{aligned} |\Psi> &= \alpha_1|\phi_1>+\alpha_2|\phi_2>+\ \cdots\cdots \\ &\qquad \text{(with expansion coefficients } \alpha_n;\ n=1,2,\cdots\cdots) \\ &= \sum_n \alpha_n|\phi_n> \end{aligned} \tag{6.4.1}$$

For examplc, $\hat{x}$, $\hat{y}$, and $\hat{z}$ are the orthonormal basis and any position vectors in the rectangular coordinate system are expressed by these orthonormal basis set.

The expansion coefficients α_m are determined by multiplying $<\phi_m|$ on the both sides of the equation,

$$\begin{aligned} <\phi_m|\Psi> &= \sum_n \alpha_n <\phi_m|\phi_n> \\ &= \sum_n \alpha_n \delta_{mn} = \alpha_m \end{aligned} \tag{6.4.2}$$

By substituting this into Eq. (6.4.1), we obtain

$$|\Psi>=\sum_n <\phi_n|\Psi>|\phi_n>=\sum_n |\phi_n><\phi_n|\Psi>, \tag{6.4.3}$$

where we change the order of the constant and vector.

For the equality of the equation, we have

$$\sum_n |\phi_n><\phi_n| = 1$$

This is called **completeness relation.**

For instance, $\sum_n |\phi_n><\phi_n|$ in the rectangular coordinate system gives

$$\begin{aligned} |\hat{x}><\hat{x}| + |\hat{y}><\hat{y}| + |\hat{z}><\hat{z}| &= \begin{pmatrix}1\\0\\0\end{pmatrix}(1\ 0\ 0) + \begin{pmatrix}0\\1\\0\end{pmatrix}(0\ 1\ 0) + \begin{pmatrix}0\\0\\1\end{pmatrix}(0\ 0\ 1) \\ &= \begin{pmatrix}1 & 0 & 0\\0 & 1 & 0\\0 & 0 & 1\end{pmatrix} \end{aligned}$$

The $|\phi_n><\phi_n|$ is called the **outer product** in contrast to the inner product. In the operator concept, the outer product $|\phi_n><\phi_n|$ is called a **projection operator** P.

We have

$$P^2 = PP = |\phi_m><\phi_m|\phi_n><\phi_n| = \delta_{mn}|\phi_m><\phi_n| = |\phi_n><\phi_n| = P$$

so that the projection operator has the property of $P^2 = P$.

If we define $<\phi_n|\Psi>=\Psi_n$, Eq. (6.4.3) is given as

$$|\Psi>=\sum_n \Psi_n|\phi_n>$$

Since we can write the orthonormal basis, $|\phi_1>=\begin{pmatrix}1\\0\\0\\ \vdots\\ \vdots\end{pmatrix}$, $|\phi_2>=\begin{pmatrix}0\\1\\0\\ \vdots\\ \vdots\end{pmatrix}$, ……,

etc, we get

$$|\Psi> \ = \ \Psi_1|\phi_1>+\Psi_2|\phi_2>+\Psi_3|\phi_3>+ \ \cdots\cdots \ = \begin{pmatrix}\Psi_1\\ \Psi_2\\ \Psi_3\\ \vdots\\ \vdots\end{pmatrix}$$

Consider an operation A acts on a state $|\Psi>$ and results in another state $|\varphi>$,

$$A|\Psi> \ = \ |\varphi>$$

By using the completeness relation, the left-hand side of the equation is given as

$$A|\Psi>=A \quad |\phi_n><\phi_n|\Psi>= \quad A|\phi_n><\phi_n|\Psi>$$

If we multiply $<\phi_m|$ on the both sides of the equation $A|\Psi>=|\varphi>$, we obtain

$$\sum_n <\phi_m|A|\phi_n><\phi_n|\Psi>=<\phi_m|\varphi>=\varphi_m$$

As we define $<\phi_n|\Psi>=\Psi_n$, we can write $<\phi_m|\varphi>=\varphi_m$.

To find a meaning of $<\phi_m|A|\phi_n>$ in the equation, let's calculate $<\phi_2|A|\phi_3>$:

$$<\phi_2|A|\phi_3> \ = (0 \ 1 \ 0 \cdots\cdots 0)\begin{pmatrix}a_{11} & a_{12} & a_{13} & \cdots & \cdots\\ a_{21} & a_{22} & a_{23} & \cdots & \cdots\\ \vdots & \vdots & \ddots & \vdots & \vdots\\ \vdots & \vdots & \vdots & \ddots & \vdots\\ a_{m1} & a_{m2} & a_{m3} & \cdots & \cdots\end{pmatrix}\begin{pmatrix}0\\0\\1\\ \vdots\\ \vdots\\0\end{pmatrix}$$

$$= (0 \ 1 \ 0 \cdots\cdots 0)\begin{pmatrix}a_{13}\\ a_{23}\\ a_{33}\\ \vdots\\ \vdots\\ a_{m3}\end{pmatrix} = a_{23}$$

Therefore, the $<\phi_m|A|\phi_n>$ on the left-hand side of the equation gives an element at m^{th} row and n^{th} column of the matrix A.

Therefore, we can write the equation

$$\varphi_m = \sum_n A_{mn} \Psi_n$$

Thus, the operator equation is expressed in the matrix representation as:

$$\begin{pmatrix} \varphi_1 \\ \varphi_1 \\ \vdots \\ \vdots \end{pmatrix} = \begin{pmatrix} a_{11} & a_{12} & \cdots & \cdots \\ a_{21} & a_{22} & \cdots & \cdots \\ \vdots & \vdots & \ddots & \vdots \\ a_{m1} & a_{m2} & \cdots & \cdots \end{pmatrix} \begin{pmatrix} \Psi_1 \\ \Psi_2 \\ \vdots \\ \vdots \end{pmatrix}$$

Let's take a look at some of important properties of Hermitian and unitary operators.

(1) Consider the Hermitian operator A and eigenvector X. Using Dirac notation the eigenvalue equation can be written as

$$A|X> \;=\; \lambda|X>$$

By multiplying $<X|$ on the both sides of the equation, we get

$$<X|A|X> \;=\; \lambda <X|X>$$

If we take their complex conjugates on the both sides of the equation, we have

$$<X|A|X>^* = \lambda^* <X|X>^* = \lambda^* <X|X> \tag{6.4.4}$$

Since A is the Hermitian operator, we have

$$<X|A|X>^* \;=\; <AX|X> \;=\; \lambda <X|X> \tag{6.4.5}$$

From Eqs. (6.4.4) and (6.4.5), we obtain

$$\lambda <X|X> \;=\; \lambda^* <X|X>$$

Thus, **the eigenvalues of the Hermitian operator are real numbers**. This is a very important concept in Quantum Mechanics because physical quantities we measure give real numbers.

(2) Consider **unitary transformation** in which the unitary operator is used to change a vector in one orthonormal basis set into a vector in another orthonormal basis set:

$$|X^{'}> = U|X>,$$

where U is the unitary operator, $|X>$ is the vector in unprimed orthonormal basis set and $|X^{'}>$ is the vector in primed orthonormal basis set.
If we apply $U^{\dagger}$ on both sides of the equation, we get

$$|X> = U^{\dagger}|X^{'}>$$

For instance, consider a vector $\overrightarrow{X} = x\hat{x} + y\hat{y} + z\hat{z}$ in the rectangular coordinate system. It is expressed in terms of the matrix

$$|X> = \begin{pmatrix} x \\ y \\ z \end{pmatrix}$$

in the orthonormal basis $\hat{x}, \hat{y}, \hat{z}$.
We know that the rotation matrix is the unitary and gives

$$|X^{'}> = R_z(\theta)|X> = \begin{pmatrix} \cos\theta & \sin\theta & 0 \\ -\sin\theta & \cos\theta & 0 \\ 0 & 0 & 1 \end{pmatrix}\begin{pmatrix} x \\ y \\ z \end{pmatrix} = \begin{pmatrix} x\cos\theta + y\sin\theta \\ -x\sin\theta + y\cos\theta \\ z \end{pmatrix}$$

This is the same vector but in different orthonormal basis.

Consider the following linear transformation,

$$|Y> = A|X>$$

Then, we have

$$|Y^{'}> = U|Y> = UA|X> = UAU^{\dagger}|X^{'}>$$

By defining $UAU^{\dagger} = A^{'}$ called **similarity transformation**, we get

$$|Y^{'}> = A^{'}|X^{'}>$$

Therefore, the unitary transformation changes the matrix (or operator) in one orthonormal basis set into the matrix in another orthonormal basis set.

Then, we get for the unitary transformation

$$< X^{'}|X^{'}> = < X|U^{+}U|X> = < X|U^{-1}U|X> = < X|X>,$$

which gives $||X^{'}|| = ||X||$.
Therefore, the norm is not changed in the unitary transformation. We conclude that unitary transformation preserves the inner product.

(3) The eigenvalue equation for the unitary operator U is

$$U|X> \;=\; \lambda|X>$$

This gives

$$< UX|UX> \;=\; < \lambda X|\lambda X>$$

The left- and right-hand sides of the equation are

$$< U^{+}UX|X> \;=\; < X|X> \text{ and } \lambda^{*}\lambda < X|X> \;=\; |\lambda|^{2} < X|X>,$$

respectively.

Thus, it gives

$$< X|X> \;=\; |\lambda|^{2} < X|X>$$

so that we have

$$|\lambda|^{2} = 1$$

Thus, the eigenvalue of the unitary operator is in the form of the function $e^{\pm ix}$[12].

Example

6.4.1 Calculate $< A|B>$ for $A = \begin{pmatrix} a_1 \\ a_2 \end{pmatrix}$ and $B = \begin{pmatrix} b_1 \\ b_2 \end{pmatrix}$ in the complex vector space.

Ans. Since the bra and ket vectors are represented by the row and column matrices, respectively, we have

$$\begin{aligned} < A|B> &= \begin{pmatrix} a_1^{*} & a_2^{*} \end{pmatrix}\begin{pmatrix} b_1 \\ b_2 \end{pmatrix} \\ &= a_1^{*}b_1 + a_2^{*}b_2 \end{aligned}$$

6.4.2 Let's assume that two level system described by the orthonormal basis vectors $|1>$ and $|2>$ has Hamiltonian H given by

$$H = m\{|1><1| + |2><2|\} + g\{|1><2| + |2><1|\},$$

where m and g are constants.

Find eigenvalues and their corresponding eigenvectors.

[12] An example is given in [Ex. 6.4.6].

Ans. First, let's express the Hamiltonian in the matrix representation:

$$|1><1| = \begin{pmatrix}1\\0\end{pmatrix}(1 \quad 0) = \begin{pmatrix}1 & 0\\0 & 0\end{pmatrix}$$

$$|2><2| = \begin{pmatrix}0\\1\end{pmatrix}(0 \quad 1) = \begin{pmatrix}0 & 0\\0 & 1\end{pmatrix}$$

$$|1><2| = \begin{pmatrix}1\\0\end{pmatrix}(0 \quad 1) = \begin{pmatrix}0 & 1\\0 & 0\end{pmatrix}$$

$$|2><1| = \begin{pmatrix}0\\1\end{pmatrix}(1 \quad 0) = \begin{pmatrix}0 & 0\\1 & 0\end{pmatrix}$$

The Hamiltonian H is then expressed in the matrix,

$$\begin{aligned} H &= m\left\{\begin{pmatrix}1 & 0\\0 & 0\end{pmatrix}+\begin{pmatrix}0 & 0\\0 & 1\end{pmatrix}\right\} + g\left\{\begin{pmatrix}0 & 1\\0 & 0\end{pmatrix}+\begin{pmatrix}0 & 0\\1 & 0\end{pmatrix}\right\} \\ &= \begin{pmatrix}m & g\\g & m\end{pmatrix}\end{aligned}$$

Second, let's find eigenvalues from the characteristic equation:

$$\begin{vmatrix}m-\lambda & g\\ g & m-\lambda\end{vmatrix} = 0$$

This gives

$$\lambda = m+g \text{ or } \lambda = m-g$$

i) For $\lambda_1 = m+g$,

$$\begin{pmatrix}-g & g\\ g & -g\end{pmatrix}\begin{pmatrix}x_1\\x_2\end{pmatrix} = 0$$

This equation yields

$$-gx_1 + gx_2 = 0 \Rightarrow x_1 = x_2$$

so that we have the eigenvector $\begin{pmatrix}1\\1\end{pmatrix}$.

Then, the normalized eigenvector is

$$|\Psi_1> = \frac{1}{\sqrt{2}}\begin{pmatrix}1\\1\end{pmatrix}$$

ii) For $\lambda_2 = m-g$, we have

$$\begin{pmatrix}g & g\\ g & g\end{pmatrix}\begin{pmatrix}x_1\\x_2\end{pmatrix} = 0,$$

which gives

$$x_1 = -x_2$$

It gives the eigenvector $\begin{pmatrix}1\\-1\end{pmatrix}$,

which gives the normalized eigenvector

$$|\Psi_2> = \frac{1}{\sqrt{2}}\begin{pmatrix}1\\-1\end{pmatrix}$$

6.4.3 We know that the rotation operator is the unitary from [Ex. 6.2.2]. Consider the clockwise rotation by θ about the $z-$axis.

Ans. For $X = x\hat{x} + y\hat{y}$, we have

$$< X|X> = (x \quad y)\begin{pmatrix} x \\ y \end{pmatrix} = x^2 + y^2$$

before the rotation.

The length is then

$$\sqrt{< A_i|A_i >} = \sqrt{x^2 + y^2}$$

Since $|X^{'}> = R_z(-\theta)|X>$ for the rotation, we get

$$|X^{'}> = \begin{pmatrix} \cos\theta & -\sin\theta \\ \sin\theta & \cos\theta \end{pmatrix}\begin{pmatrix} x \\ y \end{pmatrix} = \begin{pmatrix} x\cos\theta - y\sin\theta \\ x\sin\theta + y\cos\theta \end{pmatrix}$$

so that

$$\begin{aligned} < X^{'}|X^{'}> &= (x\cos\theta - y\sin\theta \quad x\sin\theta + y\cos\theta)\begin{pmatrix} x\cos\theta - y\sin\theta \\ x\sin\theta + y\cos\theta \end{pmatrix} \\ &= x^2 + y^2 \end{aligned}$$

The length of the vector in the rotated system is then

$$\sqrt{< X^{'}|X^{'}>} = \sqrt{x^2 + y^2}$$

Thus, we obtain

$$||X|| = ||X^{'}||$$

As expected, the length of the position vector is preserved under the unitary transformation.

6.4.4 The time-evolution operator is given by $e^{-\frac{i}{\hbar}Ht}$, where $H = -\frac{\hbar^2}{2m}\nabla^2$, considering a free particle with mass m for simplicity. Show that the time-evolution operator is the unitary.

Ans. First, let's show that the Hamiltonian is the Hermitian operator.

For a free particle moving in $[0,\ a]$ in one dimension, the Hamiltonian is

$$H = -\frac{\hbar^2}{2m}\frac{d^2}{dx^2}$$

For any functions $f(x)$ and $g(x)$ satisfying the same boundary conditions as the Hamiltonian, we have

$$\begin{aligned} < f(x)|Hg(x)> &= -\frac{\hbar^2}{2m}\int_0^a f^*(x)\frac{d}{dx}[\frac{dg(x)}{dx}]dx \\ &= -\frac{\hbar^2}{2m}[f^*(x)\frac{dg(x)}{dx}]_{x=0}^{x=a} + \frac{\hbar^2}{2m}\int_0^a \frac{df^*(x)}{dx}\frac{dg(x)}{dx}dx \end{aligned}$$

The first term of the right-hand side of the equation is zero because $f^*(0) = f^*(a) = 0$ by the boundary conditions.

If we apply integration by parts again to the equation, we get

$$< f(x)|Hg(x) > = \frac{\hbar^2}{2m}[\frac{df^*(x)}{dx}g(x)]_{x=0}^{x=a} - \frac{\hbar^2}{2m}\int_0^a \frac{d^2f^*(x)}{dx^2}g(x)dx$$

The first term of the right-hand side of the equation is zero because $g(0) = g(a) = 0$ by the boundary conditions.

The equation is then

$$\begin{aligned} < f(x)|Hg(x) > &= \int_0^a \left\{-\frac{\hbar^2}{2m}\frac{d^2f^*(x)}{dx^2}\right\}g(x)dx \\ &= \int_0^a Hf^*(x)g(x)dx = < H^*f(x)|g(x) > \\ &= < f(x)|H^\dagger g(x) > \end{aligned}$$

Thus, the Hamiltonian H is Hermitian operator, $H^\dagger = H$.

The Hermitian conjugate of the time-evolution operator $e^{-\frac{i}{\hbar}Ht}$ is then

$$e^{+\frac{i}{\hbar}H^+t} = e^{+\frac{i}{\hbar}Ht}$$

Since $e^{-\frac{i}{\hbar}Ht}e^{+\frac{i}{\hbar}Ht} = 1$, $e^{+\frac{i}{\hbar}Ht}$ is the inverse of the time-evolution operator so that the Hermitian conjugate is the same as the inverse for the time-evolution operator. Thus, the time-evolution operator is the unitary.

6.4.5 When a state at time $t = 0$ in [Ex. 6.4.2] is given by $|\varphi(0)> = |1>$, find a state $|\varphi(t)>$.

Ans. We already obtained the normalized eigenstates corresponding to eigenvalues $\lambda_1 = m + g$ and $\lambda_2 = m - g$ are $|\Psi_1>$ and $|\Psi_2>$, respectively, where

$$|\Psi_1> = \frac{1}{\sqrt{2}}\begin{pmatrix}1\\1\end{pmatrix} = \frac{1}{\sqrt{2}}\{|1>+|2>\} \tag{6.4.5.1}$$

and

$$|\Psi_2> = \frac{1}{\sqrt{2}}\begin{pmatrix}1\\-1\end{pmatrix} = \frac{1}{\sqrt{2}}\{|1>-|2>\} \tag{6.4.5.2}$$

By adding these two equations, we can express the initial state $|\varphi(0)>$ in terms of the normalized eigenstates:

$$|1> = |\varphi(0)> = \frac{1}{\sqrt{2}}\{|\Psi_1>+|\Psi_2>\}$$

We can obtain the state at time t, $|\varphi(t)>$, by applying the time-evolution operator into the initial state. Since the time-evolution operator is given by $e^{-\frac{i}{\hbar}Ht}$ in the previous example, the state at time t is

$$
\begin{aligned}
|\varphi(t)> &= e^{-\frac{i}{\hbar}Ht}|\varphi(0)> \\
&= \frac{1}{\sqrt{2}}e^{-\frac{i}{\hbar}Ht}[|\Psi_1>+|\Psi_2>] \qquad (6.4.5.3)
\end{aligned}
$$

Since $|\Psi_1>$ and $|\Psi_2>$ are the normalized eigenstates corresponding to the eigenvalues $\lambda_1 = m+g$ and $\lambda_2 = m-g$, we have

$$H|\Psi_1> = \lambda_1|\Psi_1> = (m+g)|\Psi_1> \qquad (6.4.5.4)$$

and

$$H|\Psi_2> = \lambda_2|\Psi_2> = (m-g)|\Psi_2> \qquad (6.4.5.5)$$

Let's calculate $e^{-\frac{i}{\hbar}Ht}|\Psi_1>$ in Eq. (6.4.5.3) by using Taylor series for the exponential function:

$$
\begin{aligned}
e^{-\frac{i}{\hbar}Ht}|\Psi_1> &= \left\{1+(-\frac{i}{\hbar}Ht)+\frac{1}{2!}(-\frac{i}{\hbar}Ht)(-\frac{i}{\hbar}Ht)+\cdots\cdots\right\}|\Psi_1> \\
&= |\Psi_1>-\frac{i}{\hbar}(m+g)t|\Psi_1>+\frac{1}{2!}[-\frac{i}{\hbar}(m+g)t]^2|\Psi_1> \\
&\quad + \cdots\cdots \text{ (using Eq. (6.4.5.4))} \\
&= \left\{1-\frac{i}{\hbar}(m+g)t+\frac{1}{2!}[-\frac{i}{\hbar}(m+g)t]^2+ \cdots\cdots\right\}|\Psi_1> \\
&= e^{-\frac{i}{\hbar}(m+g)t}|\Psi_1> \qquad (6.4.5.6)
\end{aligned}
$$

Similarly, we get

$$e^{-\frac{i}{\hbar}Ht}|\Psi_2> = e^{-\frac{i}{\hbar}(m-g)t}|\Psi_2> \qquad (6.4.5.7)$$

By substituting these into (6.4.5.3), we obtain the state at time t as

$$|\varphi(t)> = \frac{1}{\sqrt{2}}e^{-\frac{i}{\hbar}(m+g)t}|\Psi_1>+\frac{1}{\sqrt{2}}e^{-\frac{i}{\hbar}(m-g)t}|\Psi_2>$$

6.4.6 It was proved in [Ex. 6.2.2] that the rotation operator is the unitary. Let's find eigenvalues of this rotation operator.

Ans. For the system rotated clockwise by θ about the $z-$axis, the characteristic equation is

$$\begin{vmatrix} \cos\theta - \lambda & -\sin\theta \\ \sin\theta & \cos\theta - \lambda \end{vmatrix} = 0,$$

which gives

$$\begin{aligned} \lambda &= \cos\theta \pm \sqrt{\cos^2\theta - 1} \\ &= \cos\theta \pm i\sin\theta = e^{\pm i\theta} \end{aligned}$$

Thus, the eigenvalue of the unitary operator is in the form of the exponential function.

6.4.7 A general quadratic equation for a conic sections is $ax^2 + bxy + cy^2 + dx + ey + f = 0$. When the discriminant $b^2 - 4ac$ is negative, zero, or positive, the quadratic equation corresponds to the parabolic, ellipse or hyperbola, respectively.

Ans. This can be proved by removing the cross-product term in the quadratic equation. It is achieved by rotating the coordinates.

If we define

$$X = \begin{pmatrix} x \\ y \end{pmatrix}, \; Q = \begin{pmatrix} a & \frac{b}{2} \\ \frac{b}{2} & c \end{pmatrix} \text{ and } K = (d \quad e),$$

the quadratic equation is expressed in terms of the matrices,

$$X^T Q X + K X + f = 0 \tag{6.4.7.1}$$

Let's find a matrix P which diagonalizes the matrix Q:

$$QP = \lambda P,$$

where λ is the diagonal matrix.

For this purpose, we need to find eigenvalues and their corresponding eigenvectors. The characteristic equation for $(Q - \lambda)P = 0$ is then

$$\begin{vmatrix} a - \lambda & \frac{b}{2} \\ \frac{b}{2} & c - \lambda \end{vmatrix} = 0,$$

which gives the eigenvalues

$$\lambda_1 = \frac{(a+c) + \sqrt{(a-c)^2 + b^2}}{2}$$

and

$$\lambda_2 = \frac{(a+c) - \sqrt{(a-c)^2 + b^2}}{2}$$

Let's find the eigenvectors corresponding to the eigenvalues, $\lambda_{1,2}$.

(i) For λ_1,

we have

$$[(a-c)-\sqrt{(a-c)^2+b^2}\,]p_1=-bp_2$$

Therefore, we get

$$p_1=(a-c)+\sqrt{(a-c)^2+b^2} \text{ and } p_2=b$$

This gives the normalized eigenvector

$$\frac{1}{\sqrt{\{(a-c)+\sqrt{(a-c)^2+b^2}\}^2+b^2}}\begin{pmatrix}(a-c)+\sqrt{(a-c)^2+b^2}\\ b\end{pmatrix}\equiv\begin{pmatrix}p_{11}\\ p_{21}\end{pmatrix},$$

where p_{11} and p_{21} are the elements of the column matrix for the eigenvector corresponding to the eigenvalue λ_1.

(ii) For λ_2,

similarly, we can obtain the normalized eigenvector

$$\frac{1}{\sqrt{\{(a-c)+\sqrt{(a-c)^2+b^2}\}^2+b^2}}\begin{pmatrix}-b\\ (a-c)+\sqrt{(a-c)^2+b^2}\end{pmatrix}\equiv\begin{pmatrix}p_{12}\\ p_{22}\end{pmatrix},$$

where p_{12} and p_{22} are the elements of the column matrix for the eigenvector corresponding to the eigenvalue λ_2.

From (i) and (ii), we obtain the matrix P which diagonalizes the matrix Q is

$$P=\begin{pmatrix}p_{11} & p_{12}\\ p_{21} & p_{22}\end{pmatrix}$$

The determinant of the matrix P is

$$\det(P)=p_{11}p_{22}-p_{12}p_{21}=1$$

We already know that the determinant of the rotation matrix is 1. This makes for us to consider the following rotational transformation, $X=PX'$.

If we substitute this transformation into Eq. (6.4.7.1), we obtain

$$(PX')^TQ(PX')+K(PX')+f=0,$$

which is

$$(X')^TP^TQPX'+KPX'+f=0 \tag{6.4.7.2}$$

Let's calculate P^TQP in the equation:

$$P^T=\begin{pmatrix}p_{11} & p_{21}\\ p_{12} & p_{22}\end{pmatrix}$$

$$=\frac{1}{\sqrt{\{(a-c)+\sqrt{(a-c)^2+b^2}\}^2+b^2}}\begin{pmatrix}(a-c)+\sqrt{(a-c)^2+b^2} & b\\ -b & (a-c)+\sqrt{(a-c)^2+b^2}\end{pmatrix}$$

On the other hand,

$$P^{-1}=\frac{1}{|P|}C^T=\begin{pmatrix}C_{11} & C_{21}\\ C_{12} & C_{22}\end{pmatrix}=\begin{pmatrix}|M_{11}| & -|M_{21}|\\ -|M_{12}| & |M_{22}|\end{pmatrix}$$

$$=\frac{1}{\sqrt{\{(a-c)+\sqrt{(a-c)^2+b^2}\}^2+b^2}}\begin{pmatrix}(a-c)+\sqrt{(a-c)^2+b^2} & b\\ -b & (a-c)+\sqrt{(a-c)^2+b^2}\end{pmatrix}$$

Therefore, we get $P^T=P^{-1}$.

Since the matrix P diagonalizes the matrix Q, we have

$$P^TQP=P^{-1}QP=\begin{pmatrix}\lambda_1 & 0\\ 0 & \lambda_2\end{pmatrix}$$

If we substitute this into Eq. (6.4.7.2), the equation is now expressed in the rotated system

$$(x' \quad y')\begin{pmatrix}\lambda_1 & 0\\ 0 & \lambda_2\end{pmatrix}\begin{pmatrix}x'\\ y'\end{pmatrix}+(d \quad e)\begin{pmatrix}p_{11} & p_{12}\\ p_{21} & p_{22}\end{pmatrix}\begin{pmatrix}x'\\ y'\end{pmatrix}+f=0$$

This gives

$$\lambda_1 x'^2+\lambda_2 y'^2+(dp_{11}+ep_{21})x'+(dp_{12}+ep_{22})y'+f=0$$

Therefore, we remove the cross-product term in the equation for the conic section by introducing rotational transformation of coordinates.

This equation can be written as

$$\lambda_1(x'+\frac{dp_{11}+ep_{21}}{2\lambda_1})^2+\lambda_2(y'+\frac{dp_{12}+ep_{22}}{2\lambda_2})^2$$
$$=(\frac{dp_{11}+ep_{21}}{2\lambda_1})^2+(\frac{dp_{12}+ep_{22}}{2\lambda_2})^2-f$$

We know that this equation is for the ellipse, hyperbola, or parabola, respectively, when $\lambda_1\lambda_2=\frac{4ac-b^2}{4}$ is positive, negative or zero.

Thus, the equation is the ellipse, hyperbola, and parabola when the discriminant b^2-4ac is negative, positive and zero, respectively, as shown in Figure.

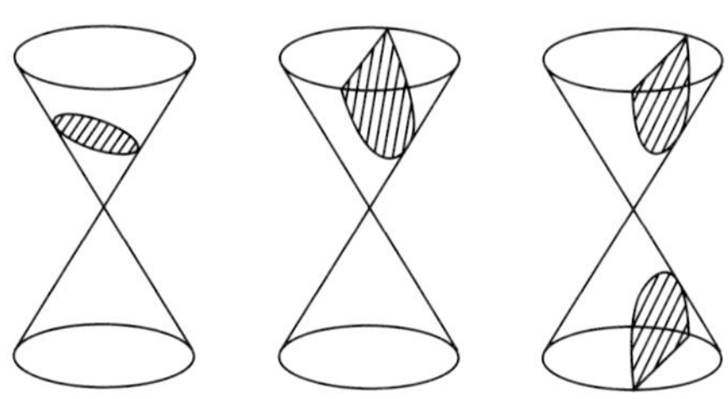

The conic sections are defined as a quadratic equation with two variables. Types of the conic section are the ellipse, parabola and hyperbola depending on values of the discriminant of the quadratic equation.

Classification of the curve's type by the discriminant for the quadratic equation $ax^2+bxy+cy^2+dx+ey+f=0$:

- $b^2-4ac<0$, ellipse (circle for $a=c$)
- $b^2-4ac>0$, hyperbola
- $b^2-4ac=0$, parabola

6.4.8 We have the quadratic equation, $\frac{x^2}{A^2}-2\frac{xy}{AB}\cos\Delta+\frac{y^2}{B^2}=\sin^2\Delta$, for two-dimensional isotropic oscillator. Prove that this equation gives the ellipse.

Ans. The discriminant of this equation is

$$(-2\frac{\cos\Delta}{AB})^2-4\frac{1}{A^2}\frac{1}{B^2} \quad = \quad -(\frac{2\sin\Delta}{AB})^2,$$

which is always a negative.

From the obtained fact of the previous example, we know that this quadratic equation is for the ellipse.

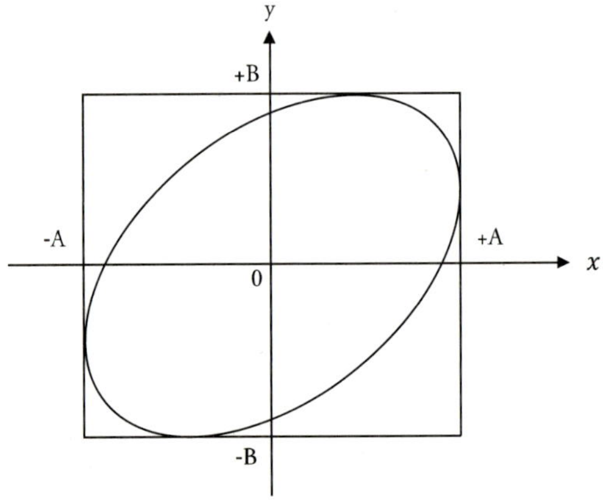

An ellipse in the rectangular for trajectory of a two-dimensional isotropic oscillator.

Chapter 07

Special Functions

In Section 7.1 we study Beta and Gamma functions which are useful in calculations of integrals in physics problems. Legendre polynomials are solutions of Legendre differential equation, and their properties and some special results are described in Section 7.2. Solutions of the Bessel differential equation and different kinds of Bessel functions are described in Section 7.3. Asymptotic behaviors of spherical Bessel functions are also discussed and their examples are presented.

7.1 Beta and Gamma Functions

· Gamma function

$$\Gamma(x) = \int_0^\infty t^{x-1} e^{-t} dt$$

· Beta function

$$B(p,q) = \int_0^1 t^{p-1}(1-t)^{q-1} dt$$

· Relationship between Gamma and Beta functions

$$B(p,q) = \frac{\Gamma(p)\Gamma(q)}{\Gamma(p+q)}$$

· Factorial function

$$\int_0^\infty x^n e^{-\alpha x} dx = \frac{n!}{\alpha^{n+1}}$$

- $\Gamma(n+1)=n\Gamma(n)=n!$
- Digamma function

$$\Psi(x)=\frac{d}{dx}ln\Gamma(x)=\frac{\Gamma'(x)}{\Gamma(x)}$$

- Stirling's formula for large n

$$n!=n^n e^{-n}\sqrt{2\pi n}$$

Gamma and **Beta functions** are defined as

$$\Gamma(x)=\int_0^\infty t^{x-1}e^{-t}dt \text{ with } x>0$$

and

$$B(p,q)=\int_0^1 t^{p-1}(1-t)^{q-1}dt \text{ with } p>0 \text{ and } q>0,$$

respectively.

(a) If we insert y^2 instead of t in the Gamma function, it gives

$$\begin{aligned}\Gamma(x)&=\int_0^\infty t^{x-1}e^{-t}dt\\&=\int_0^\infty y^{2(x-1)}e^{-y^2}2ydy=2\int_0^\infty y^{2x-1}e^{-y^2}dy\end{aligned}$$

This is another expression for the Gamma function.

(b) If we insert $\sin^2\theta$ instead of t in the Beta function,

$$\begin{aligned}B(p,q)&=\int_0^1 t^{p-1}(1-t)^{q-1}dt\\&=\int_0^{\frac{\pi}{2}}(\sin\theta)^{2(p-1)}(1-\sin^2\theta)^{q-1}\sin2\theta d\theta\\&=2\int_0^{\frac{\pi}{2}}(\sin\theta)^{2(p-1)}(\cos\theta)^{2(q-1)}\sin\theta\cos\theta d\theta\\&=2\int_0^{\frac{\pi}{2}}(\sin\theta)^{2p-1}(\cos\theta)^{2q-1}d\theta\end{aligned}$$

This is another expression for the Beta function.

There is a close relationship between the Beta function and the Gamma function. From the definition of the Gamma function, $\Gamma(x)=2\int_0^{\infty} y^{2x-1}e^{-y^2}dy$, we have

$$\Gamma(p)\Gamma(q)=(2\int_0^{\infty} y^{2p-1}e^{-y^2}dy)(2\int_0^{\infty} x^{2q-1}e^{-x^2}dx)$$
$$=4\int_0^{\infty}\int_0^{\infty} x^{2q-1}y^{2p-1}e^{-(x^2+y^2)}dxdy$$

Since $x=r\cos\theta$ and $y=r\sin\theta$ in the plane polar coordinates, we have

$$dxdy=rdrd\theta \text{ and } x^2+y^2=r^2$$

Then,

$$\Gamma(p)\Gamma(q)=4\int_0^{\infty}\int_0^{\frac{\pi}{2}}(r\cos\theta)^{2q-1}(r\sin\theta)^{2p-1}e^{-r^2}rdrd\theta$$
$$=4\int_0^{\infty} r^{2(p+q)-1}e^{-r^2}dr\int_0^{\frac{\pi}{2}}(\cos\theta)^{2q-1}(\sin\theta)^{2p-1}d\theta,$$

where an upper limit of the integral for θ is $\frac{\pi}{2}$ because $0<x<\infty$ and $0<y<\infty$ cover the first quadrant. Let's take a look at the integrals on the right-hand side of the equation.

From the definition of the Gamma function, $\Gamma(x)=2\int_0^{\infty} y^{2x-1}e^{-y^2}dy$, we get

$$\int_0^{\infty} r^{2(p+q)-1}e^{-r^2}dr=\frac{1}{2}\Gamma(p+q)$$

From the definition of the Beta function, $B(p,q)=2\int_0^{\frac{\pi}{2}}(\sin\theta)^{2p-1}(\cos\theta)^{2q-1}d\theta$,

$$\int_0^{\frac{\pi}{2}}(\cos\theta)^{2q-1}(\sin\theta)^{2p-1}d\theta=\frac{1}{2}B(p,q)$$

Therefore, we obtain

$$\Gamma(p)\Gamma(q)=4\frac{1}{2}\Gamma(p+q)\frac{1}{2}B(p,q)=\Gamma(p+q)B(p,q),$$

which gives

$$B(p,q)=\frac{\Gamma(p)\Gamma(q)}{\Gamma(p+q)}$$

This is the relation between the Beta and Gamma functions.

From the definition of the Gamma function, $\Gamma(x) = \int_0^\infty t^{x-1} e^{-t} dt$, we have

$$\Gamma(x+1) = \int_0^\infty t^x e^{-t} dt = \int_0^\infty t^x (-e^{-t})' dt$$

Using integration by parts, the right-hand side of the equation is

$$\begin{aligned}\int_0^\infty t^x (-e^{-t})' dt &= -e^{-t} t^x \Big|_{t=0}^{t=\infty} - \int_0^\infty x t^{x-1} (-e^{-t}) dt \\ &= \int_0^\infty x t^{x-1} e^{-t} dt \\ &= x \int_0^\infty t^{x-1} e^{-t} dt = x\Gamma(x)\end{aligned}$$

Therefore, we obtain the following equation:

$$\Gamma(x+1) = \int_0^\infty t^x e^{-t} dt = x\Gamma(x) \tag{7.1.1}$$

This equation satisfies all integer x values as long as functions are well-defined and finite for x.

Let's consider $\int_0^\infty e^{-\alpha x} dx$ with a positive constant α.

Then,

$$\int_0^\infty e^{-\alpha x} dx = -\frac{1}{\alpha} e^{-\alpha x} \Big|_{x=0}^{x=\infty} = \frac{1}{\alpha}$$

If we differentiate both sides of the equation with respect to α, it gives

$$-\int_0^\infty x e^{-\alpha x} dx = -\frac{1}{\alpha^2}$$

By differentiating the equation again with respect to α, we get

$$\int_0^\infty x^2 e^{-\alpha x} dx = 2\frac{1}{\alpha^3}$$

If we differentiate the equation again with respect to α, we get

$$\int_0^\infty x^3 e^{-\alpha x} dx = (3 \cdot 2)\frac{1}{\alpha^4}$$

$$\vdots$$

From this equation, we can write in general the following equation:

$$\int_0^{\infty} x^n e^{-\alpha x} dx = [n \cdot (n-1) \cdots 3 \cdot 2] \frac{1}{\alpha^{n+1}} = \frac{n!}{\alpha^{n+1}} \text{ with a positive } n$$

This is the equation for a **factorial function**:

$$\int_0^{\infty} x^n e^{-\alpha x} dx = \frac{n!}{\alpha^{n+1}}$$

From Eq. (7.1.1), we know that the left-hand side of the equation for $\alpha = 1$

corresponds to $\Gamma(n+1)$. On the other hand, the right-hand side of the equation for $\alpha = 1$ gives $\frac{n!}{\alpha^{n+1}} = \frac{n!}{1^{n+1}} = n!$

Therefore, we have

$$\Gamma(n+1) = n! \tag{7.1.2}$$

From Eqs. (7.1.1), and (7.1.2) we obtain

$$\Gamma(n+1) = n\Gamma(n) = n!$$

For $\alpha = 1$ and $n = 0$ in the factorial function, it gives

$$\int_0^{\infty} e^{-x} dx = 0!$$

Since the left-hand side of the equation is

$$\int_0^{\infty} e^{-x} dx = -e^{-x}\Big|_0^{\infty} = 1,$$

we obtain

$$0! = 1$$

There is another way to prove Eq. (7.1.2).

We know that $\Gamma(n+1) = n\Gamma(n) = n\Gamma(n-1+1)$. The right-hand side of the equation is $n(n-1)\Gamma(n-1)$ using Eq. (7.1.1), which can be written as $n(n-1)\Gamma(n-2+1)$.

If we keep applying Eq. (7.1.1) in this way, we obtain

$$\Gamma(n+1) = n(n-1)(n-2) \cdots\cdots 2\Gamma(1) = n!,$$

where $\Gamma(1) = 1$ and is proved in [Ex. 7.1.4].

Digamma function $\Psi(x)$ is defined as the first order derivative of the logarithm of the Gamma function $\Gamma(x)$:

$$\Psi(x) = \frac{d}{dx} ln\Gamma(x) = \frac{1}{\Gamma(x)} \frac{d}{dx} \Gamma(x) = \frac{\Gamma'(x)}{\Gamma(x)}$$

This function is useful to calculate Neumann function in approximation of small x.

In general, nth order derivative of the logarithm of the Gamma function is called the **polygamma** function which are denoted as $\Psi^{(0)}(x)$, $\Psi^{(1)}(x)$, ……, $\Psi^{(n)}(x)$.

Example

7.1.1 Calculate $\Gamma(\frac{1}{2})$.

Ans. From the definition of the Gamma function, $\Gamma(x) = 2\int_0^\infty y^{2x-1} e^{-y^2} dy$,

$$\Gamma(\frac{1}{2}) = 2\int_0^\infty y^{2(\frac{1}{2})-1} e^{-y^2} dy = 2\int_0^\infty e^{-y^2} dy$$

Since we already know $\int_{-\infty}^\infty e^{-x^2} dx = \sqrt{\pi}$ from [Ex. 3.1.1], we have

$$\int_0^\infty e^{-y^2} dy = \frac{1}{2}\int_{-\infty}^\infty e^{-y^2} dy = \frac{\sqrt{\pi}}{2}$$

Therefore, we get

$$\Gamma(\frac{1}{2}) = 2\frac{\sqrt{\pi}}{2} = \sqrt{\pi}$$

7.1.2 Express $\int_0^1 \frac{dx}{\sqrt{1-x^n}}$ in terms of the Beta function. This integral is appeared to find an oscillation period of an oscillator from the energy equation.

Ans. If we define $x^n = t$, we have

$$x = t^{\frac{1}{n}} \text{ and } dx = \frac{1}{n} t^{\frac{1}{n}-1} dt$$

By inserting these into the original integral, we have

$$\int_0^1 \frac{dx}{\sqrt{1-x^n}} = \frac{1}{n}\int_0^1 t^{\frac{1}{n}-1}(1-t)^{-\frac{1}{2}}dt$$

By comparing the right-hand side of this equation with the definition of the Beta function, we know that $p-1=\frac{1}{n}-1$ and $q-1=-\frac{1}{2}$.

This gives $p=\frac{1}{n}$, $q=\frac{1}{2}$ and the integral is then

$$\int_0^1 \frac{dx}{\sqrt{1-x^n}} = \frac{1}{n}\int_0^1 t^{\frac{1}{n}-1}(1-t)^{-\frac{1}{2}}dt$$

$$= \frac{1}{n}B\left(\frac{1}{n}, \frac{1}{2}\right)$$

7.1.3 We obtained $\int_0^1 \frac{dx}{\sqrt{1-x^n}} = \frac{1}{n}B\left(\frac{1}{n}, \frac{1}{2}\right)$ in the previous example. This can be expressed by the Gamma function by using the relationship between the Beta and Gamma functions:

Ans. $$\int_0^1 \frac{dx}{\sqrt{1-x^n}} = \frac{1}{n}B\left(\frac{1}{n}, \frac{1}{2}\right) = \frac{1}{n}\frac{\Gamma\left(\frac{1}{n}\right)\Gamma\left(\frac{1}{2}\right)}{\Gamma\left(\frac{1}{n}+\frac{1}{2}\right)}$$

$$= \frac{\sqrt{\pi}}{n}\frac{\Gamma\left(\frac{1}{n}\right)}{\Gamma\left(\frac{1}{n}+\frac{1}{2}\right)},$$

where $\Gamma\left(\frac{1}{2}\right)=\sqrt{\pi}$ is used.

7.1.4 Calculate $\Gamma(1)$.

Ans. From $\Gamma(n+1)=n!$, we obtain

$$\Gamma(1)=\Gamma(0+1)=0!=1$$

We can obtain the same result using the definition of the Gamma function, $\Gamma(x)=\int_0^\infty t^{x-1}e^{-t}dt$:

$$\Gamma(1)=\int_0^\infty t^{1-1}e^{-t}dt = \int_0^\infty e^{-t}dt = -e^{-t}|_0^\infty = 1$$

7.1.5 Calculate $B(\frac{1}{2},\frac{1}{2})$ by using the relationship between Beta and Gamma functions.

Ans. $$B(\frac{1}{2},\frac{1}{2})=\frac{\Gamma(\frac{1}{2})\Gamma(\frac{1}{2})}{\Gamma(\frac{1}{2}+\frac{1}{2})}=\frac{\Gamma(\frac{1}{2})\Gamma(\frac{1}{2})}{\Gamma(1)}=\frac{\sqrt{\pi}\sqrt{\pi}}{1}=\pi$$

7.1.6 Calculate $\Gamma(-\frac{3}{2})$, $\Gamma(-\frac{1}{2})$ and $\Gamma(\frac{3}{2})$.

Ans. From $\Gamma(x+1)=x\Gamma(x)$, we get $\Gamma(x)=\dfrac{\Gamma(x+1)}{x}$.
Then, we have

$$\Gamma(-\frac{3}{2})=-\frac{2}{3}\Gamma(-\frac{1}{2})$$

Similarly, we have

$$\Gamma(-\frac{1}{2})=\frac{\Gamma(-\frac{1}{2}+1)}{-\frac{1}{2}}=-2\Gamma(\frac{1}{2})=-2\sqrt{\pi}$$

so that

$$\Gamma(-\frac{3}{2})=-\frac{2}{3}\Gamma(-\frac{1}{2})=-\frac{2}{3}(-2)\sqrt{\pi}=\frac{4}{3}\sqrt{\pi}$$

From $\Gamma(x+1)=x\Gamma(x)$, we have

$$\Gamma(\frac{3}{2})=\frac{1}{2}\Gamma(\frac{1}{2})=\frac{1}{2}\sqrt{\pi}$$

7.1.7 From the definition of the Gamma function and Eq. (7.1.2), we know that

$$\Gamma(n+1)=n!=\int_0^\infty t^n e^{-t}dt$$

From this equation, prove **Stirling's formula** $n!\approx n^n e^{-n}\sqrt{2\pi n}$.

Ans. If we take the logarithm on the integrand,

$$\ell n(t^n e^{-t})=\ell n t^n - t = n\ell n t - t \qquad (7.1.7.1)$$

If we differentiate the equation with respect to t, it gives

$$\frac{d}{dt}\ell n(t^n e^{-t})=\frac{d}{dt}(nlnt-t)=\frac{n}{t}-1$$

When the right-hand side of the equation is zero, the logarithm of the integrand function, $ln(t^n e^{-t})$, has an extreme value. Therefore, the logarithm of the integrand function has a maximum value at $t=n$. By assuming $t=n+\alpha$ with small $\alpha \ll n$, we have $dt = d\alpha$. If we substitute $t = n + \alpha$ into Eq. (7.1.7.1), the right-hand side of the equation gives

$$n\ell n(n+\alpha) - (n+\alpha) = n\ell n[n(1+\frac{\alpha}{n})] - (n+\alpha)$$
$$= n[\ell n(n) + \ell n(1+\frac{\alpha}{n})] - (n+\alpha) \qquad (7.1.7.2)$$

We know from [Ex. 5.2.5],

$$\ell n(1+x) = x - \frac{x^2}{2} + \frac{x^3}{3} - \cdots\cdots \text{, for small } x$$

If we expand $\ell n(1+\frac{\alpha}{n})$ for large n, we get

$$\ell n(1+\frac{\alpha}{n}) = \frac{\alpha}{n} - \frac{1}{2}(\frac{\alpha}{n})^2 + \frac{1}{3}(\frac{\alpha}{n})^3 - \cdots\cdots$$

Then, the right-hand side of Eq. (7.1.7.2) is written as

$$n[\ell n(n) + \frac{\alpha}{n} - \frac{1}{2}(\frac{\alpha}{n})^2] - (n+\alpha)$$

neglecting higher order terms,
which gives

$$n\ell n(n) - \frac{1}{2n}\alpha^2 - n$$

Therefore, Eq. (7.1.7.1) gives

$$\ell n(t^n e^{-t}) = n\ell n(n) - \frac{1}{2n}\alpha^2 - n,$$

which gives the integrand

$$t^n e^{-t} = e^{n\ell n(n) - \frac{1}{2n}\alpha^2 - n} = e^{\ell n(n^n)} e^{-n} e^{-\frac{1}{2n}\alpha^2} = n^n e^{-n} e^{-\frac{1}{2n}\alpha^2} \qquad (7.1.7.3)$$

If we substitute Eq. (7.1.7.3) and $dt = d\alpha$ into the original equation, we obtain

$$n! = \int_{-n}^{\infty} n^n e^{-n} e^{-\frac{1}{2n}\alpha^2} d\alpha = \int_{-\infty}^{\infty} n^n e^{-n} e^{-\frac{1}{2n}\alpha^2} d\alpha \qquad (7.1.7.4)$$

The lower limit of the integral in the equation is changed from $-n$ to $-\infty$ without affecting the result because the integrand is very small for $\alpha < -n$.
The right-hand side of the equation is then

$$n^n e^{-n} \int_{-\infty}^{\infty} e^{-\frac{1}{2n}\alpha^2} d\alpha \qquad (7.1.7.5)$$

This integral is calculated by defining $\frac{1}{2n}\alpha^2 = x^2$, which gives $d\alpha = \sqrt{2n}\,dx$.
Eq. (7.1.7.5) is then

$$n^n e^{-n} \int_{-\infty}^{\infty} e^{-x^2} \sqrt{2n}\,dx = n^n e^{-n} \sqrt{2n} \int_{-\infty}^{\infty} e^{-x^2} dx = n^n e^{-n} \sqrt{2\pi n}$$

Thus, we obtain the Stirling's formula

$$n! = n^n e^{-n}\sqrt{2\pi n} \quad \text{for large } n$$

This is also expressed in terms of another form by taking the logarithm on both sides of the equation:

$$\ell n(n!) = n\ell n(n) - n + \frac{1}{2}\ell n(2\pi n)$$

7.2 Legendre Differential Equation

7.2.1 Legendre Polynomials

· Legendre polynomials of order ℓ

$$\begin{aligned} P_\ell(x) &= \frac{1}{2^\ell}\sum_{k=0}^{[\frac{\ell}{2}]}\frac{(-1)^k(2\ell-2k)!}{k!(\ell-k)!(\ell-2k)!}x^{\ell-2k} \\ &= \frac{1}{2^\ell}\sum_{k=0}^{[\frac{\ell}{2}]}\frac{(-1)^k}{k!(\ell-k)!}\frac{d^\ell}{dx^\ell}x^{2\ell-2k} \end{aligned}$$

· Rodrigues' formula

$$P_\ell(x) = \frac{1}{2^\ell \ell!}\frac{d^\ell}{dx^\ell}(x^2-1)^\ell$$

· Associated Legendre functions

$$P_\ell^m(x) = (1-x^2)^{\frac{m}{2}}\frac{d^m}{dx^m}P_\ell(x)$$

· Relationship between associated Legendre functions with a positive and negative m

$$\frac{P_\ell^{-m}(x)}{P_\ell^m(x)} = (-1)^m\frac{(\ell-m)!}{(\ell+m)!}$$

When a differential equation is as follows, it is called the **Legendre differential equation**:

$$[(1-x^2)\frac{d^2}{dx^2}-2x\frac{d}{dx}-\frac{m^2}{1-x^2}+\lambda]P(x)=0, \quad (7.2.1.1)$$

where m and λ are constants.

To find solutions of the Legendre differential equation, first let's consider the case of $m=0$ and then consider the case of $m\neq 0$:

(a) **For $m=0$**

The differential equation is then

$$[(1-x^2)\frac{d^2}{dx^2}-2x\frac{d}{dx}+\lambda]P(x)=0 \quad (7.2.1.2)$$

Let's define a trial solution in the power series:

$$P(x)=\sum_{n=0}^{\infty}a_n x^n \text{ with constant } a_n$$

Then, we get

$$\frac{d}{dx}P(x)=\sum_{n=1}^{\infty}na_n x^{n-1}$$

and

$$\frac{d^2}{dx^2}P(x)=\frac{d}{dx}[\frac{d}{dx}P(x)]=\frac{d}{dx}\sum_{n=1}^{\infty}na_n x^{n-1}=\sum_{n=2}^{\infty}n(n-1)a_n x^{n-2}$$

If we substitute these differentiations into Eq. (7.2.1.2), it gives

$$\sum_{n=2}^{\infty}n(n-1)a_n x^{n-2}-\sum_{n=2}^{\infty}n(n-1)a_n x^n$$
$$-2\sum_{n=1}^{\infty}na_n x^n+\lambda\sum_{n=0}^{\infty}a_n x^n=0 \quad (7.2.1.3)$$

The first term on the equation by substituting $n+2$ instead of n can be written as

$$\sum_{n=0}^{\infty}(n+2)(n+1)a_{n+2}x^n$$

The equation is then

$$\sum_{n=0}^{\infty}(n+2)(n+1)a_{n+2}x^n - \sum_{n=2}^{\infty}n(n-1)a_n x^n$$

$$-2\sum_{n=1}^{\infty}na_n x^n + \lambda\sum_{n=0}^{\infty}a_n x^n = 0 \qquad (7.2.1.4)$$

Let's take a look at the first few powers of x in the equation:

(i) for x^0 (i.e. $n=0$)

$$(2a_2 + \lambda a_0)x^0 = 0 \text{ gives}$$

$$a_2 = -\frac{\lambda}{2}a_0$$

(ii) for x^1 (i.e. $n=1$)

$$[3\cdot 2a_3 + (\lambda-2)a_1]x^1 = 0 \text{ gives}$$

$$a_3 = -\frac{\lambda-2}{3\cdot 2}a_1$$

(iii) for x^2 (i.e. $n=2$)

$$[4\cdot 3a_4 + (\lambda-6)a_2]x^2 = 0 \text{ gives}$$

$$\begin{aligned} a_4 &= -\frac{\lambda-6}{4\cdot 3}a_2 \\ &= \frac{\lambda(\lambda-6)}{4\cdot 3\cdot 2}a_0 = \frac{\lambda(\lambda-6)}{4!}a_0 \end{aligned}$$

The Legendre differential equation that appears in Electromagnetism and Quantum Mechanics has $\lambda=\ell(\ell+1)$ as a value with an integer ℓ. Then, it can be written as

$$a_4 = \frac{\ell(\ell+1)[\ell(\ell+1)-6]}{4!}a_0 = \frac{\ell(\ell+1)(\ell-2)(\ell+3)}{4!}a_0$$

Similarly, we can get the following equation for the coefficient of x^n from Eq. (7.2.1.4),

$$(n+2)(n+1)a_{n+2} = [n(n-1)+2n-\ell(\ell+1)]a_n = [n^2+n-\ell(\ell+1)]a_n,$$

which gives a **recurrence relation**

$$a_{n+2} = \frac{n^2+n-\ell(\ell+1)}{(n+1)(n+2)} a_n \qquad (7.2.1.5)$$

From the recurrence relation, we obtain

$$a_2 = -\frac{\ell(\ell+1)}{2} a_0 = -\frac{\ell(\ell+1)}{2!} a_0,$$

$$a_3 = \frac{2-\ell(\ell+1)}{2 \cdot 3} a_1 = -\frac{(\ell-1)(\ell+2)}{3!} a_1,$$

$$\begin{aligned} a_4 &= \frac{6-\ell(\ell+1)}{4 \cdot 3} a_2 \\ &= \frac{6-\ell(\ell+1)}{4 \cdot 3}\left[-\frac{\ell(\ell+1)}{2!}\right] a_0 = \frac{[\ell(\ell+1)-6]\ell(\ell+1)}{4!} a_0 \\ &= \frac{\ell(\ell+1)(\ell-2)(\ell+3)}{4!} a_0, \end{aligned}$$

and

$$\begin{aligned} a_5 &= \frac{12-\ell(\ell+1)}{4 \cdot 5} a_3 = -\frac{(\ell+4)(\ell-3)}{5 \cdot 4}\left[-\frac{(\ell+2)(\ell-1)}{3!}\right] a_1 \\ &= \frac{(\ell-1)(\ell+2)(\ell-3)(\ell+4)}{5!} a_1 \end{aligned}$$

$$\vdots$$

$$\vdots$$

We notice that the coefficients for even and odd n are derived from a_0 and a_1, respectively.

Therefore, the power series solution is the sum of a_0 series and a_1 series:

$$\begin{aligned} P(x) &= a_0 + a_1 x + a_2 x^2 + a_3 x^3 + a_4 x^4 + a_5 x^5 + \cdots\cdots \\ &= \left[1 - \frac{\ell(\ell+1)}{2!} x^2 + \frac{\ell(\ell+1)(\ell-2)(\ell+3)}{4!} x^4 - \cdots\cdots\right] a_0 \qquad (7.2.1.6) \\ &\quad + \left[x - \frac{(\ell-1)(\ell+2)}{3!} x^3 + \frac{(\ell-1)(\ell+2)(\ell-3)(\ell+4)}{5!} x^5 - \cdots\cdots\right] a_1, \end{aligned}$$

where a_0 and a_1 are arbitrary constants.

For large n, Eq. (7.2.1.5) gives

$$\frac{a_{n+2}}{a_n} = \frac{n^2+n-\ell(\ell+1)}{n^2+3n+2} \approx 1$$

This means that the power series solution $P(x)=\sum_{n=0}^{\infty} a_n x^n$ can be written as

$$P(x) = a_0 + a_1 x + a_2 x^2 + a_3 x^3 + \cdots\cdots = 1 + x + x^2 + x^3 + \cdots\cdots$$
$$= \frac{1}{1-x}$$

This blows up (or diverges) at $x=1$, meaning of **singular**. This problem can be avoided by making the numerator in Eq. (7.2.1.5) zero. **This is achieved by choosing $n=\ell$. Solutions for negative ℓ values can be obtained easily from solutions for positive ℓ values.** For instance, a value of $\ell(\ell+1)$ in the recurrence relation for $\ell=-2$ is the same as that for $\ell=1$.

Let's choose $a_0=1$ and $a_1=1$ in Eq. (7.2.1.6) to give $P(1)=1$.
This power series solution $P(x)$ is the polynomial of order ℓ in variable x. It is expressed as $P_\ell(x)$ and called **Legendre polynomials.**
Eq. (7.2.1.5) can be written as

$$a_n = \frac{(n+1)(n+2)}{n^2+n-\ell(\ell+1)} a_{n+2} = -\frac{(n+2)(n+1)}{(\ell-n)(n+\ell+1)} a_{n+2} \tag{7.2.1.7}$$

If we substitute $\ell-2k$ instead of n in the equation, we get

$$a_{\ell-2k} = -\frac{(\ell-2k+2)(\ell-2k+1)}{2k(2\ell-2k+1)} a_{\ell-2k+2}, \tag{7.2.1.8}$$

where $k=1,\ 2,\ \cdots\cdots,\ [\frac{\ell}{2}]$. The meaning of $[\frac{\ell}{2}]$ is

$$[\frac{\ell}{2}] = \begin{cases} \frac{\ell}{2} & \text{for even } \ell \\ \frac{\ell-1}{2} & \text{for odd } \ell \end{cases}$$

For $k=1$, $k=2$, and $k=3$, Eq. (7.2.1.8) gives

$$a_{\ell-2} = -\frac{(\ell-2\cdot 1+2)(\ell-2\cdot 1+1)}{2\cdot 1(2\ell-2\cdot 1+1)} a_\ell = -\frac{\ell(\ell-1)}{2(2\ell-1)} a_\ell,$$

$$a_{\ell-4} = -\frac{(\ell-2\cdot 2+2)(\ell-2\cdot 2+1)}{2\cdot 2(2\ell-2\cdot 2+1)} a_{\ell-2}$$

$$= (-1)^2 \frac{\ell(\ell-1)(\ell-2)(\ell-3)}{2^2 2!(2\ell-1)(2\ell-3)} a_\ell,$$

$$a_{\ell-6} = -\frac{(\ell-2\cdot 3+2)(\ell-2\cdot 3+1)}{2\cdot 3(2\ell-2\cdot 3+1)} a_{\ell-4}$$

$$= (-1)^3 \frac{\ell(\ell-1)(\ell-2)(\ell-3)(\ell-4)(\ell-5)}{2^3 3!(2\ell-1)(2\ell-3)(2\ell-5)} a_\ell,$$

respectively.

$$\vdots$$

$$\vdots$$

From these, we can generalize $a_{\ell-2k}$ to be

$$a_{\ell-2k} = (-1)^k \frac{\ell(\ell-1)(\ell-2)(\ell-3)\cdots\cdots(\ell-2k+1)}{2^k k!(2\ell-1)(2\ell-3)\cdots\cdots(2\ell-2k+1)} a_\ell$$

Thus, the power series solution $P(x) = \sum_{n=0}^{\infty} a_n x^n$ can be expressed as

$$P_\ell(x) = \sum_{k=0}^{[\frac{\ell}{2}]} a_{\ell-2k} x^{\ell-2k}$$

$$= \sum_{k=0}^{[\frac{\ell}{2}]} (-1)^k \frac{\ell(\ell-1)(\ell-2)(\ell-3)\cdots\cdots(\ell-2k+1)}{2^k k!(2\ell-1)(2\ell-3)\cdots\cdots(2\ell-2k+1)} a_\ell x^{\ell-2k}, \quad (7.2.1.9)$$

where a_ℓ is an arbitrary constant.

If we choose a_ℓ to make $P_\ell(1)=1$, we get

$$a_\ell = \frac{(2\ell)!}{2^\ell (\ell!)^2} \quad (7.2.1.10)$$

By substituting Eqs. (7.2.1.10) into (7.2.1.9), we get

$$P_\ell(x) = \sum_{k=0}^{[\frac{\ell}{2}]} (-1)^k \frac{\ell(\ell-1)(\ell-2)(\ell-3)\cdots\cdots(\ell-2k+1)}{2^k k!(2\ell-1)(2\ell-3)\cdots\cdots(2\ell-2k+1)} \frac{(2\ell)!}{2^\ell(\ell!)(\ell!)} x^{\ell-2k}$$

$$= \frac{1}{2^\ell} \sum_{k=0}^{[\frac{\ell}{2}]} \frac{(-1)^k}{k!} \frac{1}{2^k} \frac{\ell(\ell-1)(\ell-2)(\ell-3)\cdots\cdots(\ell-2k+1)}{(2\ell-1)(2\ell-3)\cdots\cdots(2\ell-2k+1)}$$

$$\times \frac{(2\ell)(2\ell-1)(2\ell-2)(2\ell-3)\cdots\cdots 2\cdot 1}{\ell(\ell-1)\cdots\cdots(\ell-2k+1)(\ell-2k)!\ell!} x^{\ell-2k}$$

$$= \frac{1}{2^\ell} \sum_{k=0}^{[\frac{\ell}{2}]} \frac{(-1)^k}{k!} \frac{1}{2^k} \frac{1}{(2\ell-1)(2\ell-3) \cdots\cdots (2\ell-2k+1)}$$

$$\times \frac{(2\ell-1)(2\ell-3) \cdots\cdots (2\ell-2k+1)(2\ell)(2\ell-2) \cdots\cdots (2\ell-2k+2)(2\ell-2k)!}{(\ell-2k)!\ell!} x^{\ell-2k}$$

$$= \frac{1}{2^\ell} \sum_{k=0}^{[\frac{\ell}{2}]} \frac{(-1)^k}{k!} \frac{1}{2^k} \frac{2^k \ell(\ell-1) \cdots\cdots (\ell-k+1)(2\ell-2k)!}{(\ell-2k)!\ell(\ell-1) \cdots\cdots (\ell-k+1)(\ell-k)!} x^{\ell-2k}$$

$$= \frac{1}{2^\ell} \sum_{k=0}^{[\frac{\ell}{2}]} \frac{(-1)^k (2\ell-2k)!}{k!(\ell-2k)!(\ell-k)!} x^{\ell-2k}$$

Therefore, we obtain the polynomials of order ℓ

$$P_\ell(x) = \frac{1}{2^\ell} \sum_{k=0}^{[\frac{\ell}{2}]} \frac{(-1)^k (2\ell-2k)!}{k!(\ell-k)!(\ell-2k)!} x^{\ell-2k} \tag{7.2.1.11}$$

Using Eq. (7.2.1.11), we obtain the following results:

$$P_0(x) = \frac{1}{2^0} \sum_{k=0}^{0} (-1)^k \frac{(-2k)!}{k!(-k)!(-2k)!} x^{-2k} = 1$$

$$P_1(x) = \frac{1}{2^1} \sum_{k=0}^{0} (-1)^k \frac{(2-2k)!}{k!(1-k)!(1-2k)!} x^{1-2k} = \frac{2 \cdot x}{2} = x$$

$$P_2(x) = \frac{1}{2^2} \sum_{k=0}^{1} (-1)^k \frac{(4-2k)!}{k!(2-k)!(2-2k)!} x^{2-2k}$$

$$= \frac{4!}{2^2 \cdot 2 \cdot 2} x^2 - \frac{2}{2^2} x^0 = \frac{1}{2}(3x^2 - 1)$$

Similarly, we can obtain

$$P_3(x) = \frac{1}{2}(5x^3 - 3x)$$

$$P_4(x) = \frac{1}{8}(35x^4 - 30x^2 + 3)$$

$$\vdots$$

$$\vdots$$

We can see that $P_\ell(1) = 1$.

Table 7.2.1 Legendre polynomials of order ℓ, $P_\ell(x)$.

- $P_0(x)=1$
- $P_1(x)=x$
- $P_2(x)=\frac{1}{2}(3x^2-1)$
- $P_3(x)=\frac{1}{2}(5x^3-3x)$
- $P_4(x)=\frac{1}{8}(35x^4-30x^2+3)$

Let's find another useful expression for the Legendre polynomials. If we differentiate $x^{2\ell-2k}$ with respect to x by ℓ times, we get

$$\begin{aligned}\frac{d^\ell}{dx^\ell}x^{2\ell-2k} &= (2\ell-2k)(2\ell-2k-1)\cdots\cdots[2\ell-2k-(\ell-1)]x^{\ell-2k} \\ &= (2\ell-2k)(2\ell-2k-1)\cdots\cdots(\ell-2k+1)x^{\ell-2k} \\ &= \frac{(2\ell-2k)!}{(\ell-2k)!}x^{\ell-2k},\end{aligned}$$

which gives

$$\frac{(2\ell-2k)!}{(\ell-2k)!}x^{\ell-2k} = \frac{d^\ell}{dx^\ell}x^{2\ell-2k}$$

If we insert this into Eq. (7.2.1.11), the Legendre polynomials of order ℓ is expressed as

$$P_\ell(x) = \frac{1}{2^\ell}\sum_{k=0}^{[\frac{\ell}{2}]}\frac{(-1)^k}{k!(\ell-k)!}\frac{d^\ell}{dx^\ell}x^{2\ell-2k} \tag{7.2.1.12}$$

Since the power term of x in the equation is less than ℓ for $k>[\frac{\ell}{2}]$, its differentiation by ℓ times yields to zero. Thus, we can extend the sum from $[\frac{\ell}{2}]$ to ℓ without affecting the result because this change just adds zeros.

Therefore, we can write Eq. (7.2.1.12) as

$$P_\ell(x) = \frac{1}{2^\ell \ell!}\frac{d^\ell}{dx^\ell}\sum_{k=0}^{\ell}\frac{(-1)^k \ell!}{k!(\ell-k)!}x^{2\ell-2k} \tag{7.2.1.13}$$

We multiply a factor $\ell!$ in the numerator and divide it in the denominator. As we will see, it makes to express the Legendre polynomials in a compact form.

If we substitute $a = x^2$, $b = -1$, and $n = \ell$ in the binomial theorem of Eq. (5.2.7), we get

$$(x^2-1)^\ell = \sum_{k=0}^{\ell}\frac{(-1)^k \ell!}{k!(\ell-k)!}x^{2\ell-2k}$$

Thus, Eq. (7.2.1.13) yields

$$P_\ell(x) = \frac{1}{2^\ell \ell!}\frac{d^\ell}{dx^\ell}(x^2-1)^\ell \tag{7.2.1.14}$$

This is called **Rodrigues' formula** which is also the solution of the Legendre differential equation with $m = 0$.

(b) **For $m \neq 0$**

The original equation, Eq. (7.2.1.1), with $\lambda = \ell(\ell+1)$ gives

$$[(1-x^2)\frac{d^2}{dx^2} - 2x\frac{d}{dx} - \frac{m^2}{1-x^2} + \ell(\ell+1)]P(x) = 0$$

Since the power series solution $P(x)$ for the equation is polynomials of order ℓ with non-zero m in variable x, let's write $P(x) = P_\ell^m(x)$.

Then, we have

$$\frac{d}{dx}[(1-x^2)\frac{dP_\ell^m(x)}{dx}] + [\ell(\ell+1) - \frac{m^2}{1-x^2}]P_\ell^m(x) = 0 \tag{7.2.1.15}$$

If we define $P_\ell^m(x) = (1-x^2)^{\frac{m}{2}}u(x)$, we have

$$\frac{dP_\ell^m(x)}{dx} = -mx(1-x^2)^{\frac{m}{2}-1}u(x) + (1-x^2)^{\frac{m}{2}}\frac{du(x)}{dx}$$

By substituting this into the first term in Eq. (7.2.1.15), it gives

$$\frac{d}{dx}[(1-x^2)\frac{dP_\ell^m(x)}{dx}]$$

$$=(1-x^2)^{\frac{m}{2}+1}u''(x)-2(m+1)x(1-x^2)^{\frac{m}{2}}u'(x)$$

$$+[m(m+1)x^2-m](1-x^2)^{\frac{m}{2}-1}u(x)=0,$$

which yields Eq. (7.2.1.15) to be

$$(1-x^2)^{\frac{m}{2}+1}u''(x)-2(m+1)x(1-x^2)^{\frac{m}{2}}u'(x)$$

$$+[\{m(m+1)x^2-m\}(1-x^2)^{\frac{m}{2}-1}$$

$$+\left\{\ell(\ell+1)-\frac{m^2}{1-x^2}\right\}(1-x^2)^{\frac{m}{2}}]u(x)=0 \quad (7.2.1.16)$$

By dividing the equation with $(1-x^2)^{\frac{m}{2}}$, we have

$$(1-x^2)u''(x)-2(m+1)xu'(x)$$

$$+[\{m(m+1)x^2-m\}(1-x^2)^{-1}+\left\{\ell(\ell+1)-\frac{m^2}{1-x^2}\right\}]u(x)=0 \quad (7.2.1.17)$$

The last term on the equation is simplified as follows:

$$\frac{\ell(\ell+1)(1-x^2)-m^2(1-x^2)-m(1-x^2)}{1-x^2} = \ell(\ell+1)-m^2-m$$

$$= \ell(\ell+1)-m(m+1)$$

The equation is then

$$(1-x^2)\frac{d^2u(x)}{dx^2}-2(m+1)x\frac{du(x)}{dx}+[\ell(\ell+1)-m(m+1)]u(x)=0 \quad (7.2.1.18)$$

For $m=0$, we know that this is just the Legendre equation with the solution $u(x)$ by comparing with Eq. (7.2.1.2).

If we differentiate Eq. (7.2.1.18) with respect to x, it gives

$$(1-x^2)\frac{d}{dx}(\frac{d^2u}{dx^2})-2[(m+1)+1]x\frac{d^2u}{dx^2}$$

$$+[(\ell-m)(\ell+m+1)-2(m+1)]\frac{du}{dx}=0 \quad (7.2.1.19)$$

The coefficient of the last term on the equation is $\ell(\ell+1)-(m+1)(m+2)$. The equation is then

$$(1-x^2)\frac{d}{dx}(\frac{d^2u}{dx^2})-2[(m+1)+1]x\frac{d^2u}{dx^2}$$

$$+[\ell(\ell+1)-(m+1)(m+2)]\frac{du}{dx}=0 \tag{7.2.1.20}$$

If we differentiate the equation with respect to x, it gives

$$(1-x^2)\frac{d^2}{dx^2}(\frac{d^2u}{dx^2})-2[(m+2)+1]x\frac{d}{dx}(\frac{d^2u}{dx^2})$$

$$+[\ell(\ell+1)-(m+2)(m+3)]\frac{d^2u}{dx^2}=0 \tag{7.2.1.21}$$

If we substitute $\frac{du(x)}{dx}$ for $u(x)$ and $m+1$ for m in Eq. (7.2.1.18), it gives Eq. (7.2.1.20). On the other hand, we get Eq. (7.2.1.21) by substituting $\frac{d^2u(x)}{dx^2}$ for $u(x)$ and $m+2$ for m in Eq. (7.2.1.18).

Therefore, if $P_\ell(x)$ is the solution for $m=0$, $\frac{d}{dx}P_\ell(x)$ is the solution for $m=1$ and $\frac{d^2}{dx^2}P_\ell(x)$ is the solution for $m=2$. In this way, we can know that $u(x)=\frac{d^m}{dx^m}P_\ell(x)$.

Since $P_\ell^m(x)=(1-x^2)^{\frac{m}{2}}u(x)$, we obtain

$$P_\ell^m(x) = (1-x^2)^{\frac{m}{2}}\frac{d^m}{dx^m}P_\ell(x). \tag{7.2.1.22}$$

This is called the **associated Legendre functions** (or **generalized Legendre polynomials**), related to solutions for $m\neq 0$.

Since the highest power of x in the Legendre polynomials $P_\ell(x)$ is x^ℓ, we know from Eq. (7.2.1.22) that the associated Legendre function has a non-zero value for $m\leq\ell$, remembering $\ell\geq 0$.

Let's calculate some of the associated Legendre functions using Eq. (7.2.1.22).

$$P_1^1(x) = (1-x^2)^{\frac{1}{2}} \frac{d}{dx} P_1(x) = (1-x^2)^{\frac{1}{2}} \frac{d}{dx} x = (1-x^2)^{\frac{1}{2}}$$

$$P_2^1(x) = (1-x^2)^{\frac{1}{2}} \frac{d}{dx} P_2(x) = (1-x^2)^{\frac{1}{2}} \frac{d}{dx} [\frac{1}{2}(3x^2-1)] = 3x(1-x^2)^{\frac{1}{2}}$$

$$P_2^2(x) = (1-x^2) \frac{d^2}{dx^2} P_2(x) = (1-x^2) \frac{d^2}{dx^2} [\frac{1}{2}(3x^2-1)] = 3(1-x^2)$$

$$P_3^1(x) = (1-x^2)^{\frac{1}{2}} \frac{d}{dx} P_3(x) = (1-x^2)^{\frac{1}{2}} \frac{d}{dx} [\frac{1}{2}(5x^3-3x)] = \frac{3}{2}(5x^2-1)(1-x^2)^{\frac{1}{2}}$$

$$P_3^2(x) = (1-x^2) \frac{d^2}{dx^2} P_3(x) = 15x(1-x^2)$$

$$P_3^3(x) = (1-x^2)^{\frac{3}{2}} \frac{d^3}{dx^3} P_3(x) = 15(1-x^2)^{\frac{3}{2}}$$

$$\vdots$$

Table 7.2.2 Associated Legendre polynomials, $P_\ell^m(x)$.

- $P_1^1(x) = (1-x^2)^{\frac{1}{2}}$
- $P_2^1(x) = 3x(1-x^2)^{\frac{1}{2}}$
- $P_2^2(x) = 3(1-x^2)$
- $P_3^1(x) = \frac{3}{2}(5x^2-1)(1-x^2)^{\frac{1}{2}}$
- $P_3^2(x) = 15x(1-x^2)$
- $P_3^3(x) = 15(1-x^2)^{\frac{3}{2}}$

Let's find out an equation for a negative m using Eq. (7.2.1.22) and the Rodrigues' formula. By inserting the Rodrigues' formula into (7.2.1.22), we get

$$\begin{aligned} P_\ell^m(x) &= (1-x^2)^{\frac{m}{2}} \frac{d^m}{dx^m} [\frac{1}{2^\ell \ell!} \frac{d^\ell}{dx^\ell}(x^2-1)^\ell] \\ &= \frac{1}{2^\ell \ell!}(1-x^2)^{\frac{m}{2}} \frac{d^{\ell+m}}{dx^{\ell+m}}(x^2-1)^\ell, \end{aligned} \tag{7.2.1.23}$$

where

$$\frac{d^{\ell+m}}{dx^{\ell+m}}(x^2-1)^\ell = \frac{d^{\ell+m}}{dx^{\ell+m}}[(x+1)^\ell(x-1)^\ell]$$

By using **Leibnitz's formula**[13], $\frac{d^n}{dx^n}[f(x)g(x)] = \sum_{k=0}^{n}\binom{n}{k}[\frac{d^{n-k}}{dx^{n-k}}f(x)][\frac{d^k}{dx^k}g(x)]$,
the right-hand side of the equation is

$$\begin{aligned}
&\frac{d^{\ell+m}}{dx^{\ell+m}}[(x+1)^\ell(x-1)^\ell] \\
&\quad = \sum_{k=0}^{\ell+m}\binom{\ell+m}{k}\frac{d^{\ell+m-k}}{dx^{\ell+m-k}}(x+1)^\ell\frac{d^k}{dx^k}(x-1)^\ell \\
&\quad = \sum_{k=0}^{\ell+m}\frac{(\ell+m)!}{(\ell+m-k)!k!}\frac{d^{\ell+m-k}}{dx^{\ell+m-k}}(x+1)^\ell\frac{d^k}{dx^k}(x-1)^\ell
\end{aligned} \tag{7.2.1.24}$$

We know that $\ell+m-k$ and k should be $\le \ell$, respectively, for a non-zero value of the right-hand side in Eq. (7.2.1.24).
It means that $k \ge m$ and $k \le \ell$, respectively, so that
(i) k runs from m to ℓ in the summation for a positive m.
(ii) k runs from $k=0$ to $\ell-|m|$ in the summation for a negative m.

The right-hand side of Eq. (7.2.1.24) is

$$\begin{aligned}
&\sum_{k=0}^{\ell+m}\frac{(\ell+m)!}{(\ell+m-k)k!}\frac{\ell(\ell-1)(\ell-2)\cdots(k-m+1)(k-m)!}{(k-m)!}(x+1)^{\ell-\ell-m+k} \\
&\qquad\times\frac{\ell(\ell-1)(\ell-2)\cdots(\ell-k+1)(\ell-k)!}{(\ell-k)!}(x-1)^{\ell-k} \\
&= \sum_{k=0}^{\ell+m}\frac{(\ell+m)!}{(\ell+m-k)!k!}\frac{\ell!}{(k-m)!}(x+1)^{k-m}\frac{\ell!}{(\ell-k)!}(x-1)^{\ell-k}
\end{aligned}$$

Then, Eq. (7.2.1.24) gives
(i) for a positive m,

$$\frac{d^{\ell+m}}{dx^{\ell+m}}(x^2-1)^\ell$$

[13] It is proved in Appendix 7-A.

$$= \sum_{k=m}^{\ell} \frac{(\ell+m)!}{(\ell+m-k)!k!} \frac{\ell!}{(k-m)!}(x+1)^{k-m} \frac{\ell!}{(\ell-k)!}(x-1)^{\ell-k}$$

$$= (\ell!)^2 \sum_{k=m}^{\ell} \frac{(\ell+m)!}{(\ell+m-k)!k!} \frac{1}{(k-m)!} \frac{1}{(\ell-k)!}(x+1)^{k-m}(x-1)^{\ell-k} \quad (7.2.1.25)$$

(ii) for a negative m,

$$\frac{d^{\ell-m}}{dx^{\ell-m}}(x^2-1)^{\ell}$$
$$= (\ell!)^2 \sum_{k'=0}^{\ell-m} \frac{(\ell-m)!}{(\ell-m-k')!(k')!} \frac{1}{(k'+m)!} \frac{1}{(\ell-k')!}(x+1)^{k'+m}(x-1)^{\ell-k'} \quad (7.2.1.26)$$

If we substitute $k' = k-m$ on the right-hand side of the equation, we get

$$\frac{d^{\ell-m}}{dx^{\ell-m}}(x^2-1)^{\ell}$$
$$= (\ell!)^2 \sum_{k=m}^{\ell} \frac{(\ell-m)!}{(\ell-k)!(k-m)!} \frac{1}{k!} \frac{1}{(\ell+m-k)!}(x+1)^{k}(x-1)^{\ell+m-k} \quad (7.2.1.27)$$

By comparing Eqs. (7.2.1.25) and (7.2.1.27), we obtain

$$\frac{d^{\ell-m}}{dx^{\ell-m}}(x^2-1)^{\ell} = \frac{(\ell-m)!}{(\ell+m)!}(x+1)^m(x-1)^m \frac{d^{\ell+m}}{dx^{\ell+m}}(x^2-1)^{\ell}$$
$$= \frac{(\ell-m)!}{(\ell+m)!}(x^2-1)^m \frac{d^{\ell+m}}{dx^{\ell+m}}(x^2-1)^{\ell} \quad (7.2.1.28)$$

If we substitute $-m$ instead of m in Eq. (7.2.1.23), it gives

$$P_\ell^{-m}(x) = \frac{1}{2^\ell \ell!}(1-x^2)^{-\frac{m}{2}} \frac{d^{\ell-m}}{dx^{\ell-m}}(x^2-1)^{\ell}$$
$$= \frac{1}{2^\ell \ell!}(1-x^2)^{-\frac{m}{2}} \frac{(\ell-m)!}{(\ell+m)!}(x^2-1)^m \frac{d^{\ell+m}}{dx^{\ell+m}}(x^2-1)^{\ell} \quad (7.2.1.29)$$

From Eqs. (7.2.1.23) and (7.2.1.29), we obtain the relationship between a positive and negative m in the associated Legendre functions:

$$P_\ell^m(x) = \frac{1}{2^\ell \ell!}(1-x^2)^{\frac{m}{2}} 2^\ell \ell!(1-x^2)^{\frac{m}{2}} \frac{(\ell+m)!}{(\ell-m)!}(x^2-1)^{-m} P_\ell^{-m}(x)$$
$$= (1-x^2)^m \frac{(\ell+m)!}{(\ell-m)!}(x^2-1)^{-m} P_\ell^{-m}(x),$$

which gives

$$\frac{P_\ell^{-m}(x)}{P_\ell^m(x)} = (1-x^2)^{-m}(x^2-1)^m\frac{(\ell-m)!}{(\ell+m)!}$$
$$= (-1)^m\frac{(\ell-m)!}{(\ell+m)!} \quad (7.2.1.30)$$

Example

7.2.1.1 If we express the Laplace's equation, $\nabla^2 U=0$, using Eq. (4.3.11) in the spherical coordinate system, it is

$$\nabla^2 U = \frac{1}{r^2}\frac{\partial}{\partial r}(r^2\frac{\partial U}{\partial r})+\frac{1}{r^2\sin\theta}\frac{\partial}{\partial\theta}(\sin\theta\frac{\partial U}{\partial\theta})+\frac{1}{r^2\sin^2\theta}\frac{\partial^2 U}{\partial\phi^2} = 0$$

For simplicity, let's assume that U is independent of ϕ, $U(r,\theta)=R(r)\Theta(\theta)$. Since the last term on the equation is zero, we get

$$\frac{1}{r^2}\frac{\partial}{\partial r}(r^2\frac{\partial U}{\partial r})+\frac{1}{r^2\sin\theta}\frac{\partial}{\partial\theta}(\sin\theta\frac{\partial U}{\partial\theta}) = 0,$$

which gives

$$\frac{\Theta}{r^2}\frac{d}{dr}(r^2\frac{dR}{dr})+\frac{R}{r^2\sin\theta}\frac{d}{d\theta}(\sin\theta\frac{d\Theta}{d\theta})=0$$

By dividing the equation with $R(r)\Theta(\theta)$ and multiplying r^2, we have

$$\frac{1}{R}\frac{d}{dr}(r^2\frac{dR}{dr})+\frac{1}{\Theta\sin\theta}\frac{d}{d\theta}(\sin\theta\frac{d\Theta}{d\theta})=0$$

It is convenient to define $\frac{1}{R}\frac{d}{dr}(r^2\frac{dR}{dr})=\ell(\ell+1)$.

Then, we have

$$\frac{1}{\Theta\sin\theta}\frac{d}{d\theta}(\sin\theta\frac{d\Theta}{d\theta}) = -\ell(\ell+1),$$

which gives

$$\frac{d^2\Theta}{d\theta^2}+\frac{\cos\theta}{\sin\theta}\frac{d\Theta}{d\theta}+\ell(\ell+1)\Theta=0 \quad (7.2.1.1.1)$$

If we define $x=\cos\theta$, we have

$$\frac{d}{dx}=\frac{d\theta}{dx}\frac{d}{d\theta}=-\frac{1}{\sin\theta}\frac{d}{d\theta}$$

and

$$\frac{d^2}{dx^2}=\frac{d}{dx}(\frac{d}{dx})=-\frac{1}{\sin\theta}\frac{d}{d\theta}(-\frac{1}{\sin\theta}\frac{d}{d\theta})=-\frac{\cos\theta}{\sin^3\theta}\frac{d}{d\theta}+\frac{1}{\sin^2\theta}\frac{d^2}{d\theta^2}$$

If we substitute these into Eq. (7.2.1.2), it gives

$$-\frac{\cos\theta}{\sin\theta}\frac{dP_\ell(\cos\theta)}{d\theta}+\frac{d^2P_\ell(\cos\theta)}{d\theta^2}+2\frac{\cos\theta}{\sin\theta}\frac{dP_\ell(\cos\theta)}{d\theta}$$
$$+\ell(\ell+1)P_\ell(\cos\theta)=0,$$

which gives

$$\frac{d^2P_\ell(\cos\theta)}{d^2\theta}+\frac{\cos\theta}{\sin\theta}\frac{dP_\ell(\cos\theta)}{d\theta}+\ell(\ell+1)P_\ell(\cos\theta) = 0 \qquad (7.2.1.1.2)$$

By comparing Eqs. (7.2.1.1.1) and (7.2.1.1.2), we obtain

$$\Theta(\theta)=P_\ell(\cos\theta)$$

It means that the solution for the angular part in the problem is the Legendre polynomials.

7.2.2 Properties of Legendre Polynomials

- Orthogonality of the Legendre polynomials

$$\int_{-1}^{1} P_n(x)P_m(x)dx = \frac{2}{2n+1}\delta_{nm}$$

- Useful recurrence relation

$$P'_{n+1}(x)-P'_{n-1}(x) = (2n+1)P_n(x)$$

- Useful recurrence relation

$$nP_{n-1}(x)-(2n+1)xP_n(x)+(n+1)P_{n+1}(x)=0$$

- Orthogonality of the associated Legendre functions

$$\int_{-1}^{1} P_\ell^m(x)P_{\ell'}^m(x)dx = \frac{2}{2\ell+1}\frac{(\ell+m)!}{(\ell-m)!}\delta_{\ell\ell'}$$

- Spherical harmonics

$$Y_{\ell m}(\theta,\phi) = (-1)^m\sqrt{\frac{2\ell+1}{4\pi}\frac{(\ell-m)!}{(\ell+m)!}}\,P_\ell^m(\cos\theta)e^{im\phi}$$

- Relationship between positive and negative m

$$Y_{\ell,-|m|}(\theta,\phi) = (-1)^m Y^*_{\ell|m|}(\theta,\phi)$$

(1) Let's calculate $\int_{-1}^{1} P_n(x)P_m(x)dx$, where $P_n(x)$ and $P_m(x)$ are the Legendre polynomials.

Since $P_n(x)$ and $P_m(x)$ are the Legendre polynomials, they satisfy the Legendre differential equation with $m=0$:

$$\frac{d}{dx}[(1-x^2)\frac{dP_n(x)}{dx}]+n(n+1)P_n(x)=0 \quad (7.2.2.1)$$

and

$$\frac{d}{dx}[(1-x^2)\frac{dP_m(x)}{dx}]+m(m+1)P_m(x)=0 \quad (7.2.2.2)$$

If $P_m \times$Eq. (7.2.2.1)$-P_n \times$Eq. (7.2.2.2), we get

$$P_m(x)[(1-x^2)P_n'(x)]' - P_n(x)[(1-x^2)P_m'(x)]'$$
$$+[n(n+1)-m(m+1)]P_n(x)P_m(x)=0$$

If we take the integral $\int_{-1}^{1} dx$ on the equation, it gives

$$\int_{-1}^{1}[P_m(x)\{(1-x^2)P_n'(x)\}' - P_n(x)\{(1-x^2)P_m'(x)\}']dx$$
$$+[n(n+1)-m(m+1)]\int_{-1}^{1} P_n(x)P_m(x)dx=0 \quad (7.2.2.3)$$

The first integral term of the left-hand side of the equation can be calculated using integration by parts and is then

$$[P_m(x)(1-x^2)P_n'(x)]_{x=-1}^{x=1} - [P_n(x)(1-x^2)P_m'(x)]_{x=-1}^{x=1}$$
$$-\int_{-1}^{1}[P_m'(x)(1-x^2)P_n'(x) - P_n'(x)(1-x^2)P_m'(x)]dx \quad (7.2.2.4)$$

Since the each term of the equation is zero, Eq. (7.2.2.3) gives

$$[n(n+1)-m(m+1)]\int_{-1}^{1} P_n(x)P_m(x)dx=0 \quad (7.2.2.5)$$

(i) For $n \neq m$, Eq. (7.2.2.5) gives

$$\int_{-1}^{1} P_n(x)P_m(x)dx=0 \quad (7.2.2.6)$$

(ii) For $n=m$, we need to calculate $\int_{-1}^{1}[P_n(x)]^2dx$ and it is convenient to use the Rodrigues' formula, $P_n(x)=\frac{1}{2^n n!}\frac{d^n}{dx^n}(x^2-1)^n$:

$$\begin{aligned}\int_{-1}^{1}[P_n(x)]^2dx &= \frac{1}{2^{2n}(n!)^2}\int_{-1}^{1}\frac{d^n}{dx^n}(x^2-1)^n\frac{d^n}{dx^n}(x^2-1)^n dx\\ &= \frac{1}{2^{2n}(n!)^2}\int_{-1}^{1}[\frac{d^{n-1}}{dx^{n-1}}(x^2-1)^n]'\frac{d^n}{dx^n}(x^2-1)^n dx\end{aligned}$$

The right-hand side of the equation is calculated using integration by parts and we have

$$\frac{1}{2^{2n}(n!)^2}[\frac{d^{n-1}}{dx^{n-1}}(x^2-1)^n\frac{d^n}{dx^n}(x^2-1)^n]_{x=-1}^{x=1} - \frac{1}{2^{2n}(n!)^2}\int_{-1}^{1}\frac{d^{n-1}}{dx^{n-1}}(x^2-1)^n\frac{d^{n+1}}{dx^{n+1}}(x^2-1)^n dx \quad (7.2.2.7)$$

Since $\frac{d^{n-1}}{dx^{n-1}}(x^2-1)^n$ leaves (x^2-1) as a factor, the first term of Eq. (7.2.2.7) is zero at $x=\pm 1$. By performing integration of parts by n times, the second term of Eq. (7.2.2.7) is obtained as

$$\frac{(-1)^n}{2^{2n}(n!)^2}\int_{-1}^{1}(x^2-1)^n\frac{d^{2n}}{dx^{2n}}(x^2-1)^n dx \quad (7.2.2.8)$$

Since the function $(x^2-1)^n$ in the integrand of the equation has the degree of $2n$, the $(2n)$th derivative of the function gives $(2n)!$.

Then, the equation gives

$$\frac{(-1)^n(2n)!}{2^{2n}(n!)^2}\int_{-1}^{1}(x^2-1)^n dx$$

Thus, we obtain

$$\int_{-1}^{1}[P_n(x)]^2dx = \frac{(-1)^n(2n)!}{2^{2n}(n!)^2}\int_{-1}^{1}(x^2-1)^n dx \quad (7.2.2.9)$$

The integral on the right-hand side of the equation can be calculated using integration by substitution, and let's define $t=\frac{x+1}{2}$.

Then, we have

$$\int_{-1}^{1}(x^2-1)^n dx = 2^{2n+1}(-1)^n\int_{0}^{1}t^n(1-t)^n dt \quad (7.2.2.10)$$

Using the definition of the Beta function, the integral on the right-hand side of the equation gives

$$\int_0^1 t^n(1-t)^n dt = B(n+1,n+1) = \frac{\Gamma(n+1)\Gamma(n+1)}{\Gamma(2n+1+1)}$$
$$= \frac{(n!)^2}{(2n+1)!}$$

If we insert these into Eq. (7.2.2.9), it gives

$$\int_{-1}^1 [P_n(x)]^2 dx = \frac{(-1)^n(2n)!}{2^{2n}(n!)^2} 2^{2n+1}(-1)^n \frac{(n!)^2}{(2n+1)!}$$
$$= \frac{2}{2n+1} \tag{7.2.2.11}$$

From Eq. (7.2.2.6) for $n \neq m$ and Eq. (7.2.2.11) for $n=m$, we obtain

$$\int_{-1}^1 P_n(x)P_m(x)dx = \frac{2}{2n+1}\delta_{nm}, \tag{7.2.2.12}$$

which is the orthogonality of the Legendre polynomials.

(2) If we differentiate the Rodrigues' formula with respect to x, it gives

$$\frac{d}{dx}P_\ell(x) = \frac{1}{2^\ell \ell!}\frac{d}{dx}[\frac{d^\ell}{dx^\ell}(x^2-1)^\ell] = \frac{2\ell}{2^\ell \ell!}\frac{d^\ell}{dx^\ell}[x(x^2-1)^{\ell-1}]$$
$$= \frac{1}{2^{\ell-1}(\ell-1)!}\frac{d^{\ell-1}}{dx^{\ell-1}}[(2\ell-1)x^2-1](x^2-1)^{\ell-2}$$

If we substitute $\ell = n+1$ into the equation, we have

$$P'_{n+1}(x) = \frac{1}{2^n n!}\frac{d^n}{dx^n}[(2n+1)x^2-1](x^2-1)^{n-1} \tag{7.2.2.13}$$

By substituting $\ell = n-1$ in the Rodrigues' formula, we have

$$P_{n-1}(x) = \frac{1}{2^{n-1}(n-1)!}\frac{d^{n-1}}{dx^{n-1}}(x^2-1)^{n-1}$$

If we differentiate this equation with respect to x,

$$P'_{n-1}(x) = \frac{d}{dx}[\frac{1}{2^{n-1}(n-1)!}\frac{d^{n-1}}{dx^{n-1}}(x^2-1)^{n-1}]$$
$$= \frac{2n}{2^n n!}\frac{d^n}{dx^n}[(x^2-1)^{n-1}] \tag{7.2.2.14}$$

From Eqs. (7.2.2.13) and (7.2.2.14), we get

$$\begin{aligned} & P'_{n+1}(x) - P'_{n-1}(x) \\ &= \frac{1}{2^n n!}\frac{d^n}{dx^n}[\{(2n+1)x^2-1\}(x^2-1)^{n-1} - 2n(x^2-1)^{n-1}] \\ &= \frac{1}{2^n n!}\frac{d^n}{dx^n}[(2n+1)x^2-(2n+1)](x^2-1)^{n-1} \\ &= \frac{1}{2^n n!}\frac{d^n}{dx^n}[(2n+1)(x^2-1)(x^2-1)^{n-1}] \\ &= \frac{2n+1}{2^n n!}\frac{d^n}{dx^n}(x^2-1)^n \end{aligned}$$

Since $\frac{d^n}{dx^n}(x^2-1)^n = 2^n n! P_n(x)$ from the Rodrigues' formula, we obtain **the recurrence relation**:

$$P'_{n+1}(x) - P'_{n-1}(x) = (2n+1)P_n(x)$$

Consider a function $g(x,t) = (1-2xt+t^2)^{-\frac{1}{2}}$. It is called the **generating function** because the Legendre polynomials can be obtained by expanding the generating function in t using Taylor series.

This method gives

$$g(x,t) = (1-2xt+t^2)^{-\frac{1}{2}} = \sum_{\ell=0}^{\infty} P_\ell(x)t^\ell$$

If we differentiate the equation with respect to t, we get

$$\frac{x-t}{(1-2xt+t^2)^{\frac{3}{2}}} = \sum_{\ell=0}^{\infty} \ell P_\ell(x)t^{\ell-1}$$

Since the left-hand side of the equation can be expressed as

$$\begin{aligned} \frac{x-t}{(1-2xt+t^2)^{\frac{3}{2}}} &= \frac{x-t}{1-2xt+t^2}(1-2xt+t^2)^{-\frac{1}{2}} \\ &= \frac{x-t}{1-2xt+t^2}\sum_{\ell=0}^{\infty} P_\ell(x)t^\ell, \end{aligned}$$

the equation gives

$$\frac{x-t}{1-2xt+t^2}\sum_{\ell=0}^{\infty}P_\ell(x)t^\ell = \sum_{\ell=0}^{\infty}\ell P_\ell(x)t^{\ell-1},$$

which is

$$(x-t)\sum_{\ell=0}^{\infty}P_\ell(x)t^\ell = (1-2xt+t^2)\sum_{\ell=0}^{\infty}\ell P_\ell(x)t^{\ell-1}$$

This yields

$$\sum_{\ell=0}^{\infty}[\ell P_\ell(x)t^{\ell+1}+P_\ell(x)t^{\ell+1}-2\ell xP_\ell(x)t^\ell-xP_\ell(x)t^\ell+\ell P_\ell(x)t^{\ell-1}]=0$$

Since this equation satisfies for each power of t, we have

$$(\ell-1)P_{\ell-1}(x)+P_{\ell-1}(x)-2\ell xP_\ell(x)-xP_\ell(x)+(\ell+1)P_{\ell+1}(x)=0,$$

which yields the following **recurrence relation**

$$\ell P_{\ell-1}(x)-(2\ell+1)xP_\ell(x)+(\ell+1)P_{\ell+1}(x)=0$$

This is very useful relation to calculate higher order Legendre polynomials. Let's calculate some of the Legendre polynomials by using the recurrence relation.

We know that $P_0(x)=1$ and $P_1(x)=x$. From the recurrence relation for $\ell=1$, we get

$$P_0(x)-3xP_1(x)+2P_2(x) = 0$$

which gives

$$P_2(x)=\frac{1}{2}(3x^2-1)$$

For $\ell=2$, we have

$$2P_1(x)-5xP_2(x)+3P_3(x) = 0$$

from the recurrence relation,

which gives

$$\begin{aligned}P_3(x) &= \frac{1}{3}[5xP_2(x)-2P_1(x)]\\ &= \frac{1}{3}[5x\cdot\frac{1}{2}(3x^2-1)-2x]\\ &= \frac{1}{2}(5x^3-3x)\end{aligned}$$

(3) Let's calculate orthogonality of the associated Legendre functions $P_\ell^m(x)$,

$$\int_{-1}^{1}[P_\ell^m(x)]^2 dx = \frac{2}{2\ell+1}\frac{(\ell+m)!}{(\ell-m)!}$$

From Eq. (7.2.1.28), we have

$$\frac{d^{\ell+m}}{dx^{\ell+m}}(x^2-1)^\ell = \frac{(\ell+m)!}{(\ell-m)!}(x^2-1)^{-m}\frac{d^{\ell-m}}{dx^{\ell-m}}(x^2-1)^\ell$$

By inserting this into Eq. (7.2.1.23) we get

$$P_\ell^m(x) = \frac{(-1)^m}{2^\ell \ell!}\frac{(\ell+m)!}{(\ell-m)!}(1-x^2)^{-\frac{m}{2}}\frac{d^{\ell-m}}{dx^{\ell-m}}(x^2-1)^\ell$$

Therefore, it gives

$$[P_\ell^m(x)]^2 = \frac{1}{2^{2\ell}(\ell!)^2}[\frac{(\ell+m)!}{(\ell-m)!}]^2(1-x^2)^{-m}\left[\frac{d^{\ell-m}}{dx^{\ell-m}}(x^2-1)^\ell\right]\left[\frac{d^{\ell-m}}{dx^{\ell-m}}(x^2-1)^\ell\right]$$

Using Eq. (7.2.1.28), the equation is expressed as

$$\begin{aligned}
&[P_\ell^m(x)]^2 \\
&= \frac{1}{2^{2\ell}(\ell!)^2}[\frac{(\ell+m)!}{(\ell-m)!}]^2(1-x^2)^{-m}\frac{(\ell-m)!}{(\ell+m)!}(x^2-1)^m\left[\frac{d^{\ell+m}}{dx^{\ell+m}}(x^2-1)^\ell\right]\left[\frac{d^{\ell-m}}{dx^{\ell-m}}(x^2-1)^\ell\right] \\
&= \frac{(-1)^m}{2^{2\ell}(\ell!)^2}\frac{(\ell+m)!}{(\ell-m)!}\left[\frac{d^{\ell+m}}{dx^{\ell+m}}(x^2-1)^\ell\right]\left[\frac{d^{\ell-m}}{dx^{\ell-m}}(x^2-1)^\ell\right]
\end{aligned}$$

If we take integrals on both sides of the equation, we have

$$\begin{aligned}
&\int_{-1}^{1}[P_\ell^m(x)]^2 dx \\
&= \frac{(-1)^m}{2^{2\ell}(\ell!)^2}\frac{(\ell+m)!}{(\ell-m)!}\int_{-1}^{1}\left[\frac{d^{\ell+m}}{dx^{\ell+m}}(x^2-1)^\ell\right]\left[\frac{d^{\ell-m}}{dx^{\ell-m}}(x^2-1)^\ell\right]dx \quad (7.2.2.15)
\end{aligned}$$

The right-hand side of the equation can be calculated using integration by parts:

$$\begin{aligned}
&\int_{-1}^{1}\left[\frac{d^{\ell+m}}{dx^{\ell+m}}(x^2-1)^\ell\right]\left[\frac{d^{\ell-m}}{dx^{\ell-m}}(x^2-1)^\ell\right]dx \\
&= \int_{-1}^{1}\left[\frac{d^{\ell+m-1}}{dx^{\ell+m-1}}(x^2-1)^\ell\right]'\frac{d^{\ell-m}}{dx^{\ell-m}}(x^2-1)^\ell dx
\end{aligned}$$

$$= [\frac{d^{\ell+m-1}}{dx^{\ell+m-1}}(x^2-1)^\ell \frac{d^{\ell-m}}{dx^{\ell-m}}(x^2-1)^\ell]_{x=-1}^{x=1}$$

$$-\int_{-1}^{1}\left[\frac{d^{\ell+m-1}}{dx^{\ell+m-1}}(x^2-1)^\ell\right]\frac{d^{\ell-m+1}}{dx^{\ell-m+1}}(x^2-1)^\ell dx$$

In here, $\frac{d^{\ell-m}}{dx^{\ell-m}}(x^2-1)^\ell$ leaves (x^2-1) as a factor. It makes the first term on the right-hand side of the equation is zero at $x=\pm 1$.

The second term is calculated by using integration by parts,

$$-\int_{-1}^{1}\left[\frac{d^{\ell+m-1}}{dx^{\ell+m-1}}(x^2-1)^\ell\right]\frac{d^{\ell-m+1}}{dx^{\ell-m+1}}(x^2-1)^\ell dx$$

$$= -\int_{-1}^{1}\left[\frac{d^{\ell+m-2}}{dx^{\ell+m-2}}(x^2-1)^\ell\right]'\frac{d^{\ell-m+1}}{dx^{\ell-m+1}}(x^2-1)^\ell dx$$

$$= \left[-\frac{d^{\ell+m-2}}{dx^{\ell+m-2}}(x^2-1)^\ell \frac{d^{\ell-m+1}}{dx^{\ell-m+1}}(x^2-1)^\ell\right]_{x=-1}^{x=1}$$

$$+\int_{-1}^{1}\left[\frac{d^{\ell+m-2}}{dx^{\ell+m-2}}(x^2-1)^\ell\right]\frac{d^{\ell-m+2}}{dx^{\ell-m+2}}(x^2-1)^\ell dx$$

Similarly, the first term of the right-hand side of the equation is zero at $x=\pm 1$.

If we perform integration by parts by m times for the second term of the right-hand side of the equation, we get

$$(-1)^m\int_{-1}^{1}\left[\frac{d^\ell}{dx^\ell}(x^2-1)^\ell\right]\frac{d^\ell}{dx^\ell}(x^2-1)^\ell dx$$

If we substitute this into Eq. (7.2.2.15), we obtain

$$\int_{-1}^{1}[P_\ell^m(x)]^2 dx = \frac{1}{2^{2\ell}(\ell!)^2}\frac{(\ell+m)!}{(\ell-m)!}\int_{-1}^{1}[\frac{d^\ell}{dx^\ell}(x^2-1)^\ell]^2 dx$$

Using the Rodrigues' formula, the right-hand side of the equation can be written as

$$\frac{1}{2^{2\ell}(\ell!)^2}\frac{(\ell+m)!}{(\ell-m)!}\int_{-1}^{1}[\frac{d^\ell}{dx^\ell}(x^2-1)^\ell]^2 dx$$

$$= \frac{1}{2^{2\ell}(\ell!)^2}\frac{(\ell+m)!}{(\ell-m)!}2^{2\ell}(\ell!)^2\int_{-1}^{1}[P_\ell(x)]^2 dx$$

$$= \frac{(\ell+m)!}{(\ell-m)!}\int_{-1}^{1}[P_\ell(x)]^2 dx$$

$$= \frac{(\ell+m)!}{(\ell-m)!}\frac{2}{2\ell+1}$$

Thus, we obtain the orthogonality of the associated Legendre functions

$$\int_{-1}^{1}[P_\ell^m(x)]^2 dx = \frac{2}{2\ell+1}\frac{(\ell+m)!}{(\ell-m)!}$$

This can be generalized to

$$\int_{-1}^{1}P_\ell^m(x)P_{\ell'}^m dx = \frac{2}{2\ell+1}\frac{(\ell+m)!}{(\ell-m)!}\delta_{\ell\ell'}$$

Example

7.2.2.1 Calculate $\int_0^\pi P_\ell(\cos\theta)P_{\ell'}(\cos\theta)\sin\theta d\theta$.

Ans. Let's define $\cos\theta = x$ and the integral is then

$$\int_1^{-1}P_\ell(x)P_{\ell'}(x)(-dx) = \int_{-1}^{1}P_\ell(x)P_{\ell'}(x)dx$$

$$= \frac{2}{2\ell+1}\delta_{\ell\ell'}$$

7.2.2.2 Find a solution for the Laplace's equation in the spherical coordinate system.

Ans. We know that $\nabla^2\Psi = 0$ in the spherical coordinate system is expressed by

$$\nabla^2\Psi = \frac{1}{r^2}\frac{\partial}{\partial r}(r^2\frac{\partial\Psi}{\partial r}) + \frac{1}{r^2\sin\theta}\frac{\partial}{\partial\theta}(\sin\theta\frac{\partial\Psi}{\partial\theta}) + \frac{1}{r^2\sin^2\theta}\frac{\partial^2\Psi}{\partial\phi^2} = 0$$

If we assume $\Psi(r,\theta,\phi) = R(r)\Theta(\theta)\Phi(\phi)$, the equation is then

$$\frac{\sin^2\theta}{R}\frac{d}{dr}(r^2\frac{dR}{dr}) + \frac{\sin\theta}{\Theta}\frac{d}{d\theta}(\sin\theta\frac{d\Theta}{d\theta}) + \frac{1}{\Phi}\frac{d^2\Phi}{d\phi^2} = 0$$

Let's define

$$\frac{1}{\Phi}\frac{d^2\Phi}{d\phi^2} = -m^2$$

The general solution for $\frac{d^2\Phi}{d\phi^2}+m^2\Phi=0$ is

$$\Phi(\phi) \;=\; Ae^{\pm im\phi}$$

The reason we choose the negative constant $-m^2$ is that the solution $\Phi(\phi)$ yields the same value for each $\phi=2\pi$ turn. Since $\phi=\phi+2\pi$, $e^{\pm im2\pi}=1$ should be satisfied. It means that $m = 0,\ 1,\ 2,\ \cdots\cdots$.

The constant, A, in the solution is found by requiring **orthonormal condition** (or **orthonormality**) for the solution $\Phi(\phi)$, $\int_0^{2\pi}|\Phi(\phi)|^2 d\phi \;=\; \int_0^{2\pi}\Phi^*(\phi)\Phi(\phi)d\phi \;=1$:

$$|A|^2\int_0^{2\pi}d\phi \;=\; 1$$

which gives

$$A=\frac{1}{\sqrt{2\pi}}$$

Thus, the orthonormal solution for ϕ is

$$\Phi(\phi) \;=\; \frac{1}{\sqrt{2\pi}}e^{im\phi},\ \text{where } m \text{ is an integer.}$$

With the assumption that $\frac{1}{R}\frac{d}{dr}(r^2\frac{dR}{dr})=\ell(\ell+1)$ as [Ex. 7.2.1.1], the original equation is then expressed as

$$\frac{1}{\sin\theta}\frac{d}{d\theta}(\sin\theta\frac{d\Theta}{d\theta})+[\ell(\ell+1)-\frac{m^2}{\sin^2\theta}]\Theta \;=\; 0 \qquad (7.2.2.2.1)$$

If we define $x=\cos\theta$, we have

$$\frac{d}{dx}=\frac{d\theta}{dx}\frac{d}{d\theta} \;=\; -\frac{1}{\sin\theta}\frac{d}{d\theta}$$

By substituting these into Eq. (7.2.1.15), we have

$$\frac{1}{\sin\theta}\frac{d}{d\theta}[\sin\theta\frac{dP_\ell^m(\cos\theta)}{d\theta}]+[\ell(\ell+1)-\frac{m^2}{\sin^2\theta}]P_\ell^m(\cos\theta)=0 \quad (7.2.2.2.2)$$

Since Eqs. (7.2.2.2.1) and (7.2.2.2.2) are the same, we know that

$$\Theta(\theta) \;=\; P_\ell^m(\cos\theta)$$

Let's define **spherical harmonics** $Y_{\ell m}$ for the angular parts so that

$$\Psi(r,\theta,\phi) \;=\; R(r)\Theta(\theta)\Phi(\phi)= \; R(r)Y_{\ell m}(\theta,\phi),$$

where

$$Y_{\ell m}(\theta,\phi) \;=\; C(-1)^m P_\ell^m(\cos\theta)e^{im\phi}$$

The $(-1)^m$ in the definition of the spherical harmonics, analogous to $(-1)^m$ in Eq. (7.2.1.30), is added. It does not change physics results but makes calculations more convenient in some Quantum Mechanics problems.

The normalization constant C is obtained by requiring the orthonormality $\int d\Omega |Y_{\ell m}(\theta, \phi)|^2 = 1$ of the spherical harmonics,

$$\int_0^{2\pi} d\phi \int_0^{\pi} Y_{\ell m}^*(\theta,\phi) Y_{\ell' m'}(\theta,\phi) \sin\theta d\theta = \delta_{\ell\ell'}\delta_{mm'}$$

It gives

$$|C|^2 \int_0^{2\pi} d\phi \int_0^{\pi} P_\ell^m(\cos\theta) P_{\ell'}^{m'}(\cos\theta) \sin\theta d\theta = \delta_{\ell\ell'}\delta_{mm'}$$

Using the orthogonality of the associated Legendre functions, we get

$$\frac{4\pi}{2\ell+1}\frac{(\ell+m)!}{(\ell-m)!}|C|^2 = 1,$$

which gives

$$C = \sqrt{\frac{2\ell+1}{4\pi}\frac{(\ell-m)!}{(\ell+m)!}}$$

Thus, we obtain for the spherical harmonics

$$Y_{\ell m}(\theta,\phi) = (-1)^m \sqrt{\frac{2\ell+1}{4\pi}\frac{(\ell-m)!}{(\ell+m)!}} P_\ell^m(\cos\theta) e^{im\phi} \qquad (7.2.2.2.3)$$

7.2.2.3 Find the relationship between a positive and negative m in the spherical harmonics.

Ans. We get $Y_{\ell,-m}(\theta,\phi)$ by substituting $-m$ instead of m in the spherical harmonics of Eq. (7.2.2.2.3):

$$Y_{\ell,-m}(\theta,\phi) = (-1)^{-m} \sqrt{\frac{2\ell+1}{4\pi}\frac{(\ell+m)!}{(\ell-m)!}} P_\ell^{-m}(\cos\theta) e^{-im\phi}$$

Since $P_\ell^{-m}(x) = (-1)^m \frac{(\ell-m)!}{(\ell+m)!} P_\ell^m(x)$, this equation is expressed as

$$\begin{aligned} Y_{\ell,-m}(\theta,\phi) &= (-1)^m \sqrt{\frac{2\ell+1}{4\pi}\frac{(\ell+m)!}{(\ell-m)!}} (-1)^m \frac{(\ell-m)!}{(\ell+m)!} P_\ell^m(\cos\theta) e^{-im\phi} \\ &= \sqrt{\frac{2\ell+1}{4\pi}\frac{(\ell-m)!}{(\ell+m)!}} P_\ell^m(\cos\theta) e^{-im\phi} \\ &= (-1)^m Y_{\ell m}^*(\theta,\phi) \end{aligned}$$

Thus, we obtain

$$Y_{\ell,-m}(\theta,\phi) = (-1)^m Y_{\ell m}^*(\theta,\phi)$$

Since we take the positive m, the equation is expressed as

$$Y_{\ell,-|m|}(\theta,\phi) = (-1)^m Y_{\ell |m|}^*(\theta,\phi)$$

7.2.2.4 Let's calculate some of the spherical harmonics, for $\ell = 0$ and 1, using Eq. (7.2.2.2.3).

Ans.

$$Y_{00} = \frac{1}{\sqrt{4\pi}} P_0^0 = \frac{1}{\sqrt{4\pi}} P_0 = \frac{1}{\sqrt{4\pi}}$$

$$Y_{10} = \sqrt{\frac{3}{4\pi}} P_1^0 = \sqrt{\frac{3}{4\pi}} P_1 = \sqrt{\frac{3}{4\pi}} \cos\theta$$

$$Y_{11} = -\sqrt{\frac{3}{8\pi}} P_1^1 e^{i\phi} = -\sqrt{\frac{3}{8\pi}} (1-\cos^2\theta)^{\frac{1}{2}} e^{i\phi}$$

$$= -\sqrt{\frac{3}{8\pi}} \sin\theta e^{i\phi}$$

$$Y_{1,-1} = (-1)^{-1} Y_{11}^* = \sqrt{\frac{3}{8\pi}} \sin\theta e^{-i\phi}$$

7.3 Solutions for Bessel's Equation

In this section we study many kinds of Bessel functions which are solutions of Bessel's equation. The Bessel functions of the first and second kind are covered in the first subsection, while other kinds are described in the second subsection.

7.3.1 Bessel and Neumann Functions

· Solutions of Bessel's equation:

$$y(x) = AJ_\nu(x) + BN_\nu(x),$$

where $J_\nu(x) = \sum_{k=0}^{\infty} \frac{(-1)^k}{k!\Gamma(\nu+k+1)} (\frac{x}{2})^{\nu+2k}$

(Bessel function of the first kind and order ν)

and $N_\nu(x) = Y_\nu(x) = \dfrac{J_\nu(x)\cos\nu\pi - J_{-\nu}(x)}{\sin\nu\pi}$

(Neumann (or Weber) function or the second kind and order

ν Bessel function)

· Relationship between a positive and negative integer ℓ:

$$J_{-\ell}(x) = (-1)^{\ell} J_{\ell}(x)$$

Bessel's equation is expressed as

$$x^2 \frac{d^2y(x)}{dx^2} + x\frac{dy(x)}{dx} + (x^2 - \nu^2)y(x) = 0, \tag{7.3.1.1}$$

where ν is any real numbers with $\nu \geq 0$. Solutions of the Bessel's equation are called **Bessel's functions** of order ν. For instance, the orders of the Bessel's function in the cylindrical[14] and spherical[15] problems are an integer value and half of an odd integer value, respectively.

$J_{\nu}(x)$ is called the **first kind and order ν Bessel function.** $N_{\nu}(x)$[16] is called the **second kind and order ν Bessel function or Neumann function**. When $J_{\nu}(x)$ and $J_{-\nu}(x)$ are the Bessel functions, their linear combination is also the solution of the Bessel's equation and is called as the Neumann function.

As a trial solution of the Bessel's equation, let's consider

$$y(x) = \sum_{n=0}^{\infty} c_n x^{n+m},$$

where $c_0 \neq 0$.

This gives

$$y'(x) = \sum_{n=0}^{\infty}(n+m)c_n x^{n+m-1} \text{ and } y''(x) = \sum_{n=0}^{\infty}(n+m)(n+m-1)c_n x^{n+m-2}$$

If we substitute these into the Bessel's equation, we get

[14] See [Example 7.3.1.1].

[15] See [Example 7.3.2.1].

[16] Sometimes it is called Weber function, $Y_{\nu}(x)$.

$$\sum_{n=0}^{\infty}[\{(n+m)(n+m-1)+(n+m)-\nu^2\}c_n x^{n+m}+c_n x^{n+m+2}] = 0,$$

which gives

$$x^m \sum_{n=0}^{\infty}[\{(n+m)^2-\nu^2\}c_n x^n + c_n x^{n+2}] = 0 \tag{7.3.1.2}$$

The term of the left-hand side of the equation should be zero for each power of x.

For x^0, we have

$$(m^2-\nu^2)c_0 = 0$$

Since $c_0 \neq 0$, we have $m = \pm\nu$ and let's choose the positive value:

$$m = \nu$$

For x^1, the equation gives

$$[(m+1)^2-\nu^2]c_1 = 0$$

If we substitute $m=\nu$ in the equation, we have

$$[(\nu+1)^2-\nu^2]c_1 = 0 \Rightarrow (2\nu+1)c_1 = 0$$

Since $\nu \geq 0$, the equation gives $c_1 = 0$.

Eq. (7.3.1.2) can be written as

$$\sum_{n=0}^{\infty}[(n+m)^2-\nu^2]c_n x^n + \sum_{n=2}^{\infty} c_{n-2}x^n = 0, \tag{7.3.1.3}$$

which gives the following recurrence relation

$$\begin{aligned} \frac{c_n}{c_{n-2}} &= -\frac{1}{(n+m)^2-\nu^2} \\ &= -\frac{1}{(n+\nu)^2-\nu^2} = -\frac{1}{n(n+2\nu)} \end{aligned} \tag{7.3.1.4}$$

From the recurrence relation we know that only even terms remain for $c_1 = 0$.

Then, we have

$$\frac{c_2}{c_0}=-\frac{1}{2(2+2\nu)}=-\frac{1}{4(1+\nu)}$$

$$\frac{c_4}{c_0}=\frac{c_4}{c_2}\frac{c_2}{c_0}=[-\frac{1}{4(4+2\nu)}][-\frac{1}{4(1+\nu)}]$$

$$=\frac{1}{8(2+\nu)}\frac{1}{4(1+\nu)}=\frac{1}{4\cdot 8(1+\nu)(2+\nu)}$$

$$\frac{c_6}{c_0}=\frac{c_6}{c_4}\frac{c_4}{c_0}=-\frac{1}{6(6+2\nu)}\frac{1}{4\cdot 8(1+\nu)(2+\nu)}$$

$$=-\frac{1}{4\cdot 8\cdot 12(1+\nu)(2+\nu)(3+\nu)}$$

$$\vdots$$

$$\vdots$$

Therefore, we get the solution

$$y(x)=x^m[c_0x^0+c_2x^2+c_4x^4+c_6x^6+\cdots\cdots]$$

$$=x^\nu[c_0x^0+c_2x^2+c_4x^4+c_6x^6+\cdots\cdots]$$

$$=c_0x^\nu[x^0-\frac{x^2}{4(1+\nu)}+\frac{x^4}{4\cdot 8(1+\nu)(2+\nu)}$$

$$-\frac{x^6}{4\cdot 8\cdot 12(1+\nu)(2+\nu)(3+\nu)}+\cdots\cdots]$$

$$=c_0x^\nu[\frac{x^0}{2^0}-\frac{x^2}{1\cdot 2^2(1+\nu)}+\frac{x^4}{2\cdot 2^4(1+\nu)(2+\nu)}$$

$$-\frac{x^6}{2\cdot 3\cdot 2^6(1+\nu)(2+\nu)(3+\nu)}+\cdots\cdots] \tag{7.3.1.5}$$

Since $\Gamma(n+1)=n\Gamma(n)$, we have

$$(1+\nu)=\frac{\Gamma(\nu+2)}{\Gamma(\nu+1)}$$

$$(2+\nu)=\frac{\Gamma(\nu+3)}{\Gamma(\nu+2)}=\frac{\Gamma(\nu+3)}{(1+\nu)\Gamma(\nu+1)} \Rightarrow (1+\nu)(2+\nu)=\frac{\Gamma(\nu+3)}{\Gamma(\nu+1)}$$

$$(3+\nu)=\frac{\Gamma(\nu+4)}{\Gamma(\nu+3)}=\frac{\Gamma(\nu+4)}{(1+\nu)(2+\nu)\Gamma(\nu+1)}$$

$$\Rightarrow (1+\nu)(2+\nu)(3+\nu)=\frac{\Gamma(\nu+4)}{\Gamma(\nu+1)}$$

Therefore, Eq. (7.3.1.5) can be expressed in terms of the Gamma functions.

$$\begin{aligned} y(x) &= c_0 x^{\nu}\Gamma(\nu+1)[\frac{x^0}{2^0\cdot\Gamma(\nu+1)} - \frac{x^2}{1\cdot 2^2\cdot\Gamma(\nu+2)} \\ &\quad + \frac{x^4}{(1\cdot 2)\cdot 2^4\cdot\Gamma(\nu+3)} - \frac{x^6}{(1\cdot 2\cdot 3)\cdot 2^6\cdot\Gamma(\nu+4)} + \cdots\cdots] \\ &= c_0\sum_{k=0}^{\infty}\frac{(-1)^k x^{\nu+2k}}{k!2^{2k}}\Gamma(\nu+1)\frac{1}{\Gamma(\nu+k+1)} \\ &= c_0\sum_{k=0}^{\infty}\frac{(-1)^k}{k!}(\frac{x}{2})^{\nu+2k}\frac{2^{\nu}\Gamma(\nu+1)}{\Gamma(\nu+k+1)} \end{aligned} \quad (7.3.1.6)$$

Let's take $c_0 = \dfrac{1}{2^{\nu}\Gamma(1+\nu)}$ to make this series more systematic and then Eq. (7.3.1.6) gives

$$y(x) = \sum_{k=0}^{\infty}\frac{(-1)^k}{k!\Gamma(\nu+k+1)}(\frac{x}{2})^{\nu+2k}$$

This is the Bessel function, $J_{\nu}(x)$, of the first kind and order ν.
Thus, the solution of the Bessel's equation of order ν for any real $\nu \geq 0$ (including integers and half of odd integers) is

$$J_{\nu}(x) = \sum_{k=0}^{\infty}\frac{(-1)^k}{k!\Gamma(\nu+k+1)}(\frac{x}{2})^{\nu+2k} \quad (7.3.1.7)$$

Similarly, we can obtain the solution $J_{-\nu}(x)$ by simply replacing ν with $-\nu$ in Eq. (7.3.1.7):

$$J_{-\nu}(x) = \sum_{k=0}^{\infty}\frac{(-1)^k}{k!\Gamma(-\nu+k+1)}(\frac{x}{2})^{-\nu+2k} \quad (7.3.1.8)$$

Since $J_{\nu}(x)$ and $J_{-\nu}(x)$ are linearly independent, their linear combination can be the solution of the Bessel's equation.
The Neumann function is defined as

$$N_{\nu}(x) = \frac{J_{\nu}(x)\cos\nu\pi - J_{-\nu}(x)}{\sin\nu\pi}, \quad (7.3.1.9)$$

which is singular at $x=0$.

To find the relationship between $J_\ell(x)$ and $J_{-\ell}(x)$ for an integer ℓ, let's substitute $n+\ell$ instead of k in Eq. (7.3.1.8). Since k and ℓ are integers, n is also an integer.
Then, it gives

$$J_{-\ell}(x) = \sum_{n=0}^{\infty} \frac{(-1)^{n+\ell}}{(n+\ell)!\Gamma(n+1)} \left(\frac{x}{2}\right)^{\ell+2n}, \tag{7.3.1.10}$$

where

$$\begin{aligned}(n+\ell)!\Gamma(n+1) &= n!(n+1)(n+2)(n+3)\cdots\cdots(n+\ell)\Gamma(n+1)\\ &= n!(n+2)(n+3)\cdots\cdots(n+\ell)\Gamma(n+1+1)\\ &= n!(n+3)\cdots\cdots(n+\ell)\Gamma(n+2+1)\\ &\vdots\\ &\vdots\\ &= n!\Gamma(n+\ell+1)\end{aligned}$$

Thus, Eq. (7.3.1.10) is expressed as

$$\begin{aligned}J_{-\ell}(x) &= \sum_{n=0}^{\infty} \frac{(-1)^\ell(-1)^n}{n!\Gamma(n+\ell+1)} \left(\frac{x}{2}\right)^{\ell+2n}\\ &= (-1)^\ell \sum_{k=0}^{\infty} \frac{(-1)^k}{k!\Gamma(k+\ell+1)} \left(\frac{x}{2}\right)^{\ell+2k} = (-1)^\ell J_\ell(x),\end{aligned}$$

where n is the dummy index, we can change it to k.
Thus, we obtain the relationship between $J_\ell(x)$ and $J_{-\ell}(x)$,

$$J_{-\ell}(x) = (-1)^\ell J_\ell(x), \text{ for an integer } \ell$$

It means that $J_\ell(x)$ and $J_{-\ell}(x)$ are not independent for an integer ℓ. Note that $J_0(0)=1$, while all other Bessel functions of higher orders are zeros at $x=0$, shown in Fig. 7.3.1.1.

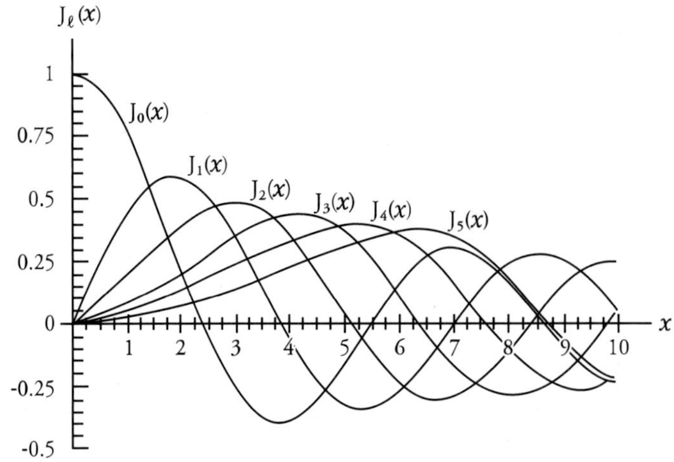

Fig. 7.3.1.1 Bessel functions.

For an integer ℓ, the Neumann function is

$$N_\ell(x) = \frac{J_\ell(x)\cos\ell\pi - J_{-\ell}(x)}{\sin\ell\pi} = \frac{J_\ell(x)(-1)^\ell - (-1)^\ell J_\ell(x)}{\sin\ell\pi} = \frac{0}{0}$$

Since $N_\ell(x)$ is $\frac{0}{0}$, it can be calculated by using L'Hopital's rule:

$$\begin{aligned} N_\ell(x) &= \frac{\frac{\partial}{\partial\ell}[J_\ell(x)\cos\ell\pi - J_{-\ell}(x)]}{\frac{\partial}{\partial\ell}(\sin\ell\pi)} \\ &= \frac{(-\pi\sin\ell\pi)J_\ell(x) + \cos\ell\pi\frac{\partial J_\ell(x)}{\partial\ell} - \frac{\partial J_{-\ell}(x)}{\partial\ell}}{\pi\cos\ell\pi} \\ &= \frac{\cos\ell\pi\frac{\partial J_\ell(x)}{\partial\ell} - \frac{\partial J_{-\ell}(x)}{\partial\ell}}{\pi\cos\ell\pi} \\ &= \frac{1}{\pi}\left[\frac{\partial J_\ell(x)}{\partial\ell} - (-1)^\ell\frac{\partial J_{-\ell}(x)}{\partial\ell}\right] \end{aligned} \tag{7.3.1.11}$$

This is well-defined for any $x \neq 0$ and is independent of $J_\ell(x)$. Some of the Neumann functions are shown in Fig. 7.3.1.2.

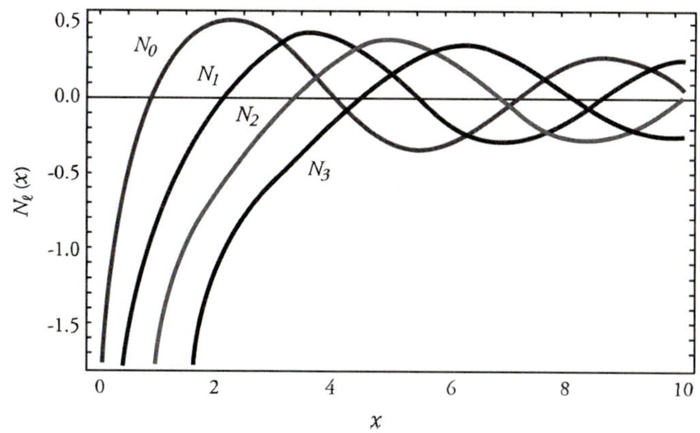

Fig. 7.3.1.2 Neumann functions.

Therefore,

(i) If ν is non-integers, $J_\nu(x)$ and $J_{-\nu}(x)$ are linearly independent so that the solution of the Bessel's equation is given as

$$y(x) = A_1 J_\nu(x) + A_2 J_{-\nu}(x),$$

where $J_\nu(x)$ and $J_{-\nu}(x)$ are given by Eqs. (7.3.1.7) and (7.3.1.8), respectively.

(ii) If $\nu = \ell$ is integers, $J_{-\ell}(x) = (-1)^\ell J_\ell(x)$ and they are not linearly independent so that we need the second linearly independent solution $N_\ell(x)$ of Eq. (7.3.1.11). Then, the solution of the Bessel's equation is given as

$$y(x) = B_1 J_\ell(x) + B_2 N_\ell(x),$$

where $J_\ell(x)$ and $N_\ell(x)$ for an integer ℓ are given by Eqs. (7.3.1.7) and (7.3.1.11), respectively.

Example

7.3.1.1 Find a general solution for the Laplace's equation in the cylindrical coordinate system.

Ans. Using Eq. (4.3.11) we get the Laplace's equation, $\nabla^2 U = 0$, in the cylindrical coordinate system:

$$\frac{1}{\rho}\frac{\partial}{\partial \rho}(\rho\frac{\partial U}{\partial \rho})+\frac{1}{\rho^2}\frac{\partial^2 U}{\partial \phi^2}+\frac{\partial^2 U}{\partial z^2}=0,$$

which gives

$$\frac{\partial^2 U}{\partial \rho^2}+\frac{1}{\rho}\frac{\partial U}{\partial \rho}+\frac{1}{\rho^2}\frac{\partial^2 U}{\partial \phi^2}+\frac{\partial^2 U}{\partial z^2}=0$$

Let's define $U(\rho,\phi,z)=R(\rho)\Phi(\phi)Z(z)$ and substitute it into the equation. We then get

$$\Phi Z\frac{d^2R}{d\rho^2}+\frac{\Phi Z}{\rho}\frac{dR}{d\rho}+\frac{RZ}{\rho^2}\frac{d^2\Phi}{d\phi^2}+R\Phi\frac{d^2Z}{dz^2}=0,$$

which gives

$$\frac{1}{R}\frac{d^2R}{d\rho^2}+\frac{1}{\rho R}\frac{dR}{d\rho}+\frac{1}{\rho^2\Phi}\frac{d^2\Phi}{d\phi^2}+\frac{1}{Z}\frac{d^2Z}{dz^2}=0$$

Then, the equation can be written as

$$\frac{1}{R}\frac{d^2R}{d\rho^2}+\frac{1}{\rho R}\frac{dR}{d\rho}+\frac{1}{\rho^2\Phi}\frac{d^2\Phi}{d\phi^2}=-\frac{1}{Z}\frac{d^2Z}{dz^2}$$

The left-hand side of the equation depends on only ρ and ϕ. On the other hand the right-hand side of the equation depends on only z. Both sides should be a constant to satisfy equality for all values of ρ, ϕ, and z.

If we define the constant to be $-k^2$, we have

$$\frac{d^2Z}{dz^2}-k^2Z=0 \tag{7.3.1.1.1}$$

and

$$\frac{1}{R}\frac{d^2R}{d\rho^2}+\frac{1}{\rho R}\frac{dR}{d\rho}+\frac{1}{\rho^2\Phi}\frac{d^2\Phi}{d\phi^2}=-k^2 \tag{7.3.1.1.2}$$

A general solution for Eq. (7.3.1.1.1) is then

$$Z(z)=A_1e^{kz}+A_2e^{-kz}$$

The negative constant, $-k^2$, is chosen so that the general solution $Z(z)$ goes to zero for $z\rightarrow\pm\infty$ for specific boundary conditions. Otherwise, a solution will be periodic trigonometric functions and will not go to zero for $z\rightarrow\pm\infty$.

Eq. (7.3.1.1.2) can be written as

$$\frac{\rho^2}{R}\frac{d^2R}{d\rho^2}+\frac{\rho}{R}\frac{dR}{d\rho}+k^2\rho^2=-\frac{1}{\Phi}\frac{d^2\Phi}{d\phi^2}$$

The left-hand side of the equation depends on only ρ, while the right-hand side of the equation depends on only ϕ. Both sides should be a constant to satisfy equality for all values of ρ and ϕ.

If we define the constant to be ν^2, it gives

$$\frac{d^2\Phi}{d\phi^2}+\nu^2\Phi=0 \tag{7.3.1.1.3}$$

and

$$\rho^2\frac{d^2R}{d\rho^2}+\rho\frac{dR}{d\rho}+(k^2\rho^2-\nu^2)R=0 \tag{7.3.1.1.4}$$

A general solution for Eq. (7.3.1.1.3) is

$$\Phi(\phi)=B_1\sin\nu\phi+B_2\cos\nu\phi, \tag{7.3.1.1.5}$$

where ν is an integer. By choosing a positive constant ν^2, the general solution $\Phi(\phi)$ can have the same value for each $\phi=2\pi$ turn.

If we define $k\rho=x$, Eq. (7.3.1.1.4) gives

$$x^2\frac{d^2R}{dx^2}+x\frac{dR}{dx}+(x^2-\nu^2)R=0,$$

which is the Bessel's equation. Therefore, a solution for $R(\rho)$ is the solution of the Bessel's equation.

Thus, the solution for the Laplace equation in the cylindrical coordinate system is

$$U(\rho,\phi,z)\propto[AJ_\nu(\rho)+BN_\nu(\rho)]e^{\pm i\nu\phi}(e^{kz}+e^{-kz}),$$

where ν is an integer.

7.3.2 Other Bessel Functions

- Hankel functions or Bessel functions of the third kind and order ν
  - $H_\nu^{(1)}(x) = J_\nu(x)+iN_\nu(x)$, Hankel function of the first kind
  - $H_\nu^{(2)}(x) = J_\nu(x)-iN_\nu(x)$, Hankel function of the second kind
- Spherical Bessel function

$$j_\ell(x) = \sqrt{\frac{\pi}{2x}}J_{\ell+\frac{1}{2}}(x)=(-x)^\ell(\frac{1}{x}\frac{d}{dx})^\ell(\frac{\sin x}{x})$$

- Spherical Neumann function

$$n_\ell(x)=(-1)^{\ell+1}\sqrt{\frac{\pi}{2x}}J_{-(\ell+\frac{1}{2})}(x)=-(-x)^\ell(\frac{1}{x}\frac{d}{dx})^\ell(\frac{\cos x}{x})$$

- Spherical Hankel function of the first kind

$$h_\ell^{(1)}(x)=\sqrt{\frac{\pi}{2x}}H_{\ell+\frac{1}{2}}^{(1)}(x)$$

- Spherical Hankel function of the second kind

$$h_\ell^{(2)}(x)=\sqrt{\frac{\pi}{2x}}H_{\ell+\frac{1}{2}}^{(2)}(x)$$

There are other types of Bessel functions which are more convenient and useful for some physics problems.

$H_\nu^{(1)}(x)$ and $H_\nu^{(2)}(x)$ are called **Hankel functions** or **Bessel functions of the third kind and order** ν, which are defined as

$$H_\nu^{(1)}(x) = J_\nu(x) + iN_\nu(x)$$

and

$$H_\nu^{(2)}(x) = J_\nu(x) - iN_\nu(x)$$

If we multiply Eq. (7.3.1.7) by x^ν, it gives

$$x^\nu J_\nu(x) = \sum_{k=0}^{\infty} \frac{(-1)^k}{k!\Gamma(\nu+k+1)} \frac{x^{2\nu+2k}}{2^{\nu+2k}}$$

If we differentiate this equation with respect to x, we have

$$\frac{d}{dx}[x^\nu J_\nu(x)] = \sum_{k=0}^{\infty} \frac{(-1)^k}{k!\Gamma(\nu+k+1)}(2\nu+2k)\frac{x^{2\nu+2k-1}}{2^{\nu+2k}}$$

$$= \sum_{k=0}^{\infty} \frac{(-1)^k}{k!\Gamma(\nu+k)} \frac{x^{2\nu+2k-1}}{2^{\nu+2k-1}}$$

By multiplying $x^{-\nu}$ on the both sides of the equation, we have

$$x^{-\nu}\frac{d}{dx}[x^\nu J_\nu(x)] = \sum_{k=0}^{\infty} \frac{(-1)^k}{k!\Gamma(\nu+k)}\left(\frac{x}{2}\right)^{\nu+2k-1}$$

The right-hand side of the equation is $J_{\nu-1}(x)$ from Eq. (7.3.1.7). Thus, we obtain the following equation

$$x^{-\nu}\frac{d}{dx}[x^\nu J_\nu(x)] = J_{\nu-1}(x) \tag{7.3.2.1}$$

Similarly, we have

$$x^{-\nu} J_\nu(x) = \sum_{k=0}^{\infty} \frac{(-1)^k}{k!\Gamma(\nu+k+1)} \frac{x^{2k}}{2^{\nu+2k}}$$

which gives

$$x^\nu \frac{d}{dx}[x^{-\nu} J_\nu(x)] = \sum_{k=0}^{\infty} \frac{(-1)^k}{(k-1)!\Gamma(\nu+k+1)}\left(\frac{x}{2}\right)^{\nu+2k-1}$$

If we replace the dummy index k with $n+1$ on the right-hand side of the equation, we get

$$\sum_{n=0}^{\infty} \frac{(-1)^{n+1}}{n!\Gamma(\nu+n+2)} (\frac{x}{2})^{\nu+2n+1} = -J_{\nu+1}(x)$$

Thus, we obtain

$$x^{\nu} \frac{d}{dx}[x^{-\nu} J_{\nu}(x)] = -J_{\nu+1}(x) \tag{7.3.2.2}$$

If we add Eqs. (7.3.2.1) and (7.3.2.2), it gives

$$\begin{aligned}
J_{\nu-1}(x) - J_{\nu+1}(x) &= x^{-\nu} \frac{d}{dx}[x^{\nu} J_{\nu}(x)] + x^{\nu} \frac{d}{dx}[x^{-\nu} J_{\nu}(x)] \\
&= \frac{d}{dx}[x^{-\nu} x^{\nu} J_{\nu}(x)] - x^{\nu} J_{\nu}(x) \frac{d}{dx} x^{-\nu} \\
&\quad + \frac{d}{dx}[x^{\nu} x^{-\nu} J_{\nu}(x)] - x^{-\nu} J_{\nu}(x) \frac{d}{dx} x^{\nu} \\
&= \frac{d}{dx} J_{\nu}(x) + \nu x^{-1} J_{\nu}(x) + \frac{d}{dx} J_{\nu}(x) - \nu x^{-1} J_{\nu}(x) \\
&= 2 \frac{d}{dx} J_{\nu}(x)
\end{aligned}$$

Thus, we obtain the following relation

$$J_{\nu-1}(x) - J_{\nu+1}(x) = 2 \frac{d}{dx} J_{\nu}(x) \tag{7.3.2.3}$$

If we subtract Eq. (7.3.2.2) from (7.3.2.1), we obtain

$$\begin{aligned}
J_{\nu-1}(x) + J_{\nu+1}(x) &= x^{-\nu} \frac{d}{dx}[x^{\nu} J_{\nu}(x)] - x^{\nu} \frac{d}{dx}[x^{-\nu} J_{\nu}(x)] \\
&= \frac{d}{dx}[x^{-\nu} x^{\nu} J_{\nu}(x)] - x^{\nu} J_{\nu}(x) \frac{d}{dx} x^{-\nu} \\
&\quad - \frac{d}{dx}[x^{\nu} x^{-\nu} J_{\nu}(x)] + x^{-\nu} J_{\nu}(x) \frac{d}{dx} x^{\nu} \\
&= \frac{d}{dx} J_{\nu}(x) + \nu x^{-1} J_{\nu}(x) - \frac{d}{dx} J_{\nu}(x) + \nu x^{-1} J_{\nu}(x) \\
&= \frac{2\nu}{x} J_{\nu}(x)
\end{aligned} \tag{7.3.2.4}$$

From Eq. (7.3.2.1), we have

$$\nu x^{\nu-1}J_\nu(x)+x^\nu J_\nu'(x) = x^\nu J_{\nu-1}(x),$$

which gives

$$J_\nu'(x) = J_{\nu-1}(x)-\frac{\nu}{x}J_\nu(x)$$

From Eq. (7.3.2.2), we have

$$-\nu x^{-\nu-1}J_\nu(x)+x^{-\nu}J_\nu'(x) = -x^{-\nu}J_{\nu+1}(x),$$

which gives

$$J_\nu'(x) = -J_{\nu+1}(x)+\frac{\nu}{x}J_\nu(x)$$

From these two equations, we have

$$J_\nu'(x) = J_{\nu-1}(x)-\frac{\nu}{x}J_\nu(x) = -J_{\nu+1}(x)+\frac{\nu}{x}J_\nu(x) \qquad (7.3.2.5)$$

From this equation, we have two results, Eqs. (7.3.2.6) and (7.3.2.7):
(i) Firstly, we have

$$J_{\nu-1}(x)=\frac{\nu}{x}J_\nu(x)+J_\nu'(x)=x^{-\nu}\frac{d}{dx}[x^\nu J_\nu(x)]$$

This gives

$$J_{\nu-1}(x)=x^{-\nu+1}(\frac{1}{x}\frac{d}{dx})[x^\nu J_\nu(x)]$$

If we calculate $(\frac{1}{x}\frac{d}{dx})^2[x^\nu J_\nu(x)]$, we get

$$\begin{aligned}
(\frac{1}{x}\frac{d}{dx})^2[x^\nu J_\nu(x)] &= (\frac{1}{x}\frac{d}{dx})(\frac{1}{x}\frac{d}{dx})[x^\nu J_\nu(x)] \\
&= \frac{1}{x}\frac{d}{dx}[x^{\nu-1}J_{\nu-1}(x)] \\
&= \frac{1}{x}[(\nu-1)x^{\nu-2}J_{\nu-1}(x)+x^{\nu-1}J_{\nu-1}'(x)] \\
&= (\nu-1)x^{\nu-3}J_{\nu-1}(x)+x^{\nu-2}[J_{\nu-2}(x)-\frac{\nu-1}{x}J_{\nu-1}(x)] \\
&\qquad [\text{using the equation } J_\nu'(x)=J_{\nu-1}(x)-\frac{\nu}{x}J_\nu(x)] \\
&= (\nu-1)x^{\nu-3}J_{\nu-1}(x)+x^{\nu-2}J_{\nu-2}(x)-(\nu-1)x^{\nu-3}J_{\nu-1}(x) \\
&= x^{\nu-2}J_{\nu-2}(x)
\end{aligned}$$

Therefore, we obtain

$$x^{\nu-2}J_{\nu-2}(x)=(\frac{1}{x}\frac{d}{dx})^2[x^{\nu}J_{\nu}(x)]$$

Similarly, let's calculate $(\frac{1}{x}\frac{d}{dx})^3[x^{\nu}J_{\nu}(x)]$ and it gives

$$\begin{aligned}(\frac{1}{x}\frac{d}{dx})^3[x^{\nu}J_{\nu}(x)] &= (\frac{1}{x}\frac{d}{dx})(\frac{1}{x}\frac{d}{dx})^2[x^{\nu}J_{\nu}(x)] \\ &= \frac{1}{x}\frac{d}{dx}[x^{\nu-2}J_{\nu-2}(x)] \\ &= \frac{1}{x}[(\nu-2)x^{\nu-3}J_{\nu-2}(x)+x^{\nu-2}J'_{\nu-2}(x)] \\ &= (\nu-2)x^{\nu-4}J_{\nu-2}(x)+x^{\nu-3}[J_{\nu-3}(x)-\frac{\nu-2}{x}J_{\nu-2}(x)] \\ &\qquad \text{[using the equation } J'_{\nu}(x)=J_{\nu-1}(x)-\frac{\nu}{x}J_{\nu}(x)] \\ &= (\nu-2)x^{\nu-4}J_{\nu-2}(x)+x^{\nu-3}J_{\nu-3}(x)-(\nu-2)x^{\nu-4}J_{\nu-2}(x) \\ &= x^{\nu-3}J_{\nu-3}(x)\end{aligned}$$

Therefore, we obtain

$$x^{\nu-3}J_{\nu-3}(x) = (\frac{1}{x}\frac{d}{dx})^3[x^{\nu}J_{\nu}(x)]$$

$$\vdots$$

$$\vdots$$

Thus, we can generalize it into

$$x^{\nu-n}J_{\nu-n}(x) = (\frac{1}{x}\frac{d}{dx})^n[x^{\nu}J_{\nu}(x)] \tag{7.3.2.6}$$

(ii) Secondly, we have

$$J_{\nu+1}(x) = \frac{\nu}{x}J_{\nu}(x)-J'_{\nu}(x) = -x^{\nu}\frac{d}{dx}[x^{-\nu}J_{\nu}(x)]$$

This gives

$$J_{\nu+1}(x) = -x^{\nu+1}(\frac{1}{x}\frac{d}{dx})[x^{-\nu}J_{\nu}(x)]$$

If we calculate $(\frac{1}{x}\frac{d}{dx})^2[x^{-\nu}J_{\nu}(x)]$, we get

$$(\frac{1}{x}\frac{d}{dx})^2[x^{-\nu}J_\nu(x)] = (\frac{1}{x}\frac{d}{dx})(\frac{1}{x}\frac{d}{dx})[x^{-\nu}J_\nu(x)]$$

$$= \frac{1}{x}\frac{d}{dx}[-x^{-\nu-1}J_{\nu+1}(x)]$$

$$= \frac{1}{x}[(\nu+1)x^{-\nu-2}J_{\nu+1}(x) - x^{-\nu-1}J'_{\nu+1}(x)]$$

$$= (\nu+1)x^{-\nu-3}J_{\nu+1}(x) - x^{-\nu-2}[-J_{\nu+2}(x) + \frac{\nu+1}{x}J_{\nu+1}(x)]$$

[using the equation $J'_\nu(x) = -J_{\nu+1}(x) + \frac{\nu}{x}J_\nu(x)$]

$$= x^{-\nu-2}J_{\nu+2}(x)$$

Therefore, we obtain

$$x^{-\nu-2}J_{\nu+2}(x) = (\frac{1}{x}\frac{d}{dx})^2[x^{-\nu}J_\nu(x)]$$

Similarly, let's calculate $(\frac{1}{x}\frac{d}{dx})^3[x^{-\nu}J_\nu(x)]$ and it gives

$$(\frac{1}{x}\frac{d}{dx})^3[x^{-\nu}J_\nu(x)]$$

$$= (\frac{1}{x}\frac{d}{dx})(\frac{1}{x}\frac{d}{dx})^2[x^{\nu}J_\nu(x)]$$

$$= (\frac{1}{x}\frac{d}{dx})[x^{-\nu-2}J_{\nu+2}(x)]$$

$$= \frac{1}{x}[-(\nu+2)x^{-\nu-3}J_{\nu+2}(x) + x^{-\nu-2}J'_{\nu+2}(x)]$$

$$= -(\nu+2)x^{-\nu-4}J_{\nu+2}(x) + x^{-\nu-3}J'_{\nu+2}(x)$$

[using the equation $J'_\nu(x) = -J_{\nu+1}(x) + \frac{\nu}{x}J_\nu(x)$]

$$= -(\nu+2)x^{-\nu-4}J_{\nu+2}(x) + x^{-\nu-3}[-J_{\nu+3}(x) + \frac{\nu+2}{x}J_{\nu+2}(x)]$$

$$= -x^{-\nu-3}J_{\nu+3}(x)$$

Therefore, we obtain

$$x^{-\nu-3}J_{\nu+3}(x) = (-\frac{1}{x}\frac{d}{dx})^3[x^{-\nu}J_\nu(x)]$$

$$\vdots$$

$$\vdots$$

Thus, we can generalize it as we did for Eq. (7.3.2.6) and have

$$x^{-\nu-n}J_{\nu+n}(x) = (-\frac{1}{x}\frac{d}{dx})^n[x^{-\nu}J_\nu(x)] \tag{7.3.2.7}$$

- $x^{-\nu}\frac{d}{dx}[x^\nu J_\nu(x)] = J_{\nu-1}(x)$ and $x^{\nu-n}J_{\nu-n}(x) = (\frac{1}{x}\frac{d}{dx})^n[x^\nu J_\nu(x)]$
- $x^\nu \frac{d}{dx}[x^{-\nu}J_\nu(x)] = -J_{\nu+1}(x)$ and $x^{-\nu-n}J_{\nu+n}(x) = (-\frac{1}{x}\frac{d}{dx})^n[x^{-\nu}J_\nu(x)]$
- $J_\nu'(x) = J_{\nu-1}(x) - \frac{\nu}{x}J_\nu(x) = -J_{\nu+1}(x) + \frac{\nu}{x}J_\nu(x)$

When the Bessel functions are of order with half of an odd integer, they have relationships with **spherical Bessel function $j_\ell(x)$ and spherical Neumann function $n_\ell(x)$**. These spherical functions are very useful in the scattering theory in Quantum Mechanics.

(i) If we substitute $\nu = \frac{1}{2}$ in Eq. (7.3.1.7), it gives

$$\begin{aligned}
J_{\frac{1}{2}}(x) &= \sum_{k=0}^{\infty}\frac{(-1)^k}{k!\Gamma(k+\frac{3}{2})}(\frac{x}{2})^{\frac{1}{2}+2k} \\
&= (\frac{x}{2})^{\frac{1}{2}}[\frac{1}{\frac{1}{2}\Gamma(\frac{1}{2})} - \frac{1}{\frac{3}{2}\Gamma(\frac{3}{2})}(\frac{x}{2})^2 + \frac{1}{2!\frac{5}{2}\Gamma(\frac{5}{2})}(\frac{x}{2})^4 - \cdots\cdots] \\
&= (\frac{x}{2})^{\frac{1}{2}}[\frac{1}{\frac{1}{2}\Gamma(\frac{1}{2})} - \frac{1}{\frac{3}{2}\frac{1}{2}\Gamma(\frac{1}{2})}(\frac{x}{2})^2 + \frac{1}{2!\frac{5}{2}\frac{3}{2}\frac{1}{2}\Gamma(\frac{1}{2})}(\frac{x}{2})^4 - \cdots\cdots] \\
&= (\frac{x}{2})^{\frac{1}{2}}\frac{1}{\Gamma(\frac{1}{2})}[2 - \frac{2}{3\cdot 2}x^2 + \frac{2}{5\cdot 4\cdot 3\cdot 2!}x^4 - \cdots\cdots] \\
&= \sqrt{\frac{2x}{\pi}}[1 - \frac{1}{3\cdot 2}x^2 + \frac{1}{5\cdot 4\cdot 3\cdot 2!}x^4 - \cdots\cdots] \\
&= \sqrt{\frac{2}{\pi x}}[x - \frac{1}{3!}x^3 + \frac{1}{5!}x^5 - \cdots\cdots] \\
&= \sqrt{\frac{2}{\pi x}}\sin x
\end{aligned} \tag{7.3.2.8}$$

From Eq. (7.3.2.7), we have

$$J_{\frac{3}{2}}(x)=x^{\frac{3}{2}}(-\frac{1}{x}\frac{d}{dx})[x^{-\frac{1}{2}}J_{\frac{1}{2}}(x)]=x^{\frac{3}{2}}(-\frac{1}{x}\frac{d}{dx})[x^{-\frac{1}{2}}\sqrt{\frac{2}{\pi x}}\sin x]$$

$$=-\sqrt{\frac{2}{\pi}}x^{\frac{3}{2}}(\frac{1}{x}\frac{d}{dx})(\frac{\sin x}{x}) \tag{7.3.2.9}$$

Similarly, we have

$$J_{\frac{5}{2}}(x)=x^{\frac{5}{2}}(-\frac{1}{x}\frac{d}{dx})[x^{-\frac{3}{2}}J_{\frac{3}{2}}(x)]$$

$$=\sqrt{\frac{2}{\pi}}x^{\frac{5}{2}}(\frac{1}{x}\frac{d}{dx})[x^{-\frac{3}{2}}x^{\frac{3}{2}}(\frac{1}{x}\frac{d}{dx})(\frac{\sin x}{x})]$$

$$=(-1)^2\sqrt{\frac{2}{\pi}}x^{\frac{5}{2}}(\frac{1}{x}\frac{d}{dx})^2(\frac{\sin x}{x})$$

$$\vdots$$

We can generalize it into the following equation

$$J_{\ell+\frac{1}{2}}(x)=(-1)^{\ell}\sqrt{\frac{2}{\pi}}x^{\ell+\frac{1}{2}}(\frac{1}{x}\frac{d}{dx})^{\ell}(\frac{\sin x}{x}) \tag{7.3.2.10}$$

(ii) To find the general expression of $J_{-(\ell+\frac{1}{2})}(x)$, let's calculate $J_{-\frac{1}{2}}(x)$ by starting with Eq. (7.3.1.8).

Then, we have

$$J_{-\frac{1}{2}}(x)=\sum_{k=0}^{\infty}\frac{(-1)^k}{k!\Gamma(k+\frac{1}{2})}(\frac{x}{2})^{-\frac{1}{2}+2k}$$

$$=(\frac{x}{2})^{-\frac{1}{2}}[\frac{1}{\Gamma(\frac{1}{2})}-\frac{1}{\frac{1}{2}\Gamma(\frac{1}{2})}(\frac{x}{2})^2+\frac{1}{2!\frac{3}{2}\frac{1}{2}\Gamma(\frac{1}{2})}(\frac{x}{2})^4-\cdots\cdots]$$

$$=\sqrt{\frac{2}{\pi x}}[1-\frac{1}{2!}x^2+\frac{1}{4!}x^4-\cdots\cdots]$$

$$=\sqrt{\frac{2}{\pi x}}\cos x$$

Using Eq. (7.3.2.6) we have

$$J_{-\frac{3}{2}}(x)=x^{\frac{3}{2}}(\frac{1}{x}\frac{d}{dx})[x^{-\frac{1}{2}}J_{-\frac{1}{2}}(x)]=x^{\frac{3}{2}}(\frac{1}{x}\frac{d}{dx})[x^{-\frac{1}{2}}\sqrt{\frac{2}{\pi x}}\cos x]$$

$$=\sqrt{\frac{2}{\pi}}x^{\frac{3}{2}}(\frac{1}{x}\frac{d}{dx})(\frac{\cos x}{x})$$

Similarly, we have

$$J_{-\frac{5}{2}}(x)=x^{\frac{5}{2}}(\frac{1}{x}\frac{d}{dx})[x^{-\frac{3}{2}}J_{-\frac{3}{2}}(x)]$$

$$=x^{\frac{5}{2}}(\frac{1}{x}\frac{d}{dx})[x^{-\frac{3}{2}}\sqrt{\frac{2}{\pi}}x^{\frac{3}{2}}(\frac{1}{x}\frac{d}{dx})(\frac{\cos x}{x})]$$

$$=\sqrt{\frac{2}{\pi}}x^{\frac{5}{2}}(\frac{1}{x}\frac{d}{dx})^2(\frac{\cos x}{x})$$

$$\vdots$$

Therefore, we can generalize into the following equation

$$J_{-(\ell+\frac{1}{2})}(x)=\sqrt{\frac{2}{\pi}}x^{\ell+\frac{1}{2}}(\frac{1}{x}\frac{d}{dx})^{\ell}(\frac{\cos x}{x}) \quad (7.3.2.11)$$

The spherical Bessel and Neumann functions, $j_\ell(x)$ and $n_\ell(x)$, are defined as

$$j_\ell(x)=\sqrt{\frac{\pi}{2x}}J_{\ell+\frac{1}{2}}(x) \quad (7.3.2.12)$$

and

$$n_\ell(x)=(-1)^{\ell+1}\sqrt{\frac{\pi}{2x}}J_{-(\ell+\frac{1}{2})}(x), \quad (7.3.2.13)$$

respectively.

Using Eqs. (7.3.2.10) and (7.3.2.11) we obtain

$$j_\ell(x)=\sqrt{\frac{\pi}{2x}}J_{\ell+\frac{1}{2}}(x)=(-x)^\ell(\frac{1}{x}\frac{d}{dx})^\ell(\frac{\sin x}{x}) \quad (7.3.2.14)$$

and

$$n_\ell(x)=(-1)^{\ell+1}\sqrt{\frac{\pi}{2x}}J_{-(\ell+\frac{1}{2})}(x)=-(-x)^\ell(\frac{1}{x}\frac{d}{dx})^\ell(\frac{\cos x}{x}) \quad (7.3.2.15)$$

Spherical Hankel functions of the first kind and second kind with order ν, $h_\nu^{(1)}(x)$ and $h_\nu^{(2)}(x)$, are defined as

$$h_\ell^{(1)}(x) = \sqrt{\frac{\pi}{2x}}\, H^{(1)}_{\ell+\frac{1}{2}}(x) \tag{7.3.2.16}$$

and

$$h_\ell^{(2)}(x) = \sqrt{\frac{\pi}{2x}}\, H^{(2)}_{\ell+\frac{1}{2}}(x), \tag{7.3.2.17}$$

respectively.

Most of the problems we are dealing in physics are for the approximate formulas of the Bessel functions for small x or $x \gg 1$.

(a) **For small** x, the contribution of the first term in Eq. (7.3.1.7) is dominant and it gives

$$J_\ell(x) \approx \frac{1}{\Gamma(\nu+1)}\left(\frac{x}{2}\right)^\ell \tag{7.3.2.18}$$

On the other hand, the Neumann function tends to $\frac{0}{0}$ so that it is calculated using L'Hopital's rule, as Eq. (7.3.1.11):

$$N_\ell(x) = \frac{1}{\pi}\left[\frac{\partial J_\ell(x)}{\partial \ell} - (-1)^\ell \frac{\partial J_{-\ell}(x)}{\partial \ell}\right]$$

If we insert Eqs. (7.3.1.7) and (7.3.1.8) for $J_\ell(x)$ and $J_{-\ell}(x)$, respectively, into the equation, we get

$$\begin{aligned} N_\ell(x) = \frac{1}{\pi}\Big[&\sum_{k=0}^{\infty} \frac{(-1)^k}{k!\,\Gamma(\ell+k+1)} \frac{\partial}{\partial \ell}\left(\frac{x}{2}\right)^{\ell+2k} \\ &+ \sum_{k=0}^{\infty}\left(\frac{x}{2}\right)^{\ell+2k} \frac{(-1)^k}{k!} \frac{\partial}{\partial \ell} \frac{1}{\Gamma(\ell+k+1)} \\ &- (-1)^\ell \sum_{k=0}^{\infty} \frac{(-1)^k}{k!\,\Gamma(-\ell+k+1)} \frac{\partial}{\partial \ell}\left(\frac{x}{2}\right)^{-\ell+2k} \\ &- (-1)^\ell \sum_{k=0}^{\infty}\left(\frac{x}{2}\right)^{-\ell+2k} \frac{(-1)^k}{k!} \frac{\partial}{\partial \ell} \frac{1}{\Gamma(-\ell+k+1)}\Big] \end{aligned}$$

Now, we need to calculate $\frac{\partial}{\partial \ell}\left(\frac{x}{2}\right)^\ell$ in the equation.

Since $(\frac{x}{2})^{\ell} = e^{\ell ln\frac{x}{2}}$, we get $(\frac{x}{2})^{\ell} = e^{\ell ln\frac{x}{2}} = e^{y}$ by defining $y = \ell ln\frac{x}{2}$.
This gives

$$\frac{\partial}{\partial \ell}(\frac{x}{2})^{\ell} = \frac{\partial}{\partial \ell}e^{\ell ln\frac{x}{2}} = \frac{dy}{d\ell}\frac{d}{dy}e^{y} = (ln\frac{x}{2})e^{y} = (ln\frac{x}{2})(\frac{x}{2})^{\ell}$$

Similarly, we also get

$$\frac{\partial}{\partial \ell}(\frac{x}{2})^{-\ell} = (-ln\frac{x}{2})(\frac{x}{2})^{-\ell}$$

If we substitute these into the equation, we get

$$\begin{aligned}N_{\ell}(x) = \frac{1}{\pi}[&\sum_{k=0}^{\infty}\frac{(-1)^k}{k!\Gamma(\ell+k+1)}(ln\frac{x}{2})(\frac{x}{2})^{\ell+2k} \\ &-\sum_{k=0}^{\infty}(\frac{x}{2})^{\ell+2k}\frac{(-1)^k\Gamma'(\ell+k+1)}{k!\{\Gamma(\ell+k+1)\}^2} \\ &+(-1)^{\ell}\sum_{k=0}^{\infty}\frac{(-1)^k}{k!\Gamma(-\ell+k+1)}(ln\frac{x}{2})(\frac{x}{2})^{-\ell+2k} \\ &-(-1)^{\ell}\sum_{k=0}^{\infty}(\frac{x}{2})^{-\ell+2k}\frac{(-1)^k\Gamma'(-\ell+k+1)}{k!\{\Gamma(-\ell+k+1)\}^2}]\end{aligned}$$

The first and third terms on the right-hand side of the equation are

$$\begin{aligned}&\sum_{k=0}^{\infty}\frac{(-1)^k}{k!\Gamma(\ell+k+1)}(ln\frac{x}{2})(\frac{x}{2})^{\ell+2k} \\ &\quad = (ln\frac{x}{2})\sum_{k=0}^{\infty}\frac{(-1)^k}{k!\Gamma(\ell+k+1)}(\frac{x}{2})^{\ell+2k} = (ln\frac{x}{2})J_{\ell}(x)\end{aligned}$$

and

$$\begin{aligned}&(-1)^{\ell}\sum_{k=0}^{\infty}\frac{(-1)^k}{k!\Gamma(-\ell+k+1)}(ln\frac{x}{2})(\frac{x}{2})^{-\ell+2k} \\ &\quad = (-1)^{\ell}(ln\frac{x}{2})\sum_{k=0}^{\infty}\frac{(-1)^k}{k!\Gamma(-\ell+k+1)}(\frac{x}{2})^{-\ell+2k} \\ &\quad = (-1)^{\ell}(ln\frac{x}{2})J_{-\ell}(x) \\ &\quad = (-1)^{2\ell}(ln\frac{x}{2})J_{\ell}(x) \\ &\quad = (ln\frac{x}{2})J_{\ell}(x),\end{aligned}$$

respectively.

Since the digamma function is defined as $\Psi(x)=\dfrac{\Gamma'(x)}{\Gamma(x)}$, the second and fourth terms on the right-hand side of the equation can be expressed as

$$\sum_{k=0}^{\infty}(\frac{x}{2})^{\ell+2k}\frac{(-1)^k\Gamma'(\ell+k+1)}{k!\{(\Gamma(\ell+k+1)\}^2} = \sum_{k=0}^{\infty}(\frac{x}{2})^{\ell+2k}\frac{(-1)^k\Psi(\ell+k+1)}{k!\Gamma(\ell+k+1)}$$

and

$$(-1)^{\ell}\sum_{k=0}^{\infty}(\frac{x}{2})^{-\ell+2k}\frac{(-1)^k\Gamma'(-\ell+k+1)}{k!\{\Gamma(-\ell+k+1)\}^2}$$
$$=(-1)^{\ell}\sum_{k=0}^{\infty}(\frac{x}{2})^{-\ell+2k}\frac{(-1)^k\Psi(-\ell+k+1)}{k!\Gamma(-\ell+k+1)},$$

respectively.

If we insert these into the equation, the Neumann function is now expressed as

$$N_{\ell}(x)=\frac{2}{\pi}(ln\frac{x}{2})J_{\ell}(x)-\frac{1}{\pi}[\sum_{k=0}^{\infty}(\frac{x}{2})^{\ell+2k}\frac{(-1)^k\Psi(\ell+k+1)}{k!\Gamma(\ell+k+1)}$$
$$+(-1)^{\ell}\sum_{k=0}^{\infty}(\frac{x}{2})^{-\ell+2k}\frac{(-1)^k\Psi(-\ell+k+1)}{k!\Gamma(-\ell+k+1)}]$$

If we replace k with $k+\ell$ on the third term of the right-hand side of the equation, it gives

$$(-1)^{\ell}\sum_{k=-\ell}^{\infty}(\frac{x}{2})^{\ell+2k}\frac{(-1)^{k+\ell}\Psi(k+1)}{(k+\ell)!\Gamma(k+1)}$$
$$=(-1)^{\ell}\sum_{k=-\ell}^{\infty}(\frac{x}{2})^{\ell+2k}\frac{(-1)^{k+\ell}\Psi(k+1)}{k!\Gamma(k+\ell+1)}$$

By substituting this into the equation, we have

$$N_{\ell}(x) = \frac{2}{\pi}(ln\frac{x}{2})J_{\ell}(x)-\frac{(-1)^{\ell}}{\pi}\sum_{k=-\ell}^{-1}(\frac{x}{2})^{\ell+2k}\frac{(-1)^{k+\ell}\Psi(k+1)}{k!\Gamma(k+\ell+1)}$$
$$-\frac{1}{\pi}\sum_{k=0}^{\infty}(\frac{x}{2})^{\ell+2k}\frac{(-1)^k}{k!\Gamma(k+\ell+1)}[\Psi(k+\ell+1)+\Psi(k+1)]$$

By replacing k with $k-\ell$ on the second term of the right-hand side of the equation, it gives

$$-\frac{(-1)^{\ell}}{\pi}\sum_{k=-\ell}^{-1}(\frac{x}{2})^{\ell+2k}\frac{(-1)^{k+\ell}\Psi(k+1)}{k!\Gamma(k+\ell+1)}$$

$$= -\frac{(-1)^{\ell}}{\pi}\sum_{k=0}^{\ell-1}(\frac{x}{2})^{-\ell+2k}\frac{(-1)^{k}\Psi(k-\ell+1)}{(k-\ell)!\Gamma(k+1)}$$

$$= -\frac{(-1)^{\ell}}{\pi}\sum_{k=0}^{\ell-1}(\frac{x}{2})^{-\ell+2k}\frac{(-1)^{k}\Psi(k-\ell+1)}{k!\Gamma(k-\ell+1)}$$

$$= -\frac{(-1)^{\ell}}{\pi}\sum_{k=0}^{\ell-1}(\frac{x}{2})^{-\ell+2k}\frac{(-1)^{k}(-1)^{\ell-k}\Gamma(\ell-k)}{k!},$$

where

$$\frac{\Psi(k-\ell+1)}{\Gamma(k-\ell+1)}=\frac{\Psi(-(\ell-k-1))}{\Gamma(-(\ell-k-1))}=(-1)^{(\ell-k-1)+1}(\ell-k-1)!$$
$$=(-1)^{\ell-k}\Gamma(\ell-k)$$

by using $\frac{\Psi(-k)}{\Gamma(-k)}=(-1)^{k+1}k!$[17].

Then, we have

$$N_{\ell}(x)=\frac{2}{\pi}(ln\frac{x}{2})J_{\ell}(x)-\frac{(-1)^{\ell}}{\pi}\sum_{k=0}^{\ell-1}(\frac{x}{2})^{-\ell+2k}\frac{(-1)^{k}(-1)^{\ell-k}\Gamma(\ell-k)}{k!}$$
$$-\frac{1}{\pi}\sum_{k=0}^{\infty}(\frac{x}{2})^{\ell+2k}\frac{(-1)^{k}}{k!\Gamma(k+\ell+1)}[\Psi(k+\ell+1)+\Psi(k+1)]$$
$$=\frac{2}{\pi}(ln\frac{x}{2})J_{\ell}(x)-\frac{1}{\pi}\sum_{k=0}^{\ell-1}(\frac{x}{2})^{-\ell+2k}\frac{\Gamma(\ell-k)}{k!}$$
$$-\frac{1}{\pi}\sum_{k=0}^{\infty}(\frac{x}{2})^{\ell+2k}\frac{(-1)^{k}}{k!\Gamma(k+\ell+1)}[\Psi(k+\ell+1)+\Psi(k+1)] \quad (7.3.2.19)$$

For $\ell=0$, the equation gives

$$N_0(x)\approx\frac{2}{\pi}ln\frac{x}{2}$$

because $J_0(x)=1$, and the contributions from the second term and the third term of the right-hand side of the equation are zero and negligible, respectively, for a small x.

---

[17] It is proved in Appendix 7-B.

For $\ell > 0$, the highest order of the second term is $x^{\ell-2}$ while the first and third terms have higher orders than x^{ℓ}. Therefore, the dominant contribution in the equation for a small x is from the second term, and the contributions from the first and third terms are negligible.
Thus, we have

$$N_\ell(x) \approx -\frac{1}{\pi}\sum_{k=0}^{\ell-1}\left(\frac{x}{2}\right)^{-\ell+2k}\frac{\Gamma(\ell-k)}{k!}$$
$$\approx -\frac{1}{\pi}\left(\frac{x}{2}\right)^{-\ell}\frac{\Gamma(\ell)}{0!}$$
$$= -\frac{\Gamma(\ell)}{\pi}\left(\frac{2}{x}\right)^{\ell} \qquad (7.3.2.20)$$

(b) **For** $x \gg 1$, Eq. (7.3.2.10) gives

$$J_{\ell+\frac{1}{2}}(x) = (-1)^\ell\sqrt{\frac{2}{\pi}}\,x^{\ell+\frac{1}{2}}\left(\frac{1}{x}\frac{d}{dx}\right)^\ell\left(\frac{\sin x}{x}\right)$$
$$= (-1)^\ell\sqrt{\frac{2}{\pi}}\,x^{\frac{1}{2}}\left(\frac{d}{dx}\right)^\ell\left(\frac{\sin x}{x}\right)$$
$$\approx (-1)^\ell\sqrt{\frac{2}{\pi}}\,x^{-\frac{1}{2}}\left(\frac{d}{dx}\right)^\ell \sin x$$
$$= \sqrt{\frac{2}{\pi x}}\left(-\frac{d}{dx}\right)^\ell \sin x$$

If we calculate $\left(-\frac{d}{dx}\right)^\ell \sin x$, it gives $\cos\left(x-\frac{\ell\pi}{2}-\frac{\pi}{2}\right)$[18] and we have

$$J_{\ell+\frac{1}{2}}(x) = \sqrt{\frac{2}{\pi x}}\cos\left(x-\frac{\ell\pi}{2}-\frac{\pi}{2}\right) \qquad (7.3.2.21)$$

By replacing ℓ with $\ell-\frac{1}{2}$ in the equation, we obtain

$$J_\ell(x) = \sqrt{\frac{2}{\pi x}}\cos\left(x-\frac{\ell\pi}{2}-\frac{\pi}{4}\right) \qquad (7.3.2.22)$$

---

[18] It is proved in Appendix 7-C.

By substituting this into Eq. (7.3.1.9), we have for the Neumann function

$$N_\ell(x) \approx \sqrt{\frac{2}{\pi x}}\frac{\cos(x-\frac{\ell\pi}{2}-\frac{\pi}{4})\cos\ell\pi-\cos(x+\frac{\ell\pi}{2}-\frac{\pi}{4})}{\sin\ell\pi}$$

The first term of the numerator in the equation is

$$\cos(x-\frac{\ell\pi}{2}-\frac{\pi}{4})\cos\ell\pi = \frac{1}{2}[\cos(x+\frac{\ell\pi}{2}-\frac{\pi}{4})+\cos(x-\frac{3\ell\pi}{2}-\frac{\pi}{4})]$$

Therefore, for the numerator we get,

$$\begin{aligned}
&\cos(x-\frac{\ell\pi}{2}-\frac{\pi}{4})\cos\ell\pi-\cos(x+\frac{\ell\pi}{2}-\frac{\pi}{4}) \\
&\qquad = -\frac{1}{2}[\cos(x+\frac{\ell\pi}{2}-\frac{\pi}{4})-\cos(x-\frac{3\ell\pi}{2}-\frac{\pi}{4})] \\
&\qquad = \sin(x-\frac{\ell\pi}{2}-\frac{\pi}{4})\sin\ell\pi
\end{aligned}$$

Thus, for the Neumann function we obtain

$$N_\ell(x) = \sqrt{\frac{2}{\pi x}}\sin(x-\frac{\ell\pi}{2}-\frac{\pi}{4}) \tag{7.3.2.23}$$

From Eqs. (7.3.2.22) and (7.3.2.23), we obtain for the Hankel functions $H_\ell^{(1)}(x)$ and $H_\ell^{(2)}(x)$, respectively:

$$\begin{aligned}
H_\ell^{(1)}(x) &= J_\ell(x)+iN_\ell(x) \approx \sqrt{\frac{2}{\pi x}}[\cos(x-\frac{\ell\pi}{2}-\frac{\pi}{4})+i\sin(x-\frac{\ell\pi}{2}-\frac{\pi}{4})] \\
&= \sqrt{\frac{2}{\pi x}}e^{i(x-\frac{\ell\pi}{2}-\frac{\pi}{4})}
\end{aligned} \tag{7.3.2.24}$$

and

$$H_\ell^{(2)}(x) = J_\ell(x)-iN_\ell(x) \approx \sqrt{\frac{2}{\pi x}}e^{-i(x-\frac{\ell\pi}{2}-\frac{\pi}{4})} \tag{7.3.2.25}$$

· For small x, the asymptotic behaviors are

- $J_\ell(x) = \frac{1}{\Gamma(\nu+1)}(\frac{x}{2})^\ell$

$$- N_\ell(x) = \begin{cases} -\dfrac{\Gamma(\ell)}{\pi}\left(\dfrac{2}{x}\right)^\ell & \text{for } \ell > 0 \\ \dfrac{2}{\pi} ln\dfrac{x}{2} & \text{for } \ell = 0 \end{cases}$$

· For large x, the asymptotic behaviors are

$$- J_\ell(x) = \sqrt{\frac{2}{\pi x}}\cos\left(x - \frac{\ell\pi}{2} - \frac{\pi}{4}\right)$$

$$- N_\ell(x) = \sqrt{\frac{2}{\pi x}}\sin\left(x - \frac{\ell\pi}{2} - \frac{\pi}{4}\right)$$

$$- H_\ell^{(1)}(x) = \sqrt{\frac{2}{\pi x}}\, e^{i\left(x - \frac{\ell\pi}{2} - \frac{\pi}{4}\right)}$$

$$- H_\ell^{(2)}(x) = \sqrt{\frac{2}{\pi x}}\, e^{-i\left(x - \frac{\ell\pi}{2} - \frac{\pi}{4}\right)}$$

Now let's look at asymptotic behaviors of the spherical Bessel and Neumann functions for small x or $x \gg 1$.

(a) **For small** x,

From Eq. (7.3.2.14), we have for the spherical Bessel function

$$j_0(x) = \frac{\sin x}{x},$$

which gives

$$j_0(x) = \frac{1}{x}\left(x - \frac{1}{3!}x^3 + \cdots\cdots\right) \approx 1$$

On the other hand, we have

$$j_1(x) = (-x)\frac{1}{x}\frac{d}{dx}\left(\frac{\sin x}{x}\right) = -\frac{x\cos x - \sin x}{x^2} = \frac{\sin x}{x^2} - \frac{\cos x}{x}$$

which gives

$$j_1(x) \approx \frac{1}{x^2}\left(x - \frac{1}{3!}x^3\right) - \frac{1}{x}\left(1 - \frac{1}{2!}x^2\right) = \frac{1}{3}x = \frac{1}{3 \cdot 1}x$$

Similarly, we have

$$j_2(x) = (-x)^2 (\frac{1}{x}\frac{d}{dx})^2 (\frac{\sin x}{x}) = x^2 (\frac{1}{x}\frac{d}{dx})[\frac{1}{x}\frac{d}{dx}(\frac{\sin x}{x})]$$

$$= x\frac{d}{dx}(\frac{\cos x}{x^2} - \frac{\sin x}{x^3})$$

$$= -\frac{\sin x}{x} - 3\frac{\cos x}{x^2} + 3\frac{\sin x}{x^3}$$

$$\approx -\frac{1}{x}(x - \frac{x^3}{3!}) - \frac{3}{x^2}(1 - \frac{x^2}{2!} + \frac{x^4}{4!}) + \frac{3}{x^3}(x - \frac{x^3}{3!} + \frac{x^5}{5!})$$

$$= (\frac{1}{3!} - \frac{3}{4!} + \frac{3}{5!})x^2 = \frac{1}{5 \cdot 3 \cdot 1}x^2$$

$$\vdots$$

$$\vdots$$

Thus, we can generalize it into

$$j_\ell(x) \approx \frac{x^\ell}{(2\ell+1)!!}, \tag{7.3.2.26}$$

where $(2\ell+1)!! = (2\ell+1)(2\ell-1) \cdots\cdots 5 \cdot 3 \cdot 1$

From Eq. (7.3.2.15), we have for the spherical Neumann function

$$n_0(x) = -\frac{\cos x}{x}$$

which gives

$$n_0(x) = -\frac{1}{x}(1 - \frac{1}{2!}x^2 + \cdots\cdots) \approx -\frac{1}{x}$$

On the other hand, we have

$$n_1(x) = x(\frac{1}{x}\frac{d}{dx})(\frac{\cos x}{x}) = -\frac{\cos x}{x^2} - \frac{\sin x}{x}$$

$$= -\frac{1}{x^2}(1 - \frac{1}{2!}x^2 + \cdots\cdots) - \frac{1}{x}(x - \frac{1}{3!}x^3 + \cdots\cdots)$$

which gives

$$n_1(x) \approx -\frac{1}{x^2}$$

Similarly, we have

$$n_2(x) = -x^2\left(\frac{1}{x}\frac{d}{dx}\right)\left[\frac{1}{x}\frac{d}{dx}\left(\frac{\cos x}{x}\right)\right] = x\frac{d}{dx}\left(\frac{\sin x}{x^2} + \frac{\cos x}{x^3}\right)$$

$$= \frac{1}{x}\cos x - 3\frac{\sin x}{x^2} - 3\frac{\cos x}{x^3}$$

$$\approx \frac{1}{x} - \frac{3}{x^2}x - \frac{3}{x^3}\left(1 - \frac{x^2}{2!}\right)$$

$$\approx -\frac{3}{x^3}$$

$$\vdots$$

$$\vdots$$

Thus, we can generalize it into

$$n_\ell(x) \approx -\frac{(2\ell-1)!!}{x^{\ell+1}} \tag{7.3.2.27}$$

(b) **For large** x,

From Eqs. (7.3.2.12) and (7.3.2.21), we obtain the spherical Bessel function

$$j_\ell(x) = \sqrt{\frac{\pi}{2x}}\, J_{\ell+\frac{1}{2}}(x) = \frac{1}{x}\cos\left(x - \frac{\ell\pi}{2} - \frac{\pi}{2}\right)$$

From Eq. (7.3.2.11) we get

$$J_{-(\ell+\frac{1}{2})}(x) = \sqrt{\frac{2}{\pi}}\, x^{\ell+\frac{1}{2}}\left(\frac{1}{x}\frac{d}{dx}\right)^\ell\left(\frac{\cos x}{x}\right) \approx \sqrt{\frac{2}{\pi x}}\left(\frac{d}{dx}\right)^\ell \cos x$$

From Eq. (7.3.2.13), we get the spherical Neumann function

$$n_\ell(x) = (-1)^{\ell+1}\sqrt{\frac{\pi}{2x}}\, J_{-(\ell+\frac{1}{2})}(x) = (-1)^{\ell+1}\sqrt{\frac{\pi}{2x}}\sqrt{\frac{2}{\pi x}}\left(\frac{d}{dx}\right)^\ell \cos x$$

$$= (-1)^{\ell+1}\frac{1}{x}\left(\frac{d}{dx}\right)^\ell \cos x$$

Since $\left(\frac{d}{dx}\right)^\ell \cos x = (-1)^{\ell+1}\sin\left(x - \frac{\ell\pi}{2} - \frac{\pi}{2}\right)$[19], the spherical Neumann function is then

$$n_\ell(x) = (-1)^{\ell+1}\frac{1}{x}(-1)^{\ell+1}\sin\left(x - \frac{\ell\pi}{2} - \frac{\pi}{2}\right) = \frac{1}{x}\sin\left(x - \frac{\ell\pi}{2} - \frac{\pi}{2}\right)$$

---

[19] It is proved in Appendix 7-C.

· For small x, the asymptotic behaviors are

- $j_\ell(x) = \dfrac{x^\ell}{(2\ell+1)!!}$

- $n_\ell(x) = -\dfrac{(2\ell-1)!!}{x^{\ell+1}}$

· For large x, the asymptotic behaviors are

- $j_\ell(x) = \dfrac{1}{x}\cos(x - \dfrac{\ell\pi}{2} - \dfrac{\pi}{2})$

- $n_\ell(x) = \dfrac{1}{x}\sin(x - \dfrac{\ell\pi}{2} - \dfrac{\pi}{2})$

Example

7.3.2.1 If we consider a free particle with mass m in Quantum Mechanics, it gives $[\frac{1}{2m}(\frac{\hbar}{i}\overrightarrow{\nabla})^2]\Psi = E\Psi$ from the eigenvalue equation $H\Psi = E\Psi$, with $\Psi(r,\theta,\phi) = R(r)\Theta(\theta)\Phi(\phi)$.

Ans. In the spherical coordinate system, the Laplacian is

$$\begin{aligned}\nabla^2 &= \frac{\partial^2}{\partial r^2} + \frac{2}{r}\frac{\partial}{\partial r} + \frac{1}{r^2}(\frac{\partial^2}{\partial \theta^2} + \cot\theta\frac{\partial}{\partial\theta} + \frac{1}{\sin^2\theta}\frac{\partial^2}{\partial\phi^2}) \\ &= \frac{\partial^2}{\partial r^2} + \frac{2}{r}\frac{\partial}{\partial r} - \frac{L^2}{r^2\hbar^2}, \text{ using [Ex. 4.3.2]}\end{aligned}$$

For the radial part, we have

$$-\frac{\hbar^2}{2m}[\frac{d^2}{dr^2} + \frac{2}{r}\frac{d}{dr} - \frac{\ell(\ell+1)}{r^2}]R(r) = ER(r),$$

which gives

$$[\frac{d^2}{dr^2} + \frac{2}{r}\frac{d}{dr} - \frac{\ell(\ell+1)}{r^2}]R(r) + k^2R(r) = 0 \tag{7.3.2.1.1}$$

with $k^2 = \dfrac{2m}{\hbar^2}E$.

If we define $\rho = kr$, we have

$$\frac{dR(r)}{dr}=\frac{d\rho}{dr}\frac{dR}{d\rho}=k\frac{dR}{d\rho}$$

and

$$\frac{d^2R(r)}{dr^2}=\frac{d}{dr}(\frac{dR}{dr})=k\frac{d}{dr}\frac{dR}{d\rho}=k\frac{d\rho}{dr}\frac{d^2R}{d\rho^2}=k^2\frac{d^2R}{d\rho^2}$$

By substituting these into Eq. (7.3.2.1.1), we get

$$\frac{d^2R(\rho)}{d\rho^2}+\frac{2}{\rho}\frac{dR(\rho)}{d\rho}-\frac{\ell(\ell+1)}{\rho^2}R(\rho)+R(\rho)=0 \tag{7.3.2.1.2}$$

Let's define $u(\rho)=\rho R(\rho)$ and then we have

$$\frac{dR}{d\rho}=\frac{d}{d\rho}(\frac{u}{\rho})=\frac{u'}{\rho}-\frac{u}{\rho^2}$$

and

$$\frac{d^2R}{d\rho^2}=\frac{d}{d\rho}(\frac{u'}{\rho}-\frac{u}{\rho^2})=\frac{u''}{\rho}-\frac{2u'}{\rho^2}+\frac{2u}{\rho^3}$$

By substituting these into Eq. (7.3.2.1.2), we obtain

$$\frac{d^2u}{d\rho^2}+[1-\frac{\ell(\ell+1)}{\rho^2}]u=0 \tag{7.3.2.1.3}$$

Let's make this equation into a familiar equation by introducing $u(\rho)=\rho^{\frac{1}{2}}y(\rho)$, which gives

$$\frac{du}{d\rho}=\frac{1}{2}\rho^{-\frac{1}{2}}y+\rho^{\frac{1}{2}}\frac{dy}{d\rho}$$

and

$$\frac{d^2u}{d\rho^2}=-\frac{1}{4}\rho^{-\frac{3}{2}}y+\rho^{-\frac{1}{2}}\frac{dy}{d\rho}+\rho^{\frac{1}{2}}\frac{d^2y}{d\rho^2}$$

By substituting these into Eq. (7.3.2.1.3), we get

$$-\frac{1}{4}\rho^{-\frac{3}{2}}y+\rho^{-\frac{1}{2}}\frac{dy}{d\rho}+\rho^{\frac{1}{2}}\frac{d^2y}{d\rho^2}+[1-\frac{\ell(\ell+1)}{\rho^2}]\rho^{\frac{1}{2}}y=0$$

If we multiply the equation by $\rho^{\frac{3}{2}}$, it gives

$$\rho^2\frac{d^2y}{d\rho^2}+\rho\frac{dy}{d\rho}+[\rho^2-(\ell+\frac{1}{2})^2]y=0 \tag{7.3.2.1.4}$$

This is the Bessel's equation corresponding to $\nu=\ell+\frac{1}{2}$. Thus, the solution of Eq. (7.3.2.1.4) is the spherical Bessel function $j_\ell(\rho)=j_\ell(kr)$.

Therefore, the general solution is

$y(\rho)=A_1J_{\ell+\frac{1}{2}}(\rho)+A_2J_{-(\ell+\frac{1}{2})}(\rho)$ with constants A_1 and A_2

This gives

$$\frac{u(\rho)}{\sqrt{\rho}} = A_1 J_{\ell+\frac{1}{2}}(\rho) + A_2 J_{-(\ell+\frac{1}{2})}(\rho) = A\sqrt{\rho}\, j_\ell(\rho) + B\sqrt{\rho}\, n_\ell(\rho),$$

which is

$$u(\rho) = A\rho j_\ell(\rho) + B\rho n_\ell(\rho) \qquad (7.3.2.1.5)$$

where A and B are constants, and $j_\ell(\rho)$ and $n_\ell(\rho)$ are the spherical Bessel function and spherical Neumann function, respectively.

Since $u(\rho) = \rho R(\rho)$, Eq. (7.3.2.1.5) gives

$$R(\rho) = A j_\ell(\rho) + B n_\ell(\rho)$$

Since the solution should be finite at the origin and the spherical Neumann function has asymptotic behavior $n_\ell(\rho) \approx -\frac{(2\ell-1)!!}{\rho^{\ell+1}}$ for small ρ, the spherical Neumann function diverges at the origin and can't be a solution.

Therefore, the general solution for the radial part is

$$R(\rho) = A j_\ell(\rho),$$

which is

$$R(kr) = A j_\ell(kr) \qquad (7.3.2.1.6)$$

We already know that the solution for the angular part in the spherical coordinate system is the spherical harmonics $Y_{\ell m}(\theta,\phi)$ so that we obtain the general solution for the eigenvalue problem to be

$$\Psi(r,\theta,\phi) = R(r)\Theta(\theta)\Phi(\phi) = \sum_{\ell=0}^{\infty}\sum_{m=-\ell}^{\ell} j_\ell(kr)\, Y_{\ell m}(\theta,\phi) \qquad (7.3.2.1.7)$$

7.3.2.2 Using Eq. (7.3.2.1.7) we know that an arbitrary plane wave propagating the z-axis can be expressed by the spherical wave,

$$e^{ikr\cos\theta} = \sum_{\ell=0}^{\infty} C_\ell j_\ell(kr) P_\ell(\cos\theta)$$

Ans. Let's find the constant C_ℓ using the orthogonality of the Legendre polynomials:

$$\int_{-1}^{1} e^{ikr\cos\theta} P_{\ell'}(\cos\theta)\, d(\cos\theta)$$

$$= \int_{-1}^{1} \sum_{\ell=0}^{\infty} C_\ell j_\ell(kr) P_\ell(\cos\theta) P_{\ell'}(\cos\theta)\, d(\cos\theta)$$

$$= \frac{2}{2\ell'+1} C_{\ell'} j_{\ell'}(kr)$$

Therefore, we have

$$C_{\ell'}j_{\ell'}(kr)=\frac{2\ell'+1}{2}\int_{-1}^{1}e^{ikr\cos\theta}P_{\ell'}(\cos\theta)d(\cos\theta)$$

$$=\frac{2\ell'+1}{2}\int_{-1}^{1}(\frac{1}{ikr}e^{ikr\cos\theta})'P_{\ell'}(\cos\theta)d(\cos\theta)$$

$$=(\frac{2\ell'+1}{2})\frac{1}{ikr}[e^{ikr\cos\theta}P_{\ell'}(\cos\theta)]_{\cos\theta=-1}^{\cos\theta=1}$$

$$-\frac{2\ell'+1}{2}\int_{-1}^{1}(\frac{1}{ikr}e^{ikr\cos\theta})P_{\ell'}'(\cos\theta)d(\cos\theta),$$

where $(\frac{1}{ikr}e^{ikr\cos\theta})'$ and $P_{\ell'}'(\cos\theta)$ are the differentiation with respect to $\cos\theta$. The first term of the right-hand side of the equation has $\frac{1}{r}$. On the other hand, we have a series of a higher order $\frac{1}{r^{\ell}}$ if we apply integration by parts by ℓ times on the second term of the right-hand side of the equation.

For large r, we can neglect the second term of the right-hand side of the equation. Thus, we get

$$C_{\ell'}j_{\ell'}(kr)=(\frac{2\ell'+1}{2})\frac{1}{ikr}[e^{ikr\cos\theta}P_{\ell'}(\cos\theta)]_{\cos\theta=-1}^{\cos\theta=1}$$

$$=(\frac{2\ell'+1}{2})\frac{1}{ikr}[e^{ikr}P_{\ell'}(1)-e^{-ikr}P_{\ell'}(-1)]$$

$$=(\frac{2\ell'+1}{2})\frac{1}{ikr}[e^{ikr}-e^{-ikr}(-1)^{\ell'}]$$

$$=(\frac{2\ell'+1}{2})\frac{1}{ikr}[e^{ikr}-e^{-ikr}e^{i\ell'\pi}]$$

$$=(\frac{2\ell'+1}{2})\frac{e^{i\frac{\ell'\pi}{2}}}{ikr}[e^{-i\frac{\ell'\pi}{2}}e^{ikr}-e^{i\frac{\ell'\pi}{2}}e^{-ikr}]$$

$$=(\frac{2\ell'+1}{2})\frac{e^{i\frac{\ell'\pi}{2}}}{ikr}[e^{i(kr-\frac{\ell'\pi}{2})}-e^{-i(kr-\frac{\ell'\pi}{2})}]$$

Since $e^{i\frac{\ell'\pi}{2}}=\cos\frac{\ell'\pi}{2}+i\sin\frac{\ell'\pi}{2} = i^{\ell'}$, the equation is then

$$C_{\ell'}j_{\ell'}(kr)=(\frac{2\ell'+1}{2})\frac{i^{\ell'}}{ikr}2i\sin(kr-\frac{\ell'\pi}{2})$$

$$=(2\ell'+1)i^{\ell'}\frac{1}{kr}\sin(kr-\frac{\ell'\pi}{2})$$

Since $j_{\ell}(x)\approx\frac{1}{x}\cos(x-\frac{\ell\pi}{2}-\frac{\pi}{2})$ for large x, we get

$$\frac{1}{kr}\sin(kr-\frac{\ell' \pi}{2})=j_{\ell'}(kr)$$

by replacing x with kr and ℓ with ℓ'.

Thus, we obtain

$$C_{\ell'}j_{\ell'}(kr)=(2\ell'+1)i^{\ell'}j_{\ell'}(kr)$$

It gives the constant

$$C_{\ell'}=(2\ell'+1)i^{\ell'}$$

By substituting it into the original equation, we obtain

$$e^{ikr\cos\theta}=\sum_{\ell=0}^{\infty}C_{\ell}j_{\ell}(kr)P_{\ell}(\cos\theta)=\sum_{\ell=0}^{\infty}(2\ell+1)i^{\ell}j_{\ell}(kr)P_{\ell}(\cos\theta)$$

which is the relationship between the plane and spherical waves.

Chapter 08 Fourier Transformation

When we express a periodic function in terms of an infinite sum of sines and cosines, it is called Fourier series and is discussed in Section 8.1. We discuss Fourier transformation in Section 8.2, which is a generalization of complex Fourier series by changing summation to the integral.

8.1 Fourier Series

· $f(x)=\dfrac{a_0}{2}+\sum_{k=1}^{\infty}(a_k\cos kx+b_k\sin kx)$,

where $a_0=\dfrac{1}{\pi}\int_{-\pi}^{\pi}f(x)dx$, $a_k=\dfrac{1}{\pi}\int_{-\pi}^{\pi}f(x)\cos kx\,dx$ and

$b_k=\dfrac{1}{\pi}\int_{-\pi}^{\pi}f(x)\sin kx\,dx$

When a function $f(x)$ satisfies $f(x)=f(-x)$ or $f(x)=-f(-x)$, it is called an **even** or **odd** function. For instance, sine and cosine functions are odd and even functions, respectively. The product of an even function and an odd function gives another odd function. On the other hand, the products of even and even functions, and odd and odd functions are even functions.

Taylor series provides a general function as an infinite power series while Fourier series provides a function as an infinite series in the elementary trigonometric functions, sine and cosine. For a function $f(x)$ with a period 2π on the closed interval $[-\pi, \pi]$ or $[0, 2\pi]$, the Fourier series is

$$f(x) = \frac{a_0}{2} + \sum_{k=1}^{\infty} (a_k \cos kx + b_k \sin kx) \tag{8.1.1}$$

Let's calculate a_0 and the Fourier coefficients, a_k and b_k, of $f(x)$ and find a reason why a factor $\frac{1}{2}$ is introduced in the coefficient a_0.

For further calculations, it is useful to know the following results:

(a) $\int_{-\pi}^{\pi} \cos nx \cos mx dx = \frac{1}{2} \int_{-\pi}^{\pi} [\cos(n+m)x + \cos(n-m)x] dx$

(i) For $n=m=0$, the right-hand side of the equation gives

$$\int_{-\pi}^{\pi} dx = 2\pi$$

(ii) For $n \neq 0$ and $n=m$, the right-hand side of the equation is

$$\frac{1}{2} \int_{-\pi}^{\pi} \cos 2nx dx + \frac{1}{2} \int_{-\pi}^{\pi} dx = \frac{1}{2} \int_{-\pi}^{\pi} dx = \pi$$

(iii) For $n \neq 0$ and $n \neq m$, the right-hand side of the equation is zero.

From (i), (ii), and (iii), we obtain

$$\int_{-\pi}^{\pi} \cos nx \cos mx dx = \begin{cases} 2\pi \text{ for } n=m=0 \\ \pi\delta_{nm} \text{ for } n \neq 0 \end{cases}$$

(b) $\int_{-\pi}^{\pi} \sin nx \sin mx dx = -\frac{1}{2} \int_{-\pi}^{\pi} [\cos(n+m)x - \cos(n-m)x] dx$

(i) For $n=0$, the right-hand side of the equation is zero.

(ii) For $n \neq 0$ and $n=m$, the right-hand side of the equation gives

$$\frac{1}{2} \int_{-\pi}^{\pi} dx = \pi$$

(iii) For $n \neq 0$ and $n \neq m$, the right-hand side of the equation is zero.

From (i), (ii), and (iii), we obtain

$$\int_{-\pi}^{\pi} \sin nx \sin mx dx = \begin{cases} 0 \text{ for } n=0 \\ \pi\delta_{nm} \text{ for } n \neq 0 \end{cases}$$

(c) $\int_{-\pi}^{\pi} \sin nx \cos mx dx = \frac{1}{2}\int_{-\pi}^{\pi} [\sin(n+m)x + \sin(n-m)x]dx = 0$

because the integral of the product of odd and even functions over the symmetric interval is zero.

If we summarize (a), (b), and (c) in Table 8.1, we have

Table 8.1
· $\int_{-\pi}^{\pi} \cos nx \cos mx dx = \begin{cases} 2\pi \text{ for } n=m=0 \\ \pi\delta_{mn} \text{ for } n \neq 0 \end{cases}$
· $\int_{-\pi}^{\pi} \sin nx \sin mx dx = \begin{cases} 0 \text{ for } n=0 \\ \pi\delta_{nm} \text{ for } n \neq 0 \end{cases}$
· $\int_{-\pi}^{\pi} \sin nx \cos mx dx = 0$

For a continuous function $f(x)$ on the interval $[a, b]$, **an average value of** $f(x)$ from $x=a$ to $x=b$ is defined as

$$\overline{f(x)} = \frac{1}{b-a}\int_{a}^{b} f(x)dx$$

(1) Let's calculate the average value of $f(x)$ on the interval $[-\pi, \pi]$ to find a_0.

$$\begin{aligned}\overline{f(x)} &= \frac{1}{\pi-(-\pi)}\int_{-\pi}^{\pi} f(x)dx \\ &= \frac{1}{2\pi}\int_{-\pi}^{\pi} [\frac{a_0}{2} + \sum_{k=1}^{\infty}(a_k\cos kx + b_k\sin kx)]dx \\ &= \frac{1}{2\pi}\frac{a_0}{2}\int_{-\pi}^{\pi} dx + \frac{a_1}{2\pi}\int_{-\pi}^{\pi} \cos x dx + \frac{a_2}{2\pi}\int_{-\pi}^{\pi} \cos 2x dx + \cdots\cdots \\ &\quad + \frac{b_1}{2\pi}\int_{-\pi}^{\pi} \sin x dx + \frac{b_2}{2\pi}\int_{-\pi}^{\pi} \sin 2x dx + \cdots\cdots\end{aligned}$$

From Table 8.1, we know that all other terms are zero except the first term on the right-hand side of the equation so that we get

$$\frac{1}{2\pi}\int_{-\pi}^{\pi} f(x)dx = \frac{a_0}{2}$$

It gives

$$a_0 = \frac{1}{\pi}\int_{-\pi}^{\pi} f(x)dx$$

We can see that the constant a_0 has a neater form by introducing the factor $\frac{1}{2}$ in the first term of the right-hand side of Eq. (8.1.1).

(2) After multiplying $\cos x$ in Eq. (8.1.1), let's calculate the average value on the interval $[-\pi, \pi]$ to find a_1.

$$\frac{1}{2\pi}\int_{-\pi}^{\pi} f(x)\cos x dx = \frac{1}{2\pi}\frac{a_0}{2}\int_{-\pi}^{\pi}\cos x dx$$

$$+\frac{a_1}{2\pi}\int_{-\pi}^{\pi}\cos^2 x dx + \frac{a_2}{2\pi}\int_{-\pi}^{\pi}\cos 2x \cos x dx + \cdots\cdots$$

$$+\frac{b_1}{2\pi}\int_{-\pi}^{\pi}\sin x \cos x dx + \frac{b_2}{2\pi}\int_{-\pi}^{\pi}\sin 2x \cos x dx + \cdots\cdots$$

From Table 8.1, we know that all other terms are zero except the second term on the right-hand side of the equation.

Since $\int_{-\pi}^{\pi}\cos nx \cos mx dx = \pi\delta_{mn}$ for $n \neq 0$, we have

$$\int_{-\pi}^{\pi}\cos x \cos x dx = \pi$$

The average value is then

$$\frac{1}{2\pi}\int_{-\pi}^{\pi} f(x)\cos x dx = \frac{a_1}{2}$$

Thus, we obtain

$$a_1 = \frac{1}{\pi}\int_{-\pi}^{\pi} f(x)\cos x dx$$

Similarly, we get the following equation by multiplying $\cos kx$ to Eq. (8.1.1) and taking the average value on the interval $[-\pi, \pi]$ to find a_k:

$$\frac{1}{2\pi}\int_{-\pi}^{\pi} f(x)\cos kx dx$$

$$=\frac{1}{2\pi}\frac{a_0}{2}\int_{-\pi}^{\pi}\cos kx dx$$

$$+\frac{a_1}{2\pi}\int_{-\pi}^{\pi}\cos x\cos kx dx+\cdots\cdots+\frac{a_k}{2\pi}\int_{-\pi}^{\pi}\cos kx\cos kx dx+\cdots\cdots$$

$$+\frac{b_1}{2\pi}\int_{-\pi}^{\pi}\sin x\cos kx dx+\frac{b_2}{2\pi}\int_{-\pi}^{\pi}\sin 2x\cos kx dx+\cdots\cdots$$

From Table 8.1, all terms except $\int_{-\pi}^{\pi}\cos kx\cos kx dx$ on the right-hand side of the equation are zero so that we have

$$\frac{1}{2\pi}\int_{-\pi}^{\pi} f(x)\cos kx dx = \frac{a_k}{2\pi}\int_{-\pi}^{\pi}\cos kx\cos kx dx=\frac{a_k}{2}$$

Thus, we obtain

$$a_k=\frac{1}{\pi}\int_{-\pi}^{\pi} f(x)\cos kx dx$$

(3) Similarly, we can calculate for b_k by multiplying $\sin kx$ to Eq. (8.1.1) and taking the average value on the interval $[-\pi, \pi]$.

$$\frac{1}{2\pi}\int_{-\pi}^{\pi} f(x)\sin kx dx$$

$$=\frac{1}{2\pi}\frac{a_0}{2}\int_{-\pi}^{\pi}\sin kx dx+\frac{a_1}{2\pi}\int_{-\pi}^{\pi}\cos x\sin kx dx+\cdots\cdots$$

$$+\frac{b_1}{2\pi}\int_{-\pi}^{\pi}\sin x\sin kx dx+\cdots\cdots+\frac{b_k}{2\pi}\int_{-\pi}^{\pi}\sin kx\sin kx dx+\cdots\cdots$$

From Table 8.1, all terms except $\int_{-\pi}^{\pi}\sin kx\sin kx dx$ on the right-hand side of the equation are zero so that we have

$$\frac{1}{2\pi}\int_{-\pi}^{\pi} f(x)\sin kx dx = \frac{b_k}{2\pi}\int_{-\pi}^{\pi}\sin kx\sin kx dx=\frac{b_k}{2}$$

Thus, we obtain

$$b_k = \frac{1}{\pi}\int_{-\pi}^{\pi} f(x)\sin kx dx$$

Example

8.1.1 We have the same result just as well as $[-\pi, \ \pi]$ by re-scaling the x coordinate on the interval $[0, \ 2\pi]$. Let's assume that a function $f(x)$ is defined on $[0, \ 2\pi]$ and has the period 2π so that $f(x) = f(x+2\pi)$ for every value of x:

$$f(x) = \begin{cases} x & \text{for } 0 \le x \le \pi \\ \pi & \text{for } \pi \le x \le 2\pi \end{cases}$$

Represent this function in the Fourier series.

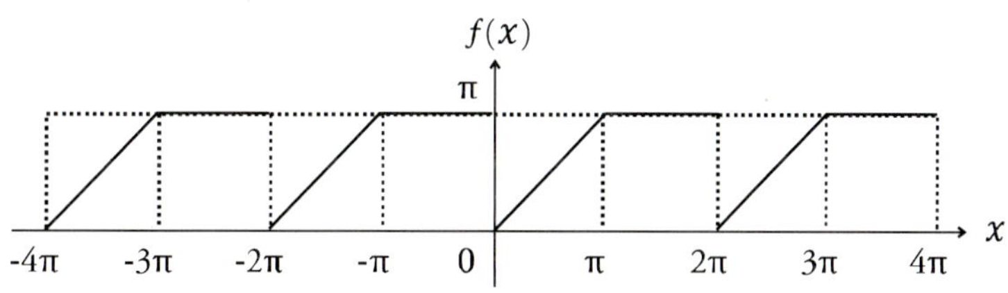

Ans. The constant a_0 and Fourier coefficients are then

$$a_0 = \frac{1}{\pi}\int_0^{2\pi} f(x)dx \ = \frac{1}{\pi}\left[\int_0^{\pi} x dx + \pi\int_{\pi}^{2\pi} dx\right] = \frac{3}{2}\pi,$$

$$a_k = \frac{1}{\pi}\int_0^{2\pi} f(x)\cos kx dx \ = \frac{1}{\pi}\left[\int_0^{\pi} x\cos kx dx + \pi\int_{\pi}^{2\pi}\cos kx dx\right]$$

$$= \frac{1}{\pi}\int_0^{\pi} x\left(\frac{1}{k}\sin kx\right)' dx$$

$$= \frac{x}{k\pi}\sin kx\Big|_{x=0}^{x=\pi} - \frac{1}{k\pi}\int_0^{\pi}\sin kx dx$$

$$= \frac{1}{k^2\pi}\cos kx\Big|_{x=0}^{x=\pi}$$

$$= \frac{1}{k^2\pi}[(-1)^k - 1] = \begin{cases} -\dfrac{2}{k^2\pi} & \text{for odd } k \\ 0 & \text{for even } k, \end{cases}$$

$$b_k = \frac{1}{\pi}\int_0^{2\pi} f(x)\sin kx dx \ = \frac{1}{\pi}\left[\int_0^{\pi} x\sin kx dx + \pi\int_{\pi}^{2\pi}\sin kx dx\right]$$

$$= \frac{1}{\pi}\int_0^{\pi} x\left(-\frac{1}{k}\cos kx\right)' dx + \int_{\pi}^{2\pi}\sin kx dx$$

$$=-\frac{x}{k\pi}\cos kx|_{x=0}^{x=\pi}+\frac{1}{k\pi}\int_0^{\pi}\cos kx dx-\frac{1}{k}\cos kx|_{x=\pi}^{x=2\pi}$$

$$=-\frac{1}{k}\cos k\pi-\frac{1}{k}\cos 2\pi k+\frac{1}{k}\cos k\pi=-\frac{1}{k}$$

By substituting $a_{0,}$ a_k, and b_k into the Fourier series, we obtain

$$f(x)=\frac{3\pi}{4}-\frac{2}{\pi}(\cos x+\frac{1}{3^2}\cos 3x+\frac{1}{5^2}\cos 5x+\cdots\cdots)$$

$$-(\sin x+\frac{1}{2}\sin 2x+\frac{1}{3}\sin 3x+\cdots\cdots)$$

Therefore, the Fourier series provides a way to represent the periodic phenomena neatly in sines and cosines.

8.1.2 Let's assume that a function $f(x)$ is defined on the interval $(-L,\ L)$ and has the period $2L$ so that $f(x)=f(x+2L)$ for every value of x:

$$f(x)=\begin{cases}1 & \text{for } -L\le x<0\\ 0 & \text{for } 0<x\le L\end{cases}$$

Represent this function in the Fourier series.

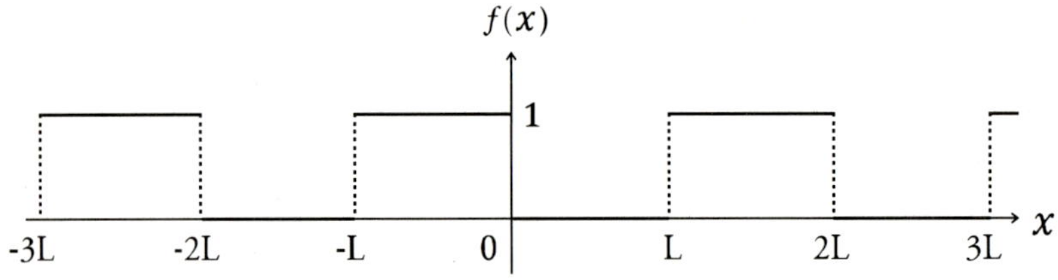

Ans. In this case the function has the form $f(x)=e^{ikx}$ and we have

$e^{-ikL}=e^{ikL}$ from $f(-L)=f(L)$

This gives

$$e^{2ikL}=1 \Rightarrow \cos 2kL+i\sin 2kL=1,$$

which yields

$$2kL=2n\pi \Rightarrow k=\frac{n\pi}{L},$$

where n is the integer.

Since its fundamental frequency is $k=\frac{n\pi}{L}$, the Fourier series corresponding to the function $f(x)$ can be expressed as

$$f(x)=\frac{a_0}{2}+a_1\cos\frac{\pi}{L}x+a_2\cos\frac{2\pi}{L}x+\cdots\cdots$$

$$+ b_1 \sin\frac{\pi}{L}x + b_2 \sin\frac{2\pi}{L}x + \cdots\cdots$$

$$= \frac{a_0}{2} + \sum_{n=1}^{\infty}\left(a_n \cos\frac{n\pi}{L}x + b_n \sin\frac{n\pi}{L}x\right)$$

The constant and Fourier coefficients are then

$$a_0 = \frac{1}{L}\int_{-L}^{L} f(x)dx = \frac{1}{L}\int_{-L}^{0} dx = 1$$

$$a_n = \frac{1}{L}\int_{-L}^{L} f(x)\cos\frac{n\pi x}{L}dx$$

$$= \frac{1}{L}\int_{-L}^{0} \cos\frac{n\pi x}{L}dx = \frac{1}{n\pi}[\sin\frac{n\pi x}{L}]_{x=-L}^{x=0}$$

$$= 0$$

$$b_n = \frac{1}{L}\int_{-L}^{L} f(x)\sin\frac{n\pi x}{L}dx$$

$$= \frac{1}{L}\int_{-L}^{0} \sin\frac{n\pi x}{L}dx = -\frac{1}{n\pi}[\cos\frac{n\pi x}{L}]_{x=-L}^{x=0}$$

$$= -\frac{1}{n\pi}[1-(-1)^n] = \begin{cases} 0 & \text{for even } n \\ -\dfrac{2}{n\pi} & \text{for odd } n \end{cases}$$

If we substitute a_0 and b_n into the Fourier series, we obtain

$$f(x) = \frac{1}{2} - \frac{2}{\pi}\sum_{n=1}^{\infty}\frac{1}{n}\sin\frac{n\pi x}{L},$$

where n is the odd integer.

8.2 Fourier Transform

- $f(x) = \int_{-\infty}^{\infty} g(\alpha)e^{i\alpha x}d\alpha$ [Inverse Fourier transform of $g(\alpha)$]
- $g(\alpha) = \frac{1}{2\pi}\int_{-\infty}^{\infty} f(x)e^{-i\alpha x}dx$ [Fourier transform of $f(x)$]
- $\delta(x-x') = \frac{1}{2\pi}\int_{-\infty}^{\infty} e^{i(x-x')\alpha}d\alpha$ (Definition of delta function)

Eq. (8.1.1) can be expressed in the complex form by using Euler formula and is written as

$$f(x) = \frac{a_0}{2} + \sum_{k=1}^{\infty} \left(a_k \frac{e^{ikx} + e^{-ikx}}{2} + b_k \frac{e^{ikx} - e^{-ikx}}{2i}\right)$$

$$= \frac{a_0}{2} + \sum_{k=1}^{\infty} \frac{a_k - ib_k}{2} e^{ikx} + \sum_{k=1}^{\infty} \frac{a_k + ib_k}{2} e^{-ikx}$$

If we replace k with $-\ell$ in the third term on the right-hand side of the equation, the equation is then

$$f(x) = \frac{a_0}{2} + \sum_{k=1}^{\infty} \frac{a_k - ib_k}{2} e^{ikx} + \sum_{\ell=-1}^{-\infty} \frac{a_{-\ell} + ib_{-\ell}}{2} e^{i\ell x},$$

which means it can be expressed in the simpler complex form as

$$f(x) = \sum_{n=-\infty}^{\infty} g(n) e^{inx}$$

For a continuous variable, this can be expressed in an integral form as

$$f(x) = \int_{-\infty}^{\infty} g(\alpha) e^{i\alpha x} d\alpha \tag{8.2.1}$$

and is called the **inverse Fourier transform** of $g(\alpha)$.

Fourier transform of $f(x)$ is then

$$g(\alpha) = \frac{1}{2\pi} \int_{-\infty}^{\infty} f(x) e^{-i\alpha x} dx \tag{8.2.2}$$

Therefore, the Fourier transform is an integral which maps one function to the other function. The Fourier transform is useful to convert a discontinuous or non-smooth function which is not easy to handle in problems into a function which can be easily solvable.

For examples, the Fourier transforms of Gaussian and step functions are calculated in [Ex. 8.2.2] and [Ex. 8.2.3], respectively, and we will show that the Fourier transform of the delta function $\delta(x)$ gives the constant function.

If we insert Eqs. (8.2.2) into (8.2.1), it gives

$$f(x)=\frac{1}{2\pi}\int_{-\infty}^{\infty}\int_{-\infty}^{\infty}f(x')e^{i(x-x')\alpha}dx'd\alpha \tag{8.2.3}$$

By defining the delta function in an integral

$$\delta(x-x')=\frac{1}{2\pi}\int_{-\infty}^{\infty}e^{i(x-x')\alpha}d\alpha, \tag{8.2.4}$$

we have the right-hand side of the equation

$$\int_{-\infty}^{\infty}f(x')\delta(x-x')dx',$$

which is equivalent to $f(x)$.

The Fourier transform of the delta function $\delta(x)$ is then given by

$$g(\alpha)=\frac{1}{2\pi}\int_{-\infty}^{\infty}\delta(x)e^{-i\alpha x}dx=\frac{1}{2\pi}$$

Therefore, the Fourier transform of the delta function $\delta(x)$ is the constant.

Let's express the Fourier series on the interval $0<x<L$. In this case an odd periodic function $f(x)$ can be written as

$$f(x)=\sum_{n=1}^{\infty}a_n\sin\frac{n\pi x}{L} \tag{8.2.5}$$

If we multiply $\sin\frac{m\pi x}{L}$ on the both sides of the equation and take integral on the interval $0\le x\le L$, it gives

$$\begin{aligned}\int_0^L f(x)\sin\frac{m\pi x}{L}dx &= \sum_{n=1}^{\infty}\int_0^L a_n\sin\frac{m\pi x}{L}\sin\frac{n\pi x}{L}dx\\ &= \sum_{n=1}^{\infty}a_n\int_0^L\sin\frac{m\pi x}{L}\sin\frac{n\pi x}{L}dx\end{aligned}$$

To calculate the right-hand side of the equation, let's substitute $\frac{\pi x}{L}=y$. Then, we have

$$\int_0^L\sin\frac{m\pi x}{L}\sin\frac{n\pi x}{L}dx=\frac{L}{\pi}\int_0^{\pi}dy\sin my\sin ny=\frac{L}{\pi}\frac{1}{2}\pi\delta_{mn}=\frac{L}{2}\delta_{mn}$$

using Table 8.1.

Thus, we obtain

$$\int_0^L f(x)\sin\frac{m\pi x}{L}dx = \frac{L}{2}\sum_{n=1}^{\infty} a_n \delta_{mn},$$

which gives

$$\int_0^L f(x)\sin\frac{n\pi x}{L}dx = \frac{L}{2}a_n$$

Thus, we get

$$a_n = \frac{2}{L}\int_0^L f(x)\sin\frac{n\pi x}{L}dx$$

If we insert a_n into Eq. (8.2.5), we have

$$f(x) = \frac{2}{L}\sum_{n=1}^{\infty}\int_0^L f(x')\sin\frac{n\pi x'}{L}\sin\frac{n\pi x}{L}dx' \tag{8.2.6}$$

Let's define $k_n = \frac{n\pi}{L}$ and the difference $\frac{\pi}{L} = \Delta k$.

Then, Eq. (8.2.6) is

$$\begin{aligned} f(x) &= \frac{2}{\pi}\Delta k\sum_{n=1}^{\infty}\sin k_n x\int_0^L f(x')\sin k_n x' dx' \\ &= \frac{2}{\pi}\Delta k\sum_{n=1}^{\infty}(\sin k_n x) g_L(k_n), \end{aligned} \tag{8.2.7}$$

where

$$g_L(k_n) = \int_0^L f(x')\sin k_n x' dx'$$

Since Δk approaches zero for large L, Eq. (8.2.7) is expressed by

$$f(x) = \frac{2}{\pi}\int_0^{\infty} g(k)\sin kx dk,$$

where

$$g(k) = \int_0^{\infty} f(x')\sin kx' dx'$$

Similarly for cosine, we can obtain

$$f(x) = \frac{2}{\pi}\int_0^{\infty} g(k)\cos kx dk,$$

where

$$g(k)=\int_0^{\infty} f(x')\cos kx' dx'$$

We can split a periodic function $f(x)$ into an even function and odd function, and the **Fourier sine and cosine transforms** are in general defined as

$$f_{\sin}(x)=\frac{2}{\pi}\int_0^{\infty} g_{\sin}(\alpha)\sin\alpha x\ d\alpha$$

and

$$f_{\cos}(x)=\frac{2}{\pi}\int_0^{\infty} g_{\cos}(\alpha)\cos\alpha x\ d\alpha,$$

respectively,

where

$$g_{\sin}(\alpha)=\int_0^{\infty} f_{\sin}(x)\sin\alpha x dx \text{ and } g_{\cos}(\alpha)=\int_0^{\infty} f_{\cos}(x)\cos\alpha x dx$$

- $f_{\sin}(x)=\frac{2}{\pi}\int_0^{\infty} g_{\sin}(\alpha)\sin\alpha x d\alpha$ (Fourier Sine Transform)
- $f_{\cos}(x)=\frac{2}{\pi}\int_0^{\infty} g_{\cos}(\alpha)\cos\alpha x d\alpha$ (Fourier Cosine Transform)

Consider two functions, $f_1(x)$ and $f_2(x)$, and let's find the Fourier transform of their product.

$$\begin{aligned}\frac{1}{2\pi}\int_{-\infty}^{\infty} f_1(x)f_2(x)e^{-i\alpha x}dx &= \frac{1}{2\pi}\int_{-\infty}^{\infty}\left[\int_{-\infty}^{\infty} g_1(\alpha')e^{i\alpha' x}d\alpha'\right]f_2(x)e^{-i\alpha x}dx \\ &= \int_{-\infty}^{\infty} d\alpha' g_1(\alpha')\left[\frac{1}{2\pi}\int_{-\infty}^{\infty} f_2(x)e^{-i(\alpha-\alpha')x}dx\right] \\ &= \int_{-\infty}^{\infty} g_1(\alpha')g_2(\alpha-\alpha')d\alpha',\end{aligned}$$

which is a convolution of g_1 and g_2.

Thus, the Fourier transform of the product of two functions is a convolution of the individual Fourier transform.

Using Eq. (8.1.1) for a function with the period 2π, we have

$$\begin{aligned}\int_{-\pi}^{\pi}|f(x)|^2dx &= \int_{-\pi}^{\pi}[\frac{a_0}{2}+\sum_{k=1}^{\infty}(a_k\cos kx+b_k\sin kx)]^2dx \\ &= \frac{a_0^2}{4}\int_{-\pi}^{\pi}dx + a_0\int_{-\pi}^{\pi}\sum_{k=1}^{\infty}(a_k\cos kx+b_k\sin kx)dx \\ &\quad + \int_{-\pi}^{\pi}\sum_{k=1}^{\infty}\sum_{\ell=1}^{\infty}[a_ka_\ell\cos kx\cos\ell x+b_kb_\ell\sin kx\sin\ell x \\ &\qquad + a_kb_\ell\cos kx\sin\ell x+a_\ell b_k\cos\ell x\sin kx]dx \\ &= \frac{\pi}{2}a_0^2+\sum_{k=1}^{\infty}\sum_{\ell=1}^{\infty}(a_ka_\ell\pi\delta_{k\ell}+b_kb_\ell\pi\delta_{k\ell}),\ \text{(using Table 8.1)} \\ &= \frac{\pi}{2}a_0^2+\pi\sum_{k=1}^{\infty}(a_k^2+b_k^2) \qquad (8.2.8)\end{aligned}$$

This gives

$$\frac{1}{2\pi}\int_{-\pi}^{\pi}|f(x)|^2dx = \frac{a_0^2}{4}+\frac{1}{2}\sum_{k=1}^{\infty}(a_k^2+b_k^2),$$

which is called as **Parseval's theorem**.

This theorem is useful to relate a function with its Fourier coefficients[20].

Example

8.2.1 Find the Fourier transform of a function

$$f(x)=\begin{cases}x & \text{for } |x|<1 \\ 0 & \text{for } |x|>1\end{cases}$$

Ans. From Eq. (8.2.2), we have

$$\begin{aligned}g(\alpha) &= \frac{1}{2\pi}\int_{-1}^{1}xe^{-i\alpha x}dx = \frac{1}{2\pi}\int_{-1}^{1}x(\frac{1}{-i\alpha}e^{-i\alpha x})'dx \\ &= \frac{1}{2\pi}[\frac{x}{-i\alpha}e^{-i\alpha x}]_{x=-1}^{x=1}+\frac{1}{2\pi}\frac{1}{i\alpha}\int_{-1}^{1}e^{-i\alpha x}dx\end{aligned}$$

[20] See [Examples 8.2.6 and 8.2.7].

$$= -\frac{1}{2\pi i\alpha}(e^{-i\alpha}+e^{i\alpha}) + \frac{1}{2\pi\alpha^2}(e^{-i\alpha}-e^{i\alpha})$$

$$= \frac{i\cos\alpha}{\pi\alpha} - \frac{i\sin\alpha}{\pi\alpha^2}$$

8.2.2 Find the Fourier transform of $f(x) = e^{-\frac{x^2}{\sigma^2}}$ which is a function appeared in wave packets problems of Quantum Mechanics. In here, σ is the constant.

Ans. We have

$$g(\alpha) = \frac{1}{2\pi}\int_{-\infty}^{\infty} e^{-\frac{x^2}{\sigma^2}} e^{-i\alpha x} dx$$

$$= \frac{1}{2\pi}\int_{-\infty}^{\infty} e^{-(\frac{x}{\sigma}+i\frac{\sigma\alpha}{2})^2} e^{(i\frac{\sigma\alpha}{2})^2} dx$$

$$= \frac{1}{2\pi} e^{-\frac{\sigma^2\alpha^2}{4}} \int_{-\infty}^{\infty} e^{-(\frac{x}{\sigma}+i\frac{\sigma\alpha}{2})^2} dx$$

If we define $\frac{x}{\sigma} + i\frac{\sigma\alpha}{2} = y$, we have $dx = \sigma dy$.
The equation is then

$$g(\alpha) = \frac{\sigma}{2\pi} e^{-\frac{\sigma^2\alpha^2}{4}} \int_{-\infty}^{\infty} e^{-y^2} dy$$

$$= \frac{\sigma}{2\pi} e^{-\frac{\sigma^2\alpha^2}{4}} \sqrt{\pi} = \frac{\sigma}{2\sqrt{\pi}} e^{-(\frac{\sigma\alpha}{2})^2}$$

Therefore, the Fourier transform of the Gaussian is the Gaussian.

8.2.3 Find the Fourier transform of a function

$$f(x) = \begin{cases} 0 & \text{for } x < -a \\ A & \text{for } -a \le x \le a, \\ 0 & \text{for } x > a \end{cases}$$

where A is the positive constant.

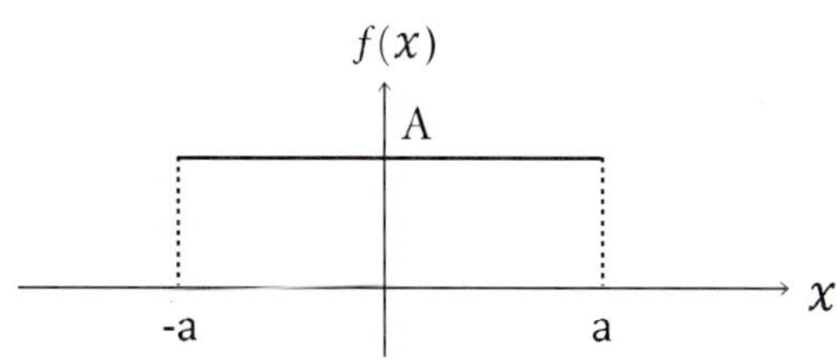

Ans. We have

$$g(\alpha)=\frac{1}{2\pi}\int_{-a}^{a}Ae^{-i\alpha x}dx \;=\frac{A}{2\pi}\frac{-1}{i\alpha}e^{-i\alpha x}\Big|_{x=-a}^{x=a}$$

$$=\frac{iA}{2\pi\alpha}(e^{-i\alpha a}-e^{i\alpha a})=\frac{A}{\pi\alpha}\sin\alpha a$$

The Fourier transform of the step function gives the sinusoidal function which has zero value at $\alpha=\frac{n\pi}{a}$, where n is the integer. Therefore, the width of this sinusoidal function is $\frac{2\pi}{a}$.

For $\alpha=0$, $g(\alpha)$ has the form $\frac{0}{0}$. Then, we have $\frac{(\sin\alpha a)'}{(\alpha)'}\Big|_{\alpha=0}=\frac{a\cos\alpha a}{1}\Big|_{\alpha=0}=a$ using L'Hopital rule. Therefore, it has the maximum value $\frac{Aa}{\pi}$ at $\alpha=0$.

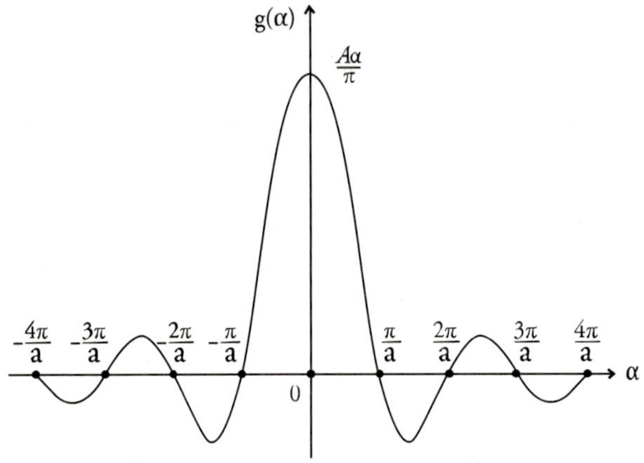

8.2.4 Find the Fourier transform of a function

$$f(x)=\begin{cases}\cos x \;\text{ for } |x|<\frac{\pi}{2}\\ 0 \;\text{ for } |x|>\frac{\pi}{2}\end{cases}$$

Ans. We have

$$g_{\cos}(\alpha)=\frac{1}{2\pi}\int_{-\frac{\pi}{2}}^{\frac{\pi}{2}}\cos x\cos\alpha x dx \;=\frac{1}{\pi}\int_{0}^{\frac{\pi}{2}}\cos x\cos\alpha x dx$$

$$=\frac{1}{\pi}\int_{0}^{\frac{\pi}{2}}(\sin x)'\cos\alpha x dx$$

$$=\frac{1}{\pi}\sin x\cos\alpha x\Big|_{x=0}^{x=\frac{\pi}{2}}+\frac{\alpha}{\pi}\int_{0}^{\frac{\pi}{2}}\sin x\sin\alpha x dx$$

$$=\frac{1}{\pi}\cos\frac{\pi\alpha}{2}-\frac{\alpha}{\pi}\int_{0}^{\frac{\pi}{2}}(\cos x)'\sin\alpha x dx$$

$$=\frac{1}{\pi}\cos\frac{\pi\alpha}{2}-\frac{\alpha}{\pi}\cos x\sin\alpha x\Big|_{x=0}^{x=\frac{\pi}{2}}+\frac{\alpha^2}{\pi}\int_{0}^{\frac{\pi}{2}}\cos x\cos\alpha x dx$$

$$= \frac{1}{\pi}\cos\frac{\pi\alpha}{2} + \frac{\alpha^2}{\pi}\int_0^{\frac{\pi}{2}} \cos x \cos\alpha x dx$$

If we let $\frac{1}{\pi}\int_0^{\frac{\pi}{2}} \cos x \cos\alpha x dx = I$, the equation is expressed as

$$I = \frac{1}{\pi}\cos\frac{\pi\alpha}{2} + \alpha^2 I,$$

which gives

$$I = \frac{1}{(1-\alpha^2)\pi}\cos\frac{\pi\alpha}{2}$$

Thus, we obtain

$$g_{\cos}(\alpha) = \frac{1}{(1-\alpha^2)\pi}\cos\frac{\pi\alpha}{2}$$

8.2.5 Calculate $\int_{-\infty}^{\infty} |f(x)|^2 dx$.

Ans.
$$\begin{aligned}\int_{-\infty}^{\infty} f^*(x) f(x) dx &= \int_{-\infty}^{\infty} f^*(x)\left[\int_{-\infty}^{\infty} g(\alpha) e^{i\alpha x} d\alpha dx\right] \\ &= \int_{-\infty}^{\infty} g(\alpha) d\alpha \int_{-\infty}^{\infty} f^*(x) e^{i\alpha x} dx \\ &= \int_{-\infty}^{\infty} g(\alpha) d\alpha [\int_{-\infty}^{\infty} f(x) e^{-i\alpha x} dx]^* \\ &= 2\pi \int_{-\infty}^{\infty} g(\alpha)[g(\alpha)]^* d\alpha\end{aligned}$$

This can be written as

$$\int_{-\infty}^{\infty} |g(\alpha)|^2 d\alpha = \frac{1}{2\pi}\int_{-\infty}^{\infty} |f(x)|^2 dx$$

This is also known as the Parseval's theorem, which is the relation between the integral of the square of a function and the integral of the square of its transform.

8.2.6 Calculate the sum $\sum_{n=1}^{\infty} \frac{1}{n^2}$ by using the Fourier series for $f(x) = 1 + x$ on the interval $[-\pi, \pi]$ and using the Parseval's theorem.

Ans. First let's find the constant and Fourier coefficients of the function $f(x)$:

$$a_0 = \frac{1}{\pi}\int_{-\pi}^{\pi} f(x) dx = \frac{1}{\pi}\int_{-\pi}^{\pi} (1+x) dx = 2$$

and

$$a_k = \frac{1}{\pi}\int_{-\pi}^{\pi} f(x)\cos kx dx = \frac{1}{\pi}\int_{-\pi}^{\pi} (1+x)\cos kx dx,$$

where

$$\int_{-\pi}^{\pi}(1+x)\cos kx dx = \int_{-\pi}^{\pi}\cos kx dx + \int_{-\pi}^{\pi} x\left(\frac{1}{k}\sin kx\right)' dx$$

$$= \frac{1}{k}\sin kx|_{-\pi}^{\pi} + \frac{x}{k}\sin kx|_{-\pi}^{\pi} - \frac{1}{k}\int_{-\pi}^{\pi}\sin kx dx$$

$$= \frac{1}{k^2}\cos kx|_{-\pi}^{\pi} = 0$$

Therefore, $a_k = 0$.

$$b_k = \frac{1}{\pi}\int_{-\pi}^{\pi} f(x)\sin kx dx = \frac{1}{\pi}\int_{-\pi}^{\pi}(1+x)\sin kx dx$$

$$= \frac{1}{\pi}\left[\int_{-\pi}^{\pi}\sin kx dx - \int_{-\pi}^{\pi} x\left(\frac{1}{k}\cos kx\right)' dx\right]$$

$$= -\frac{1}{\pi k}\cos kx|_{-\pi}^{\pi} - \frac{1}{\pi k}[x\cos kx]_{-\pi}^{\pi} + \frac{1}{\pi k^2}\sin kx|_{-\pi}^{\pi}$$

$$= -\frac{2}{k}(-1)^k = \frac{2}{k}(-1)^{k+1}$$

Substituting these coefficients into Eq. (8.2.8), we have

$$\int_{-\pi}^{\pi}(1+x)^2 dx = 2\pi + \pi\sum_{k=1}^{\infty}\frac{4}{k^2}(-1)^{2k+2} = 2\pi + \pi\sum_{k=1}^{\infty}\frac{4}{k^2}$$

Since the left-hand side of the equation is

$$\int_{-\pi}^{\pi}(1+x)^2 dx = 2\pi + \frac{2}{3}\pi^3,$$

we get

$$2\pi + \frac{2}{3}\pi^3 = 2\pi + \pi\sum_{k=1}^{\infty}\frac{4}{k^2}$$

This gives

$$\pi\sum_{k=1}^{\infty}\frac{4}{k^2} = \frac{2}{3}\pi^3,$$

which is

$$\sum_{k=1}^{\infty}\frac{1}{k^2} = \frac{1}{6}\pi^2$$

Thus, we obtain the following equation by replacing k with n

$$\sum_{n=1}^{\infty}\frac{1}{n^2}=\frac{1}{6}\pi^2$$

8.2.7 Calculate the sum $\sum_{n=1}^{\infty}\frac{1}{n^4}$ by using the Parseval's theorem with $f(x)=x^2$ on the interval $[-2,\ 2]$.

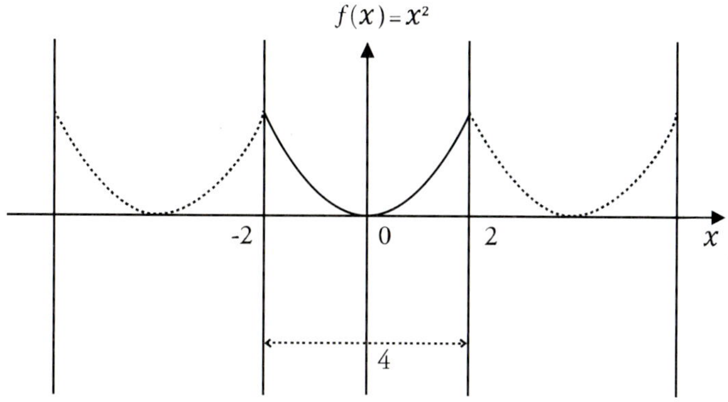

Ans. This function has a period of 4 and is even function, of which coefficient b_k is zero because the integrand is the product of even and odd functions. Now, let's calculate a_0 and the Fourier coefficient a_k:

$$a_0=\frac{1}{2}\int_{-2}^{2}x^2dx\ =\frac{8}{3}$$

$$a_k=\frac{1}{2}\int_{-2}^{2}x^2\cos kx dx$$

As we did in [Ex. 8.1.2], we have $e^{ikx}=e^{ik(x+4)}$ from $f(x)=f(x+4)$.
This gives

$$\cos 4k+i\sin 4k=1\ \Rightarrow\ 4k=2\pi n,$$

where n is the integer.
It gives

$$k=\frac{n\pi}{2}$$

Then, we get

$$a_n=\frac{1}{2}\int_{-2}^{2}x^2\cos\frac{n\pi x}{2}dx\ =\frac{1}{2}\int_{-2}^{2}x^2\left(\frac{2}{n\pi}\sin\frac{n\pi x}{2}\right)'dx$$

$$=\frac{1}{2}\left[\frac{2}{n\pi}x^2\sin\frac{n\pi x}{2}\right]_{x=-2}^{x=2}-\frac{2}{n\pi}\int_{-2}^{2}x\sin\frac{n\pi x}{2}dx$$

$$=\frac{2}{n\pi}\int_{-2}^{2}x\left(\frac{2}{n\pi}\cos\frac{n\pi x}{2}\right)'dx$$

$$= \frac{4}{n^2\pi^2}[x \cos\frac{n\pi x}{2}]_{x=-2}^{x=2} - \frac{2}{n\pi}\int_{-2}^{2}\frac{2}{n\pi}\cos\frac{n\pi x}{2}dx$$

$$= \frac{16}{n^2\pi^2}(-1)^n$$

From Eq. (8.2.8), we have

$$\int_{-2}^{2} x^4 dx = a_0^2 + 2\sum_{n=1}^{\infty} a_n^2$$

The left-hand side of the equation is

$$\int_{-2}^{2} x^4 dx = \frac{64}{5},$$

while the right-hand side of the equation is

$$a_0^2 + 2\sum_{n=1}^{\infty} a_n^2 = (\frac{8}{3})^2 + 2\sum_{n=1}^{\infty}\frac{16^2}{n^4\pi^4}$$

Thus, we obtain

$$\sum_{n=1}^{\infty}\frac{1}{n^4} = \frac{\pi^4}{2\times 16^2}[\frac{64}{5} - (\frac{8}{3})^2],$$

which gives

$$\sum_{n=1}^{\infty}\frac{1}{n^4} = \frac{\pi^4}{90}$$

8.2.8 If $\Psi(x)$ in the position representation describes the wave function for a localized particle, we have

$$\int_{-\infty}^{\infty} |\Psi(x)|^2 dx = 1$$

Express the wave function $\Psi(x)$ in terms of momentum p of a particle.

Ans. The Fourier transform maps the function $\Psi(x)$ to the function $\phi(k)$:

$$\Psi(x) = \int_{-\infty}^{\infty}\phi(k)e^{ikx}dk,$$

where k is the wave number.

Since the wave number k has the relation with the momentum p, $k = \frac{p}{\hbar}$, the equation can be expressed as

$$\Psi(x) = C\int_{-\infty}^{\infty}\phi(p)e^{i\frac{p}{\hbar}x}dp$$

where C is the complex number and is obtained from the normalization condition.

Then, we have

$$\Psi^*(x')\Psi(x) = |C|^2 \int_{-\infty}^{\infty} \phi^*(p) e^{-i\frac{p}{\hbar}x'} dp \int_{-\infty}^{\infty} \phi(p) e^{i\frac{p}{\hbar}x} dp$$

$$= |C|^2 \int_{-\infty}^{\infty}\int_{-\infty}^{\infty} |\phi(p)|^2 e^{i\frac{p}{\hbar}(x-x')} dpdp$$

$$= 2\pi |C|^2 \int_{-\infty}^{\infty} |\phi(p)|^2 \delta(\frac{x-x'}{\hbar}) dp \qquad \text{(using Eq. 8.2.4)}$$

It gives

$$\int_{-\infty}^{\infty} \Psi^*(x')\Psi(x) dx = 2\pi |C|^2 \int_{-\infty}^{\infty}\int_{-\infty}^{\infty} |\phi(p)|^2 \delta(\frac{x-x'}{\hbar}) dpdx$$

$$= 2\pi |C|^2 \int_{-\infty}^{\infty}\int_{-\infty}^{\infty} |\phi(p)|^2 \hbar \delta(x-x') dpdx,$$

where the property of the delta function, $\delta(ax) = \frac{1}{|a|}\delta(x)$ for $a \neq 0$, is used. This equation is then

$$\int_{-\infty}^{\infty} |\Psi(x)|^2 dx = 2\pi\hbar |C|^2 \int_{-\infty}^{\infty} |\phi(p)|^2 dp$$

$$= 2\pi\hbar |C|^2$$

Since the left-side of the equation is 1 by the normalization condition, we have

$$C = \frac{1}{\sqrt{2\pi\hbar}}$$

Thus, the wave function is expressed in terms of the momentum as follows:

$$\Psi(x) = \frac{1}{\sqrt{2\pi\hbar}} \int_{-\infty}^{\infty} \phi(p) e^{i\frac{p}{\hbar}x} dp$$

Chapter 09 Variational Principle

Hamilton's variational principle helps us to understand how Mother Nature determines a particular path or motion followed by a classical system. In Section 9.1 the Lagrange equations are derived from the Hamilton's principle of least action and several examples using the Lagrange equations to determine paths or motions are presented. When a motion of a system is subjected to constraints, a motion can be understood, called the Lagrange multipliers method, by introducing a new variable λ. This technique provides information of generalized forces of constraints and is described in Section 9.2.

9.1 Lagrange's Equation

· The least action principle (Hamilton's variational principle) yields Lagrange's equation

$$\frac{\partial L(q_i, \dot{q}_i, t)}{\partial q_i} - \frac{d}{dt}\left[\frac{\partial L(q_i, \dot{q}_i, t)}{\partial \dot{q}_i}\right] = 0,$$

where $L = K - U$ is the Lagrangian.

The most general principle governing motion of a mechanical system is **the least action principle** or **Hamilton's variational principle**. It means that the motion occurs when the action takes the smallest possible value. The action S is defined as

$$S=\int_{t_1}^{t_2} L(q_i,\dot{q}_i,t)dt, \tag{9.1.1}$$

where $L(q_i,\dot{q}_i,t)$ is the **Lagrangian** defined by $L(q_i,\dot{q}_i,t)=K-U$, K and U are the kinetic and potential energies, respectively, and q_i is the **generalized coordinates**.

The generalized coordinates consist of any independent coordinates which define a system uniquely and completely. Then, the number of generalized coordinates is the **number of the degree of freedom** of a system.

If there are n particles in a space and m equations of constraints (i.e. a motion is restricted by these constraints), the number of the degree of freedom is $3n-m$ which uniquely describe this n particle system[21]. When constraints are expressed in equalities (inequalities), they are called **holonomic (non-holonomic) constraints**.

For simplicity, let's assume the number of the degree of freedom for a system is $i=1$. Any results obtained with this assumption can be generalized into problems of n degrees of freedom.

Since an actual motion of a system occurring on the interval $[t_1, t_2]$ takes the smallest value of the action, this least action principle yields $\delta S=0$, which gives

$$0=\delta\int_{t_1}^{t_2} L(q,\dot{q},t)dt=\int_{t_1}^{t_2} L(q+\delta q,\dot{q}+\delta\dot{q},t)dt-\int_{t_1}^{t_2} L(q,\dot{q},t)dt \tag{9.1.2}$$

---

[21] See [Examples 9.1.1 and 9.1.2].

Since $\frac{df}{dx} = \lim_{\Delta x \to 0} \frac{f(x+\Delta x)-f(x)}{\Delta x} = \frac{f(x+\delta x)-f(x)}{\delta x}$, we can write

$$f(x+\delta x)-f(x) = \frac{df}{dx}\delta x$$

If we apply this relation to Eq. (9.1.2), the right-hand side of the equation is

$$\int_{t_1}^{t_2}\left(\frac{dL}{dq}\delta q + \frac{dL}{d\dot{q}}\delta\dot{q}\right)dt = \int_{t_1}^{t_2}\frac{dL}{dq}\delta q dt + \int_{t_1}^{t_2}\frac{dL}{d\dot{q}}\left(\frac{d}{dt}\delta q\right)dt$$

The second term on the right-hand side of the equation by using integration by part gives

$$\int_{t_1}^{t_2}\frac{dL}{d\dot{q}}\left(\frac{d}{dt}\delta q\right)dt = \frac{dL}{d\dot{q}}\delta q\Big|_{t_1}^{t_2} - \int_{t_1}^{t_2}\frac{d}{dt}\left(\frac{dL}{d\dot{q}}\right)\delta q dt = -\int_{t_1}^{t_2}\frac{d}{dt}\left(\frac{dL}{d\dot{q}}\right)\delta q dt$$

because $\delta q(t_1) = \delta q(t_2) = 0$.

Then, Eq. (9.1.2) gives

$$0 = \int_{t_1}^{t_2}\frac{dL}{dq}\delta q dt - \int_{t_1}^{t_2}\frac{d}{dt}\left(\frac{dL}{d\dot{q}}\right)\delta q dt$$

$$= \int_{t_1}^{t_2}\left[\frac{dL}{dq} - \frac{d}{dt}\left(\frac{dL}{d\dot{q}}\right)\right]\delta q dt$$

Since the equation satisfies for all δq, we obtain

$$\frac{dL(q,\dot{q},t)}{dq} - \frac{d}{dt}\left[\frac{dL(q,\dot{q},t)}{d\dot{q}}\right] = 0$$

which is called **Lagrange's equation**.

This equation can be generalized into

$$\frac{\partial L(q_i,\dot{q}_i,t)}{\partial q_i} - \frac{d}{dt}\left[\frac{\partial L(q_i,\dot{q}_i,t)}{\partial \dot{q}_i}\right] = 0,$$

where $i = 1, 2, \cdots, n$ for the degree of freedom.

Example

9.1.1 Consider a particle that moves along the surface of a sphere with radius a. Find the number of the degree of freedom of the system.

Ans. This yields an equation of constraint

$$(x^2 + y^2 + z^2) - a^2 = 0,$$

which is the holonomic constraint.

Then, the number of the degree of freedom is $(3 \times 1) - 1 = 2$. Thus, we can express the generalized coordinates as θ and ϕ for describing this particle movement.

9.1.2 Consider a dumbbell and find the number of the degree of freedom.

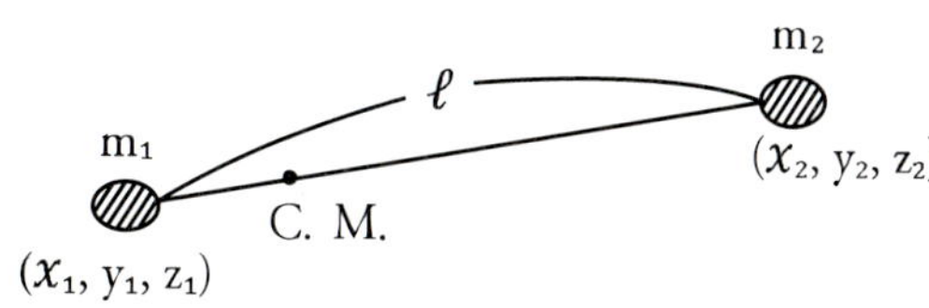

Ans. There is one constraint equation:

$$[(x_1 - x_2)^2 + (y_1 - y_2)^2 + (z_1 - z_2)^2] - \ell^2 = 0$$

We need a total of six coordinates, $(x_1, y_1, z_1, x_2, y_2, z_2)$, to describe two particles, three coordinates for each.

The number of the degree of freedom is then $(3 \times 2) - 1 = 5$. Thus, we need five generalized coordinates to uniquely describe this system and can choose them as (X, Y, Z, θ, ϕ), where (X, Y, Z) are the coordinates for the center-of-mass (C.M.) of the dumbbell and (θ, ϕ) are for its orientation.

9.1.3 Find an equation of motion for a free particle of mass m in one-dimension.

Ans. Let's start with constructing the Lagrangian

$$L = K - U = \frac{1}{2} m\dot{x}^2 - 0 = \frac{1}{2} m\dot{x}^2$$

From the Lagrange's equation

$$\frac{dL}{dx} - \frac{d}{dt}\left(\frac{dL}{d\dot{x}}\right) = 0,$$

we have

$$0 = 0 - \frac{d}{dt}\left[\frac{d}{d\dot{x}}\left(\frac{1}{2} m\dot{x}^2\right)\right] = -\frac{d}{dt}(m\dot{x})$$

Thus, we obtain the equation of motion $m\ddot{x} = 0$. It means the velocity of the particle is constant (or the momentum is the constant) throughout the motion.

9.1.4 Find the shortest path between two points x_1 and x_2.

Ans. Since $ds = \sqrt{dx^2 + dy^2}$ is the length for a small variation in the plane, a path between two points, x_1 and x_2, is

$$S = \int_{x_1}^{x_2} ds = \int_{x_1}^{x_2} \sqrt{dx^2 + dy^2} = \int_{x_1}^{x_2} \sqrt{1 + (\frac{dy}{dx})^2}\, dx = \int_{x_1}^{x_2} \sqrt{1 + y'^2}\, dx$$

By the least action principle, the actual path is for $\delta S = 0$. By comparing this equation with Eq. (9.1.1) we know that t, q, and $\dot{q}$ in Eq. (9.1.1) correspond to x, y, and y', respectively. Therefore, $L(q, \dot{q}, t)$ corresponds to $f(y, y', x)$.

If we define $f(y, y', x) = \sqrt{1 + y'^2}$, the Lagrange's equation is then expressed as

$$\frac{df}{dy} - \frac{d}{dx}(\frac{df}{dy'}) = 0$$

Then, we get

$$0 = \frac{df}{dy} - \frac{d}{dx}(\frac{df}{dy'}) = 0 - \frac{d}{dx}(\frac{y'}{\sqrt{1 + y'^2}}),$$

which gives

$$\frac{d}{dx}(\frac{y'}{\sqrt{1 + y'^2}}) = 0$$

It means that $\frac{y'}{\sqrt{1+y'^2}}$ is a constant with respect to x.

By defining this constant a, we obtain

$$y' = \frac{dy}{dx} = \pm \frac{a}{\sqrt{1 - a^2}},$$

where the left- and right-hand sides are the slope and constant, respectively. Thus, this equation corresponds to a straight line.

9.1.5 Light travels with velocity $v_1 (v_2)$ in a medium $1(2)$ from points $(x_1,\ y_1)$ to $(x_2,\ y_2)$ assuming light travels in a straight line and an arbitrary crossing point $(x,\ 0)$. Show that $\frac{\sin\theta_1}{v_1} = \frac{\sin\theta_2}{v_2}$.

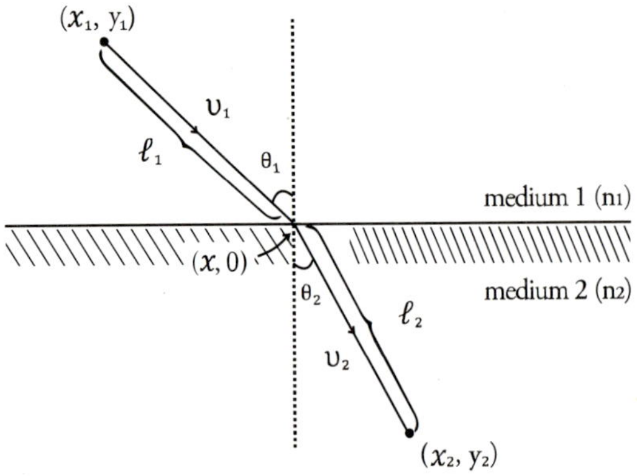

Ans. The time of traveling from (x_1, y_1) to (x_2, y_2) is

$$t_{tot}(x) = t_1 + t_2 = \frac{\ell_1}{v_1} + \frac{\ell_2}{v_2}$$
$$= \frac{\sqrt{(x-x_1)^2 + y_1^2}}{v_1} + \frac{\sqrt{(x_2-x)^2 + y_2^2}}{v_2}$$

Let's find a path which takes the smallest traveling time:

$$0 = \frac{dt_{tot}(x)}{dx} = \frac{1}{v_1}\frac{x-x_1}{\sqrt{(x-x_1)^2+y_1^2}} - \frac{1}{v_2}\frac{x_2-x}{\sqrt{(x_2-x)^2+y_2^2}}$$
$$= \frac{1}{v_1}\left(\frac{x-x_1}{\ell_1}\right) - \frac{1}{v_2}\left(\frac{x_2-x}{\ell_2}\right) = \frac{\sin\theta_1}{v_1} - \frac{\sin\theta_2}{v_2}$$

This gives

$$\frac{\sin\theta_1}{v_1} = \frac{\sin\theta_2}{v_2},$$

which is **Snell's law**.

Since an index of refraction is defined as $n_i = \frac{c}{v_i}$, the Snell's law gives

$$n_1 \sin\theta_1 = n_2 \sin\theta_2$$

9.1.6 Find an equation of motion for a particle of mass m moving under the gravitational field.

Ans. The Lagrangian is

$$L = K - U = \frac{1}{2}mv^2 - U = \frac{1}{2}m(\dot{x}^2 + \dot{y}^2 + \dot{z}^2) - mgz$$

From the Lagrange's equations

$$\frac{\partial L}{\partial x} - \frac{d}{dt}\left(\frac{\partial L}{\partial \dot{x}}\right) = 0,\ \frac{\partial L}{\partial y} - \frac{d}{dt}\left(\frac{\partial L}{\partial \dot{y}}\right) = 0,\ \text{and}\ \frac{\partial L}{\partial z} - \frac{d}{dt}\left(\frac{\partial L}{\partial \dot{z}}\right) = 0,$$

we have

$$\frac{d}{dt}(m\dot{x}) = 0 \Rightarrow \dot{x} = \text{const.},$$

$$\frac{d}{dt}(m\dot{y}) = 0 \Rightarrow \dot{y} = \text{const.},$$

and

$$\frac{d}{dt}(m\dot{z}) + mg = 0 \Rightarrow \ddot{z} = -g,$$

respectively.

9.1.7 Find an equation of motion for the one-dimensional harmonic oscillator.

Ans. The Lagrangian is

$$L = K - U = \frac{1}{2}m\dot{x}^2 - \frac{1}{2}kx^2$$

The Lagrange's equation gives

$$0 = \frac{dL}{dx} - \frac{d}{dt}\left(\frac{dL}{d\dot{x}}\right) = -kx - m\ddot{x}$$

Then, we obtain

$$m\ddot{x} + kx = 0$$

This corresponds to the equation of motion for the undamped harmonic oscillator and gives

$$\ddot{x} + \omega_0^2 x = 0,$$

where $\omega_0^2 = \sqrt{\dfrac{k}{m}}$.

The general solution of the equation is then

$$x(t) = x_0 \sin(\omega_0 t + \phi)$$

9.1.8 Consider the simple pendulum of mass m and length ℓ, and find an equation of motion.

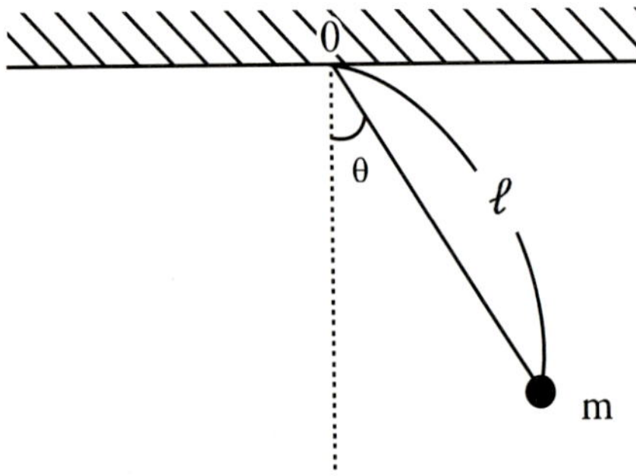

Ans. Since $x = \ell\sin\theta$ and $y = -\ell\cos\theta$, their time derivatives are $\dot{x} = \ell\dot{\theta}\cos\theta$ and $\dot{y} = \ell\dot{\theta}\sin\theta$.

The Lagrangian is then

$$L = K - U = \frac{1}{2}m(\dot{x}^2 + \dot{y}^2) - mgy = \frac{1}{2}m\ell^2\dot{\theta}^2 + mg\ell\cos\theta$$

The Lagrange's equation gives

$$0 = \frac{dL}{d\theta} - \frac{d}{dt}\left(\frac{dL}{d\dot{\theta}}\right) = -mg\ell\sin\theta - m\ell^2\ddot{\theta}$$

Then, the equation of the motion is

$$\ddot{\theta}+\frac{g}{\ell}\sin\theta=0$$

For a small angle, we have $\sin\theta \approx \theta$.
The equation of motion is then expressed as

$$\ddot{\theta}+\frac{g}{\ell}\theta=0$$

Its solution is

$$\theta(t)=\theta_0\sin(\omega t+\phi),$$

where $\omega=\sqrt{\frac{g}{\ell}}$.

For the period T of the oscillator, we have

$$\theta(t)=\theta(t+T) \Rightarrow \sin(\omega t+\phi)=\sin(\omega t+\omega T+\phi)$$

so that $\omega T=2\pi$.
Thus, the period is then

$$T=\frac{2\pi}{\omega}=2\pi\sqrt{\frac{\ell}{g}}$$

9.1.9 Find an equation of the motion of the Atwood machine with string length ℓ and pulley of radius a, ignoring the masses of the string and pulley, and assuming $m_2 > m_1$.

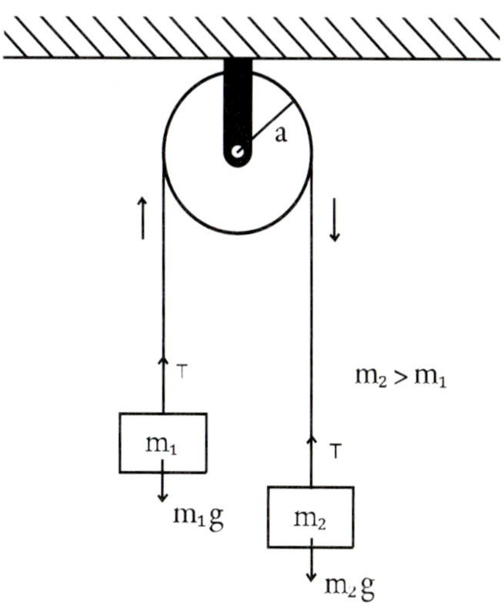

Ans. We need two coordinates x_1 and x_2 to completely describe the system. Since there is one holonomic constraint equation

$$f(x_1, x_2) = x_1 + \pi a + x_2 - \ell = 0,$$

the number of the degree of freedom is then $(1 \times 2) - 1 = 1$ and let's define x as the generalized coordinate.

The kinetic and potential energies of the system are

$$K = \frac{1}{2}m_1\dot{x}^2 + \frac{1}{2}m_2(-\dot{x})^2 \text{ and } U = m_1 g x + m_2 g(\ell - \pi a - x)$$

The Lagrangian is then

$$L = K - U = \frac{1}{2}m_1\dot{x}^2 + \frac{1}{2}m_2\dot{x}^2 - m_1 g x - m_2 g(\ell - \pi a - x)$$

It gives the Lagrange's equation

$$0 = \frac{dL}{dx} - \frac{d}{dt}\left(\frac{dL}{d\dot{x}}\right) = (m_2 - m_1)g - (m_1 + m_2)\ddot{x},$$

which yields

$$\ddot{x} = \frac{m_2 - m_1}{m_1 + m_2}g$$

Tension T has the following relation with the acceleration $\ddot{x}$:

$$T - m_1 g = m_1\ddot{x}$$

Then, we obtain the tension

$$T = \frac{2m_1 m_2}{m_1 + m_2}g$$

9.1.10 Consider a particle of mass m moving under the central force field $\vec{F} = F(r)\hat{r}$ in the $x-y$ plane. Find equations of the motion.

Ans. From [Ex. 4.1.2], we know that

$$\dot{\vec{r}} = \dot{r}\hat{r} + r\dot{\hat{r}} = \dot{r}\hat{r} + r\dot{\theta}\hat{\theta}$$

The kinetic and potential energies are then

$$K = \frac{1}{2}m\dot{r}^2 = \frac{1}{2}m\dot{\vec{r}} \cdot \dot{\vec{r}} = \frac{1}{2}m(\dot{r}^2 + r^2\dot{\theta}^2)$$

and

$$U = U(r),$$

respectively.

This gives the Lagrangian,

$$L = K - U = \frac{1}{2}m(\dot{r}^2 + r^2\dot{\theta}^2) - U(r)$$

The Lagrange's equations give

$$0 = \frac{\partial L}{\partial r} - \frac{d}{dt}\left(\frac{\partial L}{\partial \dot{r}}\right) = mr\dot{\theta}^2 - \frac{dU(r)}{dr} - m\ddot{r} \tag{9.1.10.1}$$

and

$$0 = \frac{\partial L}{\partial \theta} - \frac{d}{dt}\left(\frac{\partial L}{\partial \dot{\theta}}\right) = -\frac{d}{dt}(mr^2\dot{\theta}) \tag{9.1.10.2}$$

Eq. (9.1.10.2) for the angular component gives $mr^2\dot{\theta}$=constant. Since this constant has the unit of the angular momentum L, let's define $r^2\dot{\theta} = \frac{L}{m} = \ell$, where ℓ is the angular momentum per mass. The physical meaning of constant $mr^2\dot{\theta} = constant$ is that the angular momentum is conserved throughout the motion.
Eq. (9.1.10.1) is then

$$m\ddot{r} = mr\dot{\theta}^2 - \frac{dU}{dr} = mr\dot{\theta}^2 + F(r) = m\frac{\ell^2}{r^3} + F(r)$$

This is the equation of the motion for the radial component.

9.1.11 Consider a pendulum of mass m_1 suspended with length ℓ to support mass m_2 which freely moves in the $x-$direction only. Find equations of the motion assuming there is no frictional force on the surface.

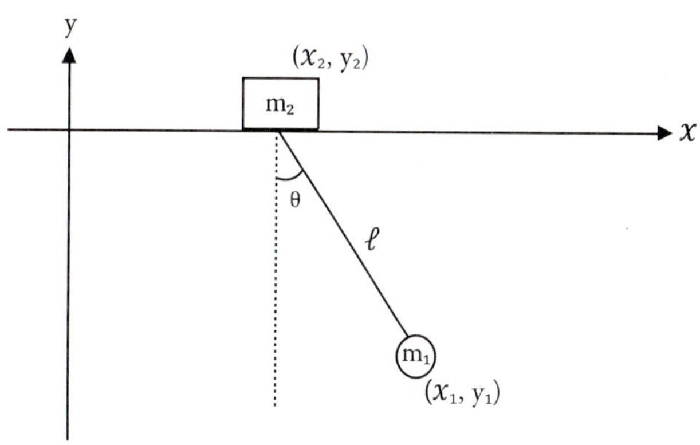

Ans. We need four coordinates, (x_1, y_1) and (x_2, y_2), to describe m_1 and m_2, respectively. Two constraint equations, $f_1 = y_2 = 0$ and $f_2 = [(x_1 - x_2)^2 + y_1^2] - \ell^2 = 0$, exist for the system.
Therefore, two generalized coordinates are needed to uniquely describe the motion of the system. We define them as x_2 and θ, which are for the support movement along the $x-$axis and for the pendulum's swing in $x-y$ plane, respectively. Then, the kinetic and potential energies are

$$K = \frac{1}{2}m_2\dot{x}_2^2 + \frac{1}{2}m_1(\dot{x}_1^2 + \dot{y}_1^2), \text{ where } \dot{x}_1 = \dot{x}_2 + \ell\dot{\theta}\cos\theta \text{ and } \dot{y}_1 = \ell\dot{\theta}\sin\theta$$

and

$$U = m_1 g y_1 = - m_1 g \ell \cos\theta,$$

respectively.

The Lagrangian is then

$$L = \frac{1}{2} m_2 \dot{x}_2^{\,2} + \frac{1}{2} m_1 (\dot{x}_2^{\,2} + \ell^2 \dot{\theta}^2 + 2\ell \dot{x}_2 \dot{\theta} \cos\theta) + m_1 g \ell \cos\theta,$$

(i) For x_2, we have

$$\frac{\partial L}{\partial x_2} = 0$$

and

$$\frac{d}{dt}\left(\frac{\partial L}{\partial \dot{x}_2}\right) = \frac{d}{dt}[(m_1 + m_2)\dot{x}_2 + m_1 \ell \dot{\theta} \cos\theta]$$

The Lagrange's equation gives

$$0 = \frac{\partial L}{\partial x_2} - \frac{d}{dt}\left(\frac{\partial L}{\partial \dot{x}_2}\right) = - \frac{d}{dt}[(m_1 + m_2)\dot{x}_2 + m_1 \ell \dot{\theta} \cos\theta]$$ [22]

which is

$$(m_1 + m_2)\dot{x}_2 + m_1 \ell \dot{\theta} \cos\theta = \text{const}.$$

(ii) For θ, we have

$$\frac{\partial L}{\partial \theta} = - m_1 \ell \dot{x}_2 \dot{\theta} \sin\theta - m_1 g \ell \sin\theta = - m_1 \ell (\dot{x}_2 \dot{\theta} + g) \sin\theta$$

and

$$\frac{d}{dt}\left(\frac{\partial L}{\partial \dot{\theta}}\right) = \frac{d}{dt}(m_1 \ell^2 \dot{\theta} + m_1 \ell \dot{x}_2 \cos\theta) = m_1 \ell^2 \ddot{\theta} + m_1 \ell \ddot{x}_2 \cos\theta - m_1 \ell \dot{x}_2 \dot{\theta} \sin\theta$$
$$= m_1 \ell (\ell \ddot{\theta} + \ddot{x}_2 \cos\theta - \dot{x}_2 \dot{\theta} \sin\theta)$$

Then, the Lagrange's equation gives

$$0 = \frac{\partial L}{\partial \theta} - \frac{d}{dt}\left(\frac{\partial L}{\partial \dot{\theta}}\right) = - m_1 \ell (\dot{x}_2 \dot{\theta} + g) \sin\theta - m_1 \ell (\ell \ddot{\theta} + \ddot{x}_2 \cos\theta - \dot{x}_2 \dot{\theta} \sin\theta)$$

so that we obtain

$$\ell \ddot{\theta} + \ddot{x}_2 \cos\theta + g \sin\theta = 0$$

9.1.12 Find the equations of motion of a falling disk with mass m and radius a from an attached string. We ignore the string mass.

---

[22] This result is used in [Example 9.2.2].

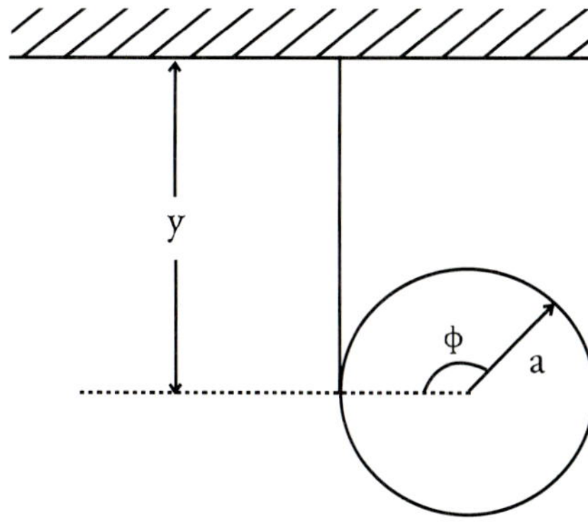

Ans. The kinetic energy of the system is

$$K = \frac{1}{2}m\dot{y}^2 + \frac{1}{2}I\dot{\phi}^2,$$

where $I = \frac{1}{2}ma^2$ is the moment of inertia of the disk.
The potential energy is

$$U = -mgy$$

Since $y = a\phi$, the kinetic and potential energies are

$$K = \frac{1}{2}ma^2\dot{\phi}^2 + \frac{1}{4}ma^2\dot{\phi}^2 \text{ and } U = -mga\phi,$$

respectively.
The Lagrangian is then

$$L = K - U = \frac{3}{4}ma^2\dot{\phi}^2 + mga\phi$$

so that the Lagrange's equation gives

$$0 = \frac{dL}{d\phi} - \frac{d}{dt}\left(\frac{dL}{d\dot{\phi}}\right) = mga - \frac{3}{2}ma^2\ddot{\phi},$$

which yields

$$\ddot{\phi} = \frac{2g}{3a} \text{ (or } \ddot{y} = a\ddot{\phi} = \frac{2}{3}g)$$

9.1.13 Derive the equation in [Ex. 6.3.1] for the coupled harmonic oscillator using the Lagrange's equations.

Ans. The kinetic and potential energies of the system are

$$K = \frac{1}{2}m\dot{x}_1^2 + \frac{1}{2}m\dot{x}_2^2$$

and

$$U=\frac{1}{2}k_1x_1^2+\frac{1}{2}k_2(x_2-x_1)^2+\frac{1}{2}k_1x_2^2,$$

respectively.

The Lagrangian is then

$$L=K-U=\frac{1}{2}m\dot{x}_1^2+\frac{1}{2}m\dot{x}_2^2-\frac{1}{2}k_1x_1^2-\frac{1}{2}k_2(x_2-x_1)^2-\frac{1}{2}k_1x_2^2$$

From this equation, we have

$$\frac{\partial L}{\partial x_1}=-k_1x_1+k_2(x_2-x_1),\ \frac{\partial L}{\partial x_2}=-k_2(x_2-x_1)-k_1x_2,$$

$$\frac{\partial L}{\partial \dot{x}_1}=m\dot{x}_1,\ \text{and}\ \frac{\partial L}{\partial \dot{x}_2}=m\dot{x}_2$$

Then, the Lagrange's equations give two equations,

$$m\ddot{x}_1+(k_1+k_2)x_1-k_2x_2=0 \tag{9.1.13.1}$$

and

$$m\ddot{x}_2-k_2x_1+(k_1+k_2)x_2=0 \tag{9.1.13.2}$$

We expect each oscillator has a motion with same frequency ω and phase ϕ but different amplitude, so that we can define $x_1=a_1\cos(\omega t+\phi)$ and $x_2=a_2\cos(\omega t+\phi)$. This gives $\ddot{x}_1=-a_1\omega^2\cos(\omega t+\phi)$ and $\ddot{x}_2=-a_2\omega^2\cos(\omega t+\phi)$.

By substituting these into Eqs. (9.1.13.1) and (9.1.13.2), we get

$$(k_1+k_2)a_1-k_2a_2=m\omega^2a_1 \tag{9.1.13.3}$$

and

$$-k_2a_1+(k_1+k_2)a_2=m\omega^2a_2, \tag{9.1.13.4}$$

respectively.

We can express these two equations in the matrix representation:

$$\begin{pmatrix} k_1+k_2 & -k_2 \\ -k_2 & k_1+k_2 \end{pmatrix}\begin{pmatrix} a_1 \\ a_2 \end{pmatrix}=m\omega^2\begin{pmatrix} a_1 \\ a_2 \end{pmatrix}$$

If we define the constant $m\omega^2=\lambda$, we have

$$\begin{pmatrix} k_1+k_2 & -k_2 \\ -k_2 & k_1+k_2 \end{pmatrix}\begin{pmatrix} a_1 \\ a_2 \end{pmatrix}=\lambda\begin{pmatrix} a_1 \\ a_2 \end{pmatrix}$$

This is the equation in [Ex. 6.3.1].

9.2 Lagrange Multipliers Method

- Lagrange undetermined multiplier method for m constraint equations gives

$$\frac{\partial L}{\partial q_i}-\frac{d}{dt}\left(\frac{\partial L}{\partial \dot{q}_i}\right)+\sum_{k=1}^{m}\lambda_k(t)\frac{\partial f_k}{\partial q_i}=0,$$

where $\lambda_k \frac{\partial f_k}{\partial q_i}$ is the generalized forces of constraint.

We have n independent Lagrange's equations for a n degrees of freedom of a system. On the other hand we can modify Lagrangian formalism to include equations of constraint. Equations derived this way are not independent but the **Lagrange undetermined multiplier method** makes these equations independent. This technique also provides information for **generalized forces of constraint** which can not be calculable from conserved potential functions.

For simplicity, consider a system which is described with generalized coordinates q_1 and q_2, and has a holonomic constraint equation $f(q_1, q_2, t)=0$.

The least action principle gives

$$0=\delta\int_{t_1}^{t_2} L(q_1,q_2,\dot{q}_1,\dot{q}_2,t)dt = \int_{t_1}^{t_2}\sum_{i=1}^{2}\left[\frac{\partial L}{\partial q_i}-\frac{d}{dt}\left(\frac{\partial L}{\partial \dot{q}_i}\right)\right]\delta q_i dt$$

$$=\int_{t_1}^{t_2}\left[\frac{\partial L}{\partial q_1}-\frac{d}{dt}\left(\frac{\partial L}{\partial \dot{q}_1}\right)\right]\delta q_1 dt+\int_{t_1}^{t_2}\left[\frac{\partial L}{\partial q_2}-\frac{d}{dt}\left(\frac{\partial L}{\partial \dot{q}_2}\right)\right]\delta q_2 dt \qquad (9.2.1)$$

If the constraint equation does not depend on the time explicitly, the total differential of the constraint equation is

$$\delta f=\frac{\partial f}{\partial q_1}\delta q_1+\frac{\partial f}{\partial q_2}\delta q_2=0$$

This equation gives

$$\delta q_2 = -\frac{\partial f/\partial q_1}{\partial f/\partial q_2}\delta q_1 \tag{9.2.2}$$

If we insert Eqs. (9.2.2) into (9.2.1), we have

$$0 = \int_{t_1}^{t_2}[\left\{\frac{\partial L}{\partial q_1} - \frac{d}{dt}(\frac{\partial L}{\partial \dot{q}_1})\right\} - \left\{\frac{\partial L}{\partial q_2} - \frac{d}{dt}(\frac{\partial L}{\partial \dot{q}_2})\right\}\frac{\partial f/\partial q_1}{\partial f/\partial q_2}]\delta q_1 dt$$

Since δq_1 is connected with δq_2 by the variation in the constraint equation and the equation should be satisfied for all δq_i, we get

$$\frac{\partial L}{\partial q_1} - \frac{d}{dt}(\frac{\partial L}{\partial \dot{q}_1}) = [\frac{\partial L}{\partial q_2} - \frac{d}{dt}(\frac{\partial L}{\partial \dot{q}_2})]\frac{\partial f/\partial q_1}{\partial f/\partial q_2}$$

It gives

$$\frac{\frac{\partial L}{\partial q_1} - \frac{d}{dt}(\frac{\partial L}{\partial \dot{q}_1})}{\partial f/\partial q_1} = \frac{\frac{\partial L}{\partial q_2} - \frac{d}{dt}(\frac{\partial L}{\partial \dot{q}_2})}{\partial f/\partial q_2}$$

The left- and right-hand sides of the equation are functions of $(q_1, \dot{q}_1)$ and $(q_2, \dot{q}_2)$, respectively. When result of each side is a function of time only, the equality satisfies for all times. Let's define this function is $-\lambda(t)$ which is called an **undetermined multiplier**.

Then, we obtain

$$\frac{\partial L}{\partial q_1} - \frac{d}{dt}(\frac{\partial L}{\partial \dot{q}_1}) + \lambda(t)\frac{\partial f}{\partial q_1} = 0 \text{ and } \frac{\partial L}{\partial q_2} - \frac{d}{dt}(\frac{\partial L}{\partial \dot{q}_2}) + \lambda(t)\frac{\partial f}{\partial q_2} = 0$$

We need three equations to solve three unknown quantities q_1, q_2, and λ. Since we have these two equations and there is one constraint equation $f(q_1, q_2, t) = 0$, the requirement for solving the problem is fulfilled. These two equations can be in general expressed as

$$\frac{\partial L}{\partial q_i} - \frac{d}{dt}(\frac{\partial L}{\partial \dot{q}_i}) + \lambda(t)\frac{\partial f}{\partial q_i} = 0, \tag{9.2.3}$$

where the Lagrange multiplier term $\lambda(t)\frac{\partial f}{\partial q_i}$ is called the **generalized forces of constraint**.

If we generalize Eq. (9.2.3) for m constraint equations, we obtain

$$\frac{\partial L}{\partial q_i}-\frac{d}{dt}\left(\frac{\partial L}{\partial \dot{q}_i}\right)+\sum_{k=1}^{m}\lambda_k(t)\frac{\partial f_k}{\partial q_i}=0 \tag{9.2.4}$$

Example

9.2.1 We considered the Atwood machine with a string length ℓ ignoring the pulley's mass in [Ex. 9.1.9]. Solve the same problem using the Lagrange undetermined multiplier method and find the generalized forces of constraint.

Ans. From Eq. (9.2.3), we have

$$\frac{\partial L}{\partial x_1}-\frac{d}{dt}\left(\frac{\partial L}{\partial \dot{x}_1}\right)+\lambda(t)\frac{\partial f}{\partial x_1}=0$$

and

$$\frac{\partial L}{\partial x_2}-\frac{d}{dt}\left(\frac{\partial L}{\partial \dot{x}_2}\right)+\lambda(t)\frac{\partial f}{\partial x_2}=0,$$

Since the Lagrangian and constraint equation are

$$L=\frac{1}{2}m_1\dot{x}_1^{\,2}+\frac{1}{2}m_2\dot{x}_2^{\,2}+(m_1gx_1+m_2gx_2)$$

and

$$f(x_1,x_2)=x_1+\pi a+x_2-\ell$$

respectively,

we get

$$m_1g-m_1\ddot{x}_1=-\lambda \tag{9.2.1.1}$$

and

$$m_2g-m_2\ddot{x}_2=-\lambda \tag{9.2.1.2}$$

Since $\ddot{x}_2=-\ddot{x}_1$, Eq. (9.2.1.2) gives

$$m_2g+m_2\ddot{x}_1=-\lambda \tag{9.2.1.3}$$

If we subtract Eqs. (9.2.1.3) from (9.2.1.1), we obtain

$$\ddot{x}_1=\frac{m_1-m_2}{m_1+m_2}g, \tag{9.2.1.4}$$

which gives

$$\ddot{x}_2 = -\ddot{x}_1 = \frac{m_2 - m_1}{m_1 + m_2} g \tag{9.2.1.5}$$

We can determine the multiplier λ by inserting Eqs. (9.2.1.4) into (9.2.1.1):

$$\lambda = -m_1 g + m_1 \frac{m_1 - m_2}{m_1 + m_2} g = -\frac{2m_1 m_2}{m_1 + m_2} g$$

The generalized forces of constraint are then

$$\lambda \frac{\partial f}{\partial x_1} = -\frac{2m_1 m_2}{m_1 + m_2} g$$

and

$$\lambda \frac{\partial f}{\partial x_2} = -\frac{2m_1 m_2}{m_1 + m_2} g$$

Since the right-hand side of the equation has the unit of mass times acceleration, it is clear that the multiplier λ has the unit of force.

Thus, we know that the generalized force of constraint is the string tension which is obtained in [Ex. 9.1.9]. The negative sign indicates that the direction of the tension is opposite to the direction of the gravitational acceleration.

9.2.2 Find equations of the motion of the system in [Ex. 9.1.11] by using the Lagrange undetermined multiplier method and confirm that this technique gives the same answer.

Ans. By using coordinates (x_1, y_1) and (x_2, y_2) to describe m_1 and m_2, the Lagrangian is given by

$$L = K - U = \frac{1}{2} m_1 (\dot{x}_1^2 + \dot{y}_1^2) + \frac{1}{2} m_2 \dot{x}_2^2 - m_1 g y_1$$

Since there are two constraint equations $f_1(x_1, y_1) = [(x_1 - x_2)^2 + y_1^2] - \ell^2$ and $f_2(x_2, y_2) = y_2$, Eq. (9.2.4) gives

$$\begin{cases} \dfrac{\partial L}{\partial x_1} - \dfrac{d}{dt}\left(\dfrac{\partial L}{\partial \dot{x}_1}\right) + \left(\lambda_1 \dfrac{\partial f_1}{\partial x_1} + \lambda_2 \dfrac{\partial f_2}{\partial x_1}\right) = 0 \\ \dfrac{\partial L}{\partial x_2} - \dfrac{d}{dt}\left(\dfrac{\partial L}{\partial \dot{x}_2}\right) + \left(\lambda_1 \dfrac{\partial f_1}{\partial x_2} + \lambda_2 \dfrac{\partial f_2}{\partial x_2}\right) = 0 \\ \dfrac{\partial L}{\partial y_1} - \dfrac{d}{dt}\left(\dfrac{\partial L}{\partial \dot{y}_1}\right) + \left(\lambda_1 \dfrac{\partial f_1}{\partial y_1} + \lambda_2 \dfrac{\partial f_2}{\partial y_1}\right) = 0 \\ \dfrac{\partial L}{\partial y_2} - \dfrac{d}{dt}\left(\dfrac{\partial L}{\partial \dot{y}_2}\right) + \left(\lambda_1 \dfrac{\partial f_1}{\partial y_2} + \lambda_2 \dfrac{\partial f_2}{\partial y_2}\right) = 0 \end{cases} \tag{9.2.2.1}$$

The last equation in Eq. (9.2.2.1) does not give any meaningful result because the left-hand side of the equation is zero.

If we calculate the rest of equations in Eq. (9.2.2.1) by inserting the Lagrangian and constraint equations, we get

$$\begin{cases} -m_1\ddot{x}_1+2\lambda_1(x_1-x_2)=0 \\ -m_2\ddot{x}_2-2\lambda_1(x_1-x_2)=0 \\ -m_1g-m_1\ddot{y}_1+2\lambda_1 y_1=0 \end{cases} \tag{9.2.2.2}$$

We can find the undetermined multiplier λ_1 by subtracting the second equation from the first equation of Eq. (9.2.2.2):

$$m_2\ddot{x}_2-m_1\ddot{x}_1+4\lambda_1(x_1-x_2)=0$$

This gives

$$\lambda_1=\frac{m_1\ddot{x}_1-m_2\ddot{x}_2}{4(x_1-x_2)} \tag{9.2.2.3}$$

By substituting Eq. (9.2.2.3) into the first equation of Eq. (9.2.2.2), we get

$$m_1\ddot{x}_1+m_2\ddot{x}_2=0,$$

which gives

$$\ddot{x}_1=-\frac{m_2}{m_1}\ddot{x}_2 \tag{9.2.2.4}$$

If we substitute Eqs. (9.2.2.3) and (9.2.2.4) into the last equation of Eq. (9.2.2.2), we obtain

$$-m_1g-m_1\ddot{y}_1-\frac{m_2\ddot{x}_2}{x_1-x_2}y_1=0 \tag{9.2.2.5}$$

Now, let's check whether this equation gives the same result as that obtained in [Ex. 9.1.11].

In this problem, we know that $x_1-x_2=\ell\sin\theta$ and $y_1=-\ell\cos\theta$.

It gives

$$\dot{y}_1=\ell\dot{\theta}\sin\theta \Rightarrow \ddot{y}_1=\ell\ddot{\theta}\sin\theta+\ell\dot{\theta}^2\cos\theta$$

If we substitute these into Eq. (9.2.2.5), we have

$$m_1g\sin\theta+m_1\ell\ddot{\theta}\sin^2\theta+m_1\ell\dot{\theta}^2\cos\theta\sin\theta-m_2\ddot{x}_2\cos\theta=0 \tag{9.2.2.6}$$

We obtained the following result for x_2 in [Ex. 9.1.11]:

$$\frac{d}{dt}[(m_1+m_2)\dot{x}_2+m_1\ell\dot{\theta}\cos\theta]=0$$

By performing the differentiation, this gives

$$m_1\ell\dot{\theta}^2\sin\theta\cos\theta=(m_1+m_2)\ddot{x}_2\cos\theta+m_1\ell\ddot{\theta}\cos^2\theta$$

If we substitute this into (9.2.2.6), we obtain

$$m_1 g\sin\theta + m_1\ell\ddot{\theta}\sin^2\theta + m_1\ddot{x}_2\cos\theta + m_1\ell\ddot{\theta}\cos^2\theta = 0,$$

which is the same equation as

$$\ell\ddot{\theta} + \ddot{x}_2\cos\theta + g\sin\theta = 0,$$

obtained in [Ex 9.1.11]

Therefore, this example shows that it is much simpler to set up independent Lagrange's equations as long as problems do not request to find the generalized forces of constraint.

9.2.3 A particle of mass m is sliding down from the top of a semi-vertical hoop of radius a. Find the height using the Lagrange undetermined multiplier method when a particle falls off.

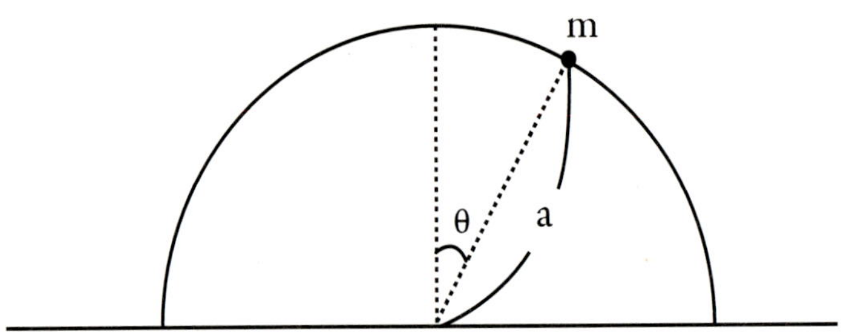

Ans. This system can be described in the plane polar coordinates. The velocity is then $v^2 = \dot{r}^2 + r^2\dot{\theta}^2$ and the constraint equation is $f(r) = r - a$.

The Lagrangian is then

$$L = \frac{1}{2}m(\dot{r}^2 + r^2\dot{\theta}^2) - mgr\cos\theta$$

From the Lagrange undetermined multiplier method

$$\frac{\partial L}{\partial r} - \frac{d}{dt}\left(\frac{\partial L}{\partial \dot{r}}\right) + \lambda\frac{\partial f}{\partial r} = 0 \text{ and } \frac{\partial L}{\partial \theta} - \frac{d}{dt}\left(\frac{\partial L}{\partial \dot{\theta}}\right) + \lambda\frac{\partial f}{\partial \theta} = 0,$$

we get

$$mr\dot{\theta}^2 - mg\cos\theta - m\ddot{r} + \lambda = 0 \qquad (9.2.3.1)$$

and

$$mgr\sin\theta - \frac{d}{dt}(mr^2\dot{\theta}) = 0 \qquad (9.2.3.2)$$

From Eq. (9.2.3.1), we get

$$\lambda = m\ddot{r} + mg\cos\theta - mr\dot{\theta}^2$$

Since we have $\dot{r} = 0$ and $\ddot{r} = 0$, we get the undetermined multiplier

$$\lambda = mg\cos\theta - ma\dot{\theta}^2 \tag{9.2.3.3}$$

because $r = a$ from the constraint equation.

We have the following relation from the energy conservation law:

$$mga = \frac{1}{2}ma^2\dot{\theta}^2 + mga\cos\theta,$$

which gives

$$\dot{\theta}^2 = \frac{2g}{a}(1 - \cos\theta)$$

By inserting this into Eq. (9.2.3.3), we get

$$\lambda = 3mg\cos\theta - 2mg$$

The generalized force of constraint is then

$$\lambda\frac{\partial f}{\partial r} = \lambda = 3mg\cos\theta - 2mg,$$

which is the constraint force to make the particle moving on the hoop. Therefore, when this force is zero, the particle leaves the hoop:

$$0 = 3mg\cos\theta - 2mg \Rightarrow \cos\theta = \frac{2}{3}$$

Thus, the height when the particle falls off is

$$a\cos\theta = \frac{2}{3}a$$

9.2.4 Find the equations of motion for a disk with radius a rolling down an incline plane and find the forces of constraints keep the disk from sliding. The constraint equation is $f = x - a\phi$. The moment of inertia of the disk is given as $I = \frac{1}{2}ma^2$.

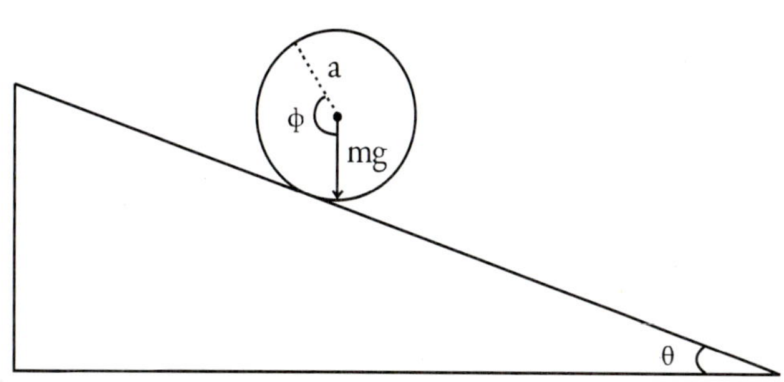

Ans. The kinetic and potential energies are

$$K=\frac{1}{2}m\dot{x}^2+\frac{1}{2}I\dot{\phi}^2=\frac{1}{2}m\dot{x}^2+\frac{1}{4}ma^2\dot{\phi}^2$$

and

$$U=-mgx\sin\theta,$$

respectively.

The Lagrangian is then

$$L=K-U=\frac{1}{2}m\dot{x}^2+\frac{1}{4}ma^2\dot{\phi}^2+mgx\sin\theta$$

Using the Lagrange undetermined multiplier method

$$\frac{\partial L}{\partial x}-\frac{d}{dt}\left(\frac{\partial L}{\partial \dot{x}}\right)+\lambda\frac{\partial f}{\partial x}=0 \text{ and } \frac{\partial L}{\partial \phi}-\frac{d}{dt}\left(\frac{\partial L}{\partial \dot{\phi}}\right)+\lambda\frac{\partial f}{\partial \phi}=0,$$

we get

$$mg\sin\theta-m\ddot{x}+\lambda=0 \tag{9.2.4.1}$$

and

$$-\frac{1}{2}ma^2\ddot{\phi}-a\lambda=0 \tag{9.2.4.2}$$

Eq. (9.2.4.2) gives $\lambda=-\frac{1}{2}ma\ddot{\phi}=-\frac{1}{2}m\ddot{x}$.

By inserting this into Eq. (9.2.4.1), we obtain

$$\ddot{x}=\frac{2}{3}g\sin\theta$$

This gives

$$\lambda=-\frac{1}{3}mg\sin\theta \text{ and } \ddot{\phi}=\frac{\ddot{x}}{a}=\frac{2g}{3a}\sin\theta$$

The generalized forces of constraints are then

$$\lambda\frac{\partial f}{\partial x}=\lambda=-\frac{1}{3}mg\sin\theta,$$

which corresponds to the force

and

$$\lambda\frac{\partial f}{\partial \phi}=\lambda(-a)=\left(\frac{1}{3}mg\sin\theta\right)a,$$

which corresponds to the torque.

9.2.5 Find the generalized forces of constraint for the problem in [Ex. 9.1.12].

Ans. The constraint equation for the system is $f=y-a\phi$.

The Lagrangian is

$$L = K - U = \frac{1}{2}m\dot{y}^2 + \frac{1}{2}I\dot{\phi}^2 + mgy = \frac{1}{2}m\dot{y}^2 + \frac{1}{4}ma^2\dot{\phi}^2 + mgy$$

From

$$\frac{\partial L}{\partial y} - \frac{d}{dt}(\frac{\partial L}{\partial \dot{y}}) + \lambda\frac{\partial f}{\partial y} = 0 \text{ and } \frac{\partial L}{\partial \phi} - \frac{d}{dt}(\frac{\partial L}{\partial \dot{\phi}}) + \lambda\frac{\partial f}{\partial \phi} = 0,$$

we obtain

$$mg - m\ddot{y} + \lambda = 0 \tag{9.2.5.1}$$

and

$$-\frac{1}{2}ma^2\ddot{\phi} - \lambda a = 0, \tag{9.2.5.2}$$

respectively.

Eq. (9.2.5.2) gives

$$\lambda = -\frac{1}{2}ma\ddot{\phi} = -\frac{1}{2}m\ddot{y}$$

By substituting this into Eq. (9.2.5.1), we get

$$\ddot{y} = \frac{2}{3}g$$

Then, we obtain the undetermined multiplier

$$\lambda = -\frac{1}{3}mg$$

so that the generalized forces of constraint are

$$\lambda\frac{\partial f}{\partial y} = \lambda = -\frac{1}{3}mg,$$

which corresponds to the tension of the string

and

$$\lambda\frac{\partial f}{\partial \phi} = \lambda(-a) = (-\frac{1}{3}mg)(-a) = (\frac{1}{3}mg)a,$$

which corresponds to the torque on the disk.

Appendix

Appendix 3-A

$$\sum_{k=1}^{3}\epsilon_{ijk}\epsilon_{k\ell m}=\sum_{k=1}^{3}\begin{vmatrix}\delta_{ik} & \delta_{i\ell} & \delta_{im}\\ \delta_{jk} & \delta_{j\ell} & \delta_{jm}\\ \delta_{kk} & \delta_{k\ell} & \delta_{km}\end{vmatrix}$$

$$=\sum_{k=1}^{3}(\delta_{ik}\delta_{j\ell}\delta_{km}-\delta_{ik}\delta_{jm}\delta_{k\ell}-\delta_{i\ell}\delta_{jk}\delta_{km}+\delta_{i\ell}\delta_{jm}\delta_{kk}-\delta_{im}\delta_{j\ell}\delta_{kk}+\delta_{im}\delta_{jk}\delta_{k\ell})$$

$$=\sum_{k=1}^{3}(\delta_{i\ell}\delta_{jm}-\delta_{im}\delta_{j\ell})\delta_{kk}+\sum_{k=1}^{3}\delta_{ik}\delta_{j\ell}\delta_{km}$$

$$-\sum_{k=1}^{3}\delta_{ik}\delta_{jm}\delta_{k\ell}-\sum_{k=1}^{3}\delta_{i\ell}\delta_{jk}\delta_{km}+\sum_{k=1}^{3}\delta_{im}\delta_{jk}\delta_{k\ell}$$

$$=3(\delta_{i\ell}\delta_{jm}-\delta_{im}\delta_{j\ell})+\delta_{im}\delta_{j\ell}-\delta_{i\ell}\delta_{jm}-\delta_{i\ell}\delta_{jm}+\delta_{im}\delta_{j\ell}$$

$$=3(\delta_{i\ell}\delta_{jm}-\delta_{im}\delta_{j\ell})+2\delta_{im}\delta_{j\ell}-2\delta_{i\ell}\delta_{jm}$$

$$=\delta_{i\ell}\delta_{jm}-\delta_{im}\delta_{j\ell}$$

Thus, we obtain

$$\sum_{k=1}^{3}\epsilon_{ijk}\epsilon_{k\ell m}=\delta_{i\ell}\delta_{jm}-\delta_{im}\delta_{j\ell}$$

Appendix 7-A

If we differentiate $f(x)g(x)$ with respect to x, we get

$$\frac{d}{dx}[f(x)g(x)] = \frac{df(x)}{dx}g(x) + f(x)\frac{dg(x)}{dx}$$

By differentiating the equation again, we have

$$\frac{d^2}{dx^2}[f(x)g(x)] = \frac{d}{dx}[\frac{d}{dx}f(x)g(x)] = \frac{d}{dx}[\frac{df}{dx}g + f\frac{dg}{dx}]$$

$$= \frac{d^2f(x)}{dx^2}g(x) + 2\frac{df(x)}{dx}\frac{dg(x)}{dx} + f(x)\frac{d^2g(x)}{dx^2}$$

By differentiating the equation again, we have

$$\frac{d^3}{dx^3}[f(x)g(x)] = \frac{d}{dx}[\frac{d^2}{dx^2}f(x)g(x)]$$

$$= \frac{d}{dx}[\frac{d^2f(x)}{dx^2}g(x) + 2\frac{df(x)}{dx}\frac{dg(x)}{dx} + f(x)\frac{d^2g(x)}{dx^2}]$$

$$= \frac{d^3f(x)}{dx^3}g(x) + 3\frac{d^2f(x)}{dx^2}\frac{dg(x)}{dx} + 3\frac{df(x)}{dx}\frac{d^2g(x)}{dx^2} + f(x)\frac{d^3g(x)}{dx^3}$$

Similarly, we get if we differentiate the equation again,

$$\frac{d^4}{dx^4}[f(x)g(x)]$$

$$= \frac{d}{dx}[\frac{d^3f(x)}{dx^3}g(x) + 3\frac{d^2f(x)}{dx^2}\frac{dg(x)}{dx} + 3\frac{df(x)}{dx}\frac{d^2g(x)}{dx^2} + f(x)\frac{d^3g(x)}{dx^3}]$$

$$= \frac{d^4f(x)}{dx^4}g(x) + 4\frac{d^3f(x)}{dx^3}\frac{dg(x)}{dx}$$

$$+ 6\frac{d^2f(x)}{dx^2}\frac{d^2g(x)}{dx^2} + 4\frac{df(x)}{dx}\frac{d^3g(x)}{dx^3} + f(x)\frac{d^4g(x)}{dx^4}$$

$$= \frac{d^4f(x)}{dx^4}g(x) + 4\frac{d^3f(x)}{dx^3}\frac{dg(x)}{dx} + \frac{4(4-1)}{2!}\frac{d^2f(x)}{dx^2}\frac{d^2g(x)}{dx^2}$$

$$+ 4\frac{df(x)}{dx}\frac{d^3g(x)}{dx^3} + f(x)\frac{d^4g(x)}{dx^4}$$

$$\vdots$$

We can generalize the equation

$$\frac{d^n}{dx^n}[f(x)g(x)] = \frac{d^n f(x)}{dx^n}g(x) + n\frac{d^{n-1}f(x)}{dx^{n-1}}\frac{dg(x)}{dx}$$

$$+\frac{n(n-1)}{2!}\frac{d^{n-2}f(x)}{dx^{n-2}}\frac{d^2g(x)}{dx^2} + \cdots\cdots + n\frac{df(x)}{dx}\frac{d^{n-1}g(x)}{dx^{n-1}} + f(x)\frac{d^n g(x)}{dx^n} \quad (1)$$

If we substitute $\frac{df(x)}{dx}$ and $\frac{dg(x)}{dx}$ instead of a and b in the binomial theorem, Eq. (5.2.7), respectively, we get

$$(\frac{df}{dx} + \frac{dg}{dx})^n = \frac{d^n f}{dx^n}\frac{d^0 g}{dx^0} + n\frac{d^{n-1}f}{dx^{n-1}}\frac{dg}{dx}$$

$$+\frac{n(n-1)}{2!}\frac{d^{n-2}f}{dx^{n-2}}\frac{d^2g}{dx^2} + \cdots\cdots + n\frac{df}{dx}\frac{d^{n-1}g}{dx^{n-1}} + \frac{d^0 f}{dx^0}\frac{dx^n g}{dx^n}$$

$$= \frac{d^n f}{dx^n}g + n\frac{d^{n-1}f}{dx^{n-1}}\frac{dg}{dx} + \frac{n(n-1)}{2!}\frac{d^{n-2}f}{dx^{n-2}}\frac{d^2g}{dx^2} + \cdots\cdots + n\frac{df}{dx}\frac{d^{n-1}g}{dx^{n-1}} + f\frac{dx^n g}{dx^n}$$

We see that the right-hand side of the equation is the same as the right-hand side of Eq. (1).

From the binomial theorem, the left-hand side of the equation can be expressed as

$$(\frac{df}{dx} + \frac{dg}{dx})^n = \sum_{k=0}^{n}\binom{n}{k}[\frac{d^{n-k}f(x)}{dx^{n-k}}][\frac{d^k g(x)}{dx^k}] \quad (2)$$

Thus, from Eqs. (1) and (2), we obtain the Leibnitz's formula

$$\frac{d^n}{dx^n}[f(x)g(x)] = \sum_{k=0}^{n}\binom{n}{k}[\frac{d^{n-k}f(x)}{dx^{n-k}}][\frac{d^k g(x)}{dx^k}]$$

Appendix 7-B

From the property of the gamma function, we have

$$\Gamma(x) = \Gamma(x-1+1) = (x-1)! = \frac{x!}{x}$$

$$= \frac{1}{x}\frac{1\cdot 2\cdot 3\cdot \cdots\cdots x[(x+1)(x+2)(x+3)\cdots\cdots(x+n)]}{[(x+1)(x+2)(x+3)\cdots\cdots(x+n)]}$$

Since n is any arbitrary, the equation can be

$$\Gamma(x) = \frac{n!(n+1)(n+2)\cdots\cdots(n+x)}{x(x+1)(x+2)\cdots\cdots(x+n)}$$

$$= \frac{n!n^x(1+\frac{1}{n})(1+\frac{2}{n})\cdots\cdots(1+\frac{x}{n})}{x(x+1)(x+2)\cdots\cdots(x+n)}$$

Then, the equation gives

$$\lim_{n\to\infty}[\frac{n!n^x(1+\frac{1}{n})(1+\frac{2}{n})(1+\frac{3}{n})\cdots\cdots(1+\frac{x}{n})}{x(x+1)(x+2)\cdots\cdots(x+n)}]$$

$$= \lim_{n\to\infty}[\frac{n!n^x}{x(x+1)(x+2)\cdots\cdots(x+n)}],$$

where $\frac{n!n^x}{x(x+1)(x+2)\cdots\cdots(x+n)}$ is defined as $\Gamma_n(x)$ and it is called **Gauss expression** for the Gamma function.

Therefore, we have

$$\lim_{n\to\infty}\Gamma_n(x) = \Gamma(x)$$

The Gauss expression then gives

$$\Gamma_n(x) = \frac{1}{x}\frac{1}{x+1}\frac{2}{x+2}\cdots\cdots\frac{n}{x+n}n^x$$

$$= \frac{e^{x[ln(n)-1-\frac{1}{2}-\cdots\cdots-\frac{1}{n}]}e^{x+\frac{x}{2}+\cdots\cdots+\frac{x}{n}}}{x(x+1)(1+\frac{x}{2})\cdots\cdots(1+\frac{x}{n})}$$

If we define $ln(n)-1-\frac{1}{2}-\cdots\cdots-\frac{1}{n} = -\gamma$ (**Euler-Mascheroni constant**), the equation is then expressed as

$$\Gamma_n(x)=e^{-\gamma x}\frac{1}{x}\frac{e^x}{1+x}\frac{e^{\frac{x}{2}}}{1+\frac{x}{2}}\cdots\cdots\frac{e^{\frac{x}{n}}}{1+\frac{x}{n}}$$

This equation gives

$$\lim_{n\to\infty}\Gamma_n(x)=\Gamma(x)=e^{-\gamma x}x^{-1}\prod_{n=1}^{\infty}\frac{n}{n+x}e^{\frac{x}{n}}$$

This is called **Weierstrass expression.**

If we take logarithm on both sides of the equation, we have

$$ln\Gamma(x)=-\gamma x+lnx^{-1}+\sum_{n=1}^{\infty}lne^{\frac{x}{n}}+\sum_{n=1}^{\infty}ln(\frac{n}{n+x})$$

By differentiating the equation with respect to x, we get

$$\frac{d}{dx}ln\Gamma(x)=-\gamma-\frac{1}{x}+\sum_{n=1}^{\infty}\frac{\frac{1}{n}e^{\frac{x}{n}}}{e^{\frac{x}{n}}}-\sum_{n=1}^{\infty}\frac{-\frac{n}{(n+x)^2}}{\frac{n}{n+x}}$$

$$=-\gamma-\frac{1}{x}+\sum_{n=1}^{\infty}\frac{1}{n}-\sum_{n=1}^{\infty}\frac{1}{n+x}$$

$$=-\gamma-\frac{1}{x}+\sum_{n=1}^{\infty}(\frac{1}{n}-\frac{1}{n+x})=-\gamma+\sum_{n=1}^{\infty}\frac{1}{n}-\sum_{n=0}^{\infty}\frac{1}{n+x}$$

$$=\lim_{n\to\infty}[ln(n)-\sum_{k=1}^{n}\frac{1}{k}]+\sum_{n=1}^{\infty}\frac{1}{n}-\sum_{n=0}^{\infty}\frac{1}{n+x}$$

$$=\lim_{n\to\infty}[ln(n)-\sum_{k=0}^{n}\frac{1}{k+x}]$$

The left-hand side equation is $\frac{d}{dx}ln\Gamma(x)=\frac{\Gamma'(x)}{\Gamma(x)}$, which is the digamma function $\Psi(x)$.

Then, the equation is

$$\Psi(x)=\lim_{n\to\infty}[ln(n)-\sum_{k=0}^{n}\frac{1}{k+x}]$$

so that

$$\frac{\Psi(x)}{\Gamma(x)}=\lim_{n\to\infty}\frac{ln(n)-(\frac{1}{x}+\frac{1}{x+1}+\frac{1}{x+2}+\cdots\cdots+\frac{1}{x+n})}{\frac{n!n^x}{x(x+1)(x+2)\cdots\cdots(x+n)}}$$

By multiplying both numerator and denominator by $x+k$, we get

$$\frac{\Psi(x)}{\Gamma(x)}=\lim_{n\to\infty}\frac{(x+k)ln(n)-(\frac{x+k}{x}+\frac{x+k}{x+1}+\cdots\cdots+\frac{x+k}{x+k-1}+1+\frac{x+k}{x+k+1}+\cdots\cdots\frac{x+k}{x+n})}{\frac{n!n^x}{x(x+1)(x+2)\cdots\cdots(x+k-1)\cdot 1\cdot(x+k+1)\cdots\cdots(x+n)}}$$

If we substitute $-k$ instead of x in the equation, we get

$$\frac{\Psi(-k)}{\Gamma(-k)}$$

$$=\lim_{n\to\infty}\frac{0\cdot ln(n)-(\frac{0}{-k}+\frac{0}{-k+1}+\cdots\cdots+\frac{0}{-1}+1+\frac{0}{1}+\cdots\cdots+\frac{0}{-k+n})}{\frac{n!n^{-k}}{-k(-k+1)(-k+2)\cdots\cdots(-1)\cdot 1\cdot[1\cdot 2\cdots\cdots(-k+n)]}}$$

$$=\lim_{n\to\infty}(-1)^{k+1}\frac{k(k-1)(k-2)\cdots\cdots(1)\cdot 1\cdot[1\cdot 2\cdots\cdots(-k+n)]}{n!n^{-k}}$$

$$=\lim_{n\to\infty}(-1)^{k+1}\frac{k!}{n!n^{-k}}\frac{n!}{(n-k+1)(n-k+2)\cdots\cdots(n-1)n}$$

$$=\lim_{n\to\infty}(-1)^{k+1}k!n^k\frac{1}{(n-k+1)(n-k+2)\cdots\cdots(n-1)n}$$

$$=\lim_{n\to\infty}(-1)^{k+1}k!\frac{1}{(1-\frac{k}{n}+\frac{1}{n})(1-\frac{k}{n}+\frac{2}{n})\cdots\cdots(1-\frac{1}{n})\cdot 1}$$

$$=(-1)^{k+1}k!$$

Thus, we obtain

$$\frac{\Psi(-k)}{\Gamma(-k)}=(-1)^{k+1}k!$$

Appendix 7-C

(i) Let's prove that $(-\frac{d}{dx})^{\ell}\sin x = \cos(x - \frac{\ell\pi}{2} - \frac{\pi}{2})$

For $\ell = 0$,

$$(-\frac{d}{dx})^{0}\sin x = \sin x = \cos(x - \frac{\pi}{2})$$

For $\ell = 1$,

$$(-\frac{d}{dx})^{1}\sin x = -\cos x = \cos(x - \pi) = \cos(x - \frac{\pi}{2} - \frac{\pi}{2})$$

For $\ell = 2$,

$$(-\frac{d}{dx})^{2}\sin x = -\frac{d}{dx}(-\frac{d}{dx}\sin x) = \frac{d}{dx}\cos x$$
$$= -\sin x = \cos(x - \frac{3}{2}\pi) = \cos(x - \frac{2\pi}{2} - \frac{\pi}{2})$$

For $\ell = 3$,

$$(-\frac{d}{dx})^{3}\sin x = -\frac{d}{dx}(-\sin x) = \cos x = \cos(x - \frac{3}{2}\pi - \frac{\pi}{2})$$

$$\vdots$$

We obtain in general

$$(-\frac{d}{dx})^{\ell}\sin x = \cos(x - \frac{\ell\pi}{2} - \frac{\pi}{2})$$

(ii) Let's prove that $(\frac{d}{dx})^{\ell}\cos x = (-1)^{\ell+1}\sin(x - \frac{\ell\pi}{2} - \frac{\pi}{2})$

For $\ell = 0$,

$$(\frac{d}{dx})^{0}\cos x = \cos x = \sin(\frac{\pi}{2} - x) = -\sin(x - \frac{\pi}{2})$$

For $\ell = 1$,

$$(\frac{d}{dx})^{1}\cos x = -\sin x = -\sin(\pi - x) = \sin(x - \pi) = \sin(x - \frac{\pi}{2} - \frac{\pi}{2})$$

For $\ell = 2$,

$$(\frac{d}{dx})^{2}\cos x = \frac{d}{dx}(-\sin x) = -\cos x = \sin(\frac{3\pi}{2} - x)$$

$$=-\sin(x-\frac{3\pi}{2})=-\sin(x-\frac{2\pi}{2}-\frac{\pi}{2})$$

For $\ell=3$,

$$(\frac{d}{dx})^3\cos x=\frac{d}{dx}(-\cos x)=\sin x=\sin(x-\frac{3\pi}{2}-\frac{\pi}{2})$$

$$\vdots$$

We obtain in general

$$(\frac{d}{dx})^\ell\cos x=(-1)^{\ell+1}\sin(x-\frac{\ell\pi}{2}-\frac{\pi}{2})$$

REFERENCES

[Chapter 1]

Alastair Buckley, "*Mathematical Methods for Physics and Astronomy*", http://www.hep.shef.ac.uk/phy226/unit1/phy250_MMblock1.htm (2012).

James Nearing, "*Mechanics*", http://www.physics.miami.edu/nearing/class/340/book/mechanics.pdf (2008).

[Chapter 2]

Mary L. Boas, "*Mathematical Methods in the Physical Sciences*", 2nd Edition, John Wiley & Sons, Inc., 1983.

Grant R. Fowles and George L. Cassiday, "*Analytical Mechanics*", 7th Edition, Thomson Learning, Inc., 2005.

James Nearing, "*Mechanics*", http://www.physics.miami.edu/nearing/class/340/book/mechanics.pdf (2008).

[Chapter 3]

Stephen T. Thornton and Jerry B. Marion, "*Classical Dynamics of Particles and Systems*", 5th Edition, TOMSON Brooks/Cole, 2004.

Roger Chen, "*Advanced Calculus; PROOF OF THE DIVERGENCE THEOREM AND STOKES' THEOREM*", http://www.math.ncku.edu.tw/~rchen/Advanced%20Calculus/divergence%20theorem.pdf.

James Nearing, "*Mechanics*", http://www.physics.miami.edu/nearing/class/340/book/mechanics.pdf (2008).

David J. Griffith, "*Introduction to Electrodynamics*", Pearson High (2013).

Joseph C. Kolecki, "*An Introduction to Tensors for Students of Physics and Engineering*", NASA/TM—2002-211716 (2002).

[Chapter 4]

James Foadi, "*Mathematical Physics: Gradient, Divergence and Curl in Curvilinear Coordinates*", Oxford 2011.

Mary L. Boas, "*Mathematical Methods in the Physical Sciences*", 2nd Edition, John Wiley & Sons, Inc., 1983.

Stephen Gasiorowicz, "*Quantum Physics*", 3rd Edition", John Wiley & Sons, Inc., 2003.

[Chapter 5]

Mary L. Boas, "*Mathematical Methods in the Physical Sciences*", 2nd Edition, John Wiley & Sons, Inc., 1983.

Hans J. Weber and George B. Arfken, "*Essential Mathematical Methods for Physicists*", Elsevier Science, 2004.

Matthias Heil, "*Lecture Notes for 2M1*", http://www.maths.manchester.ac.uk/~mheil/Lectures/2M1/Material/Summary.pdf.

[Chapter 6]

Mary L. Boas, "*Mathematical Methods in the Physical Sciences*", 2nd Edition, John Wiley & Sons, Inc., 1983.

Stephen Gasiorowicz, "*Quantum Physics*", 3rd Edition", John Wiley & Sons, Inc., 2003.

Grant R. Fowles and George L. Cassiday, "*Analytical Mechanics*", 7th Edition, Thomson Learning, Inc., 2005.

L.D. Landau and E. M. Lifshitz, "*Mechanics*", 3rd Edition, Course of Theoretical Physics Volume 1, Pergamon Press, 1976.

[Chapter 7]

Mary L. Boas, "*Mathematical Methods in the Physical Sciences*", 2nd Edition, John Wiley & Sons, Inc., 1983.

Hans J. Weber and George B. Arfken, "*Essential Mathematical Methods for Physicists*", Elsevier Science, 2004.

Jim Lambers, "*Differential Equations and Special Functions: Lecture16*", http://www.math.usm.edu/lambers/mat415/lecture16.pdf.

K. Kono, "*Alien's Mathematics*", http://fractional-calculus.com/gamma_digamma.pdf.

Joseph C. Weingartner, "*Classical Electrodynamics: Lecture Note*", Ihttp://physics.gmu.edu/~joe/PHYS685/Topic3.pdf.

송희성, "*양자역학*", 교학연구사, 1984.

Stephen Gasiorowicz, "Quantum Physics, Third Edition", John Wiley & Sons, Inc., 2003.

[Chapter 8]

Mary L. Boas, "*Mathematical Methods in the Physical Sciences*", 2nd Edition, John Wiley & Sons, Inc., 1983.

Hans J. Weber and George B. Arfken, "*Essential Mathematical Methods for Physicists*", Elsevier Science, 2004.

Stephen Gasiorowicz, "Quantum Physics, Third Edition", John Wiley & Sons, Inc., 2003.

[Chapter 9]

Grant R. Fowles and George L. Cassiday, "*Analytical Mechanics*", 7th Edition, Thomson Learning, Inc., 2005.

L.D. Landau and E. M. Lifshitz, "*Mechanics*", 3rd Edition, Course of Theoretical Physics Volume 1, Pergamon Press, 1976.

Index

[A]

"ABC"rule 64

Abelian 151

adjoint matrix 139

algebraical method 58

angular component 267

angular momentum 63, 99, 110

angular momentum operator 111

anti-commutative 59

anti-derivative 24

Argand diagram 29

arithmetic series 114

associated Legendre functions 190

Associated Legendre polynomials 191

asymptotic behaviors 229, 233

Atwood machine 265, 273

auxiliary equation 41

average value 240

azimuthal angle 53

[B]

"BAC-CAB" rule 64, 66

basis 135

Bessel function 210

Bessel functions 216

Bessel's equation 207

Beta functions 172

binomial formula 121

binomial theorem 121

bra 156

braket 156

[C]

calculus limit method 18

centripetal force 85

chain rule 22

characteristic equation 152

circle 32

closed interval 239

cofactor 139

cofactor matrix 139

column matrix 135

commutative 59, 151

complementary solution 46

completeness relation 158

complex conjugate 29

complex number 29

conic section 34

conic sections 167

conjugate transpose 146

conservative force 85

constraints 259

convergent 115

convolution 249

cosine transform 249

coupled harmonic oscillator 152, 269

Cramer's rule 140

critical damping 45

Curl 67, 69

curvilinear coordinate system 102

Cylindrical coordinate system 54

cylindrical coordinate system 100

[D]
damped harmonic oscillator 44
damping coefficient 44
"del"vector 67
De Moivre's formula 16
definite integral 25
degree of freedom 259
delta function 245, 247
determinant 61
diagonal matrix 138
diagonalizable 151
differential equation 35, 36
differential operator 41
differentiations 18
Digamma function 176
digamma function 284
Dirac delta function 31
Dirac notation 156
direction cosines 88
distribution 56
Divergence 68
divergence 77
Divergence theorem 77
Divergent 67
divergent 115
Double-angle Formulas 14
dyad 86
d'Alembert ratio test 115

[E]
eigenvalue 151
eigenvalue equation 151
eigenvector 151
ellipse 33
equations of constraint 259
equilibrium 130
equilibrium position 130
Euler-Mascheroni constant 283
Euler's formula 16
even 238

[F]
factorial 116
factorial function 175
falling disk 268
first order differential equation 37
forced (or driven) oscillator 46, 48
forces of constraint 271
Fourier coefficients 239
Fourier Cosine Transform 249
Fourier Series 238
Fourier sine 249
Fourier Sine Transform 249
Fourier Transform 245
Fourier transform 246

[G]
Gamma 172
Gauss expression 283
Gauss theorem 77
Gaussian 56, 251
generalized coordinates 259
generalized forces of constraint 272
generalized Legendre polynomials 190
generating function 199
geometric series 114
Gradient 67, 68

[H]
Half-angle formulas 15
Hamiltonian operator 73
Hamilton's variational principle 259
Hankel functions 216
harmonic oscillator 264
Hermitian 146
Hermitian onjugate 146
Hermitian operator 160
Hessian square matrix 130
hird kind 216
holonomic 259
homogeneous 36
hoop 276

horizontal 34
horizontal ellipse 33
horizontal parabolas 33
hyperbola 34
Hyperbolic Functions 16

[I]
identity 138
inconclusive 115
indefinite integral 24
indeterminate 115
inhomogeneous 36
inhomogeneous second order differential equation 46
inner product 156, 158
integrand 25
integration by parts 27
integration by substitution 25
inverse Fourier transform 246
inverse matrix 138
inverse matrix method 140
invertible 146
invertible matrix theorem 152
isotropic oscillator 170

[K]
ket 156
Kronecker delta symbol 64

[L]
Lagrange undetermined multiplier method 271
Lagrange's equation 260
Lagrangian 259
Laplace's equation 69, 203, 213
Laplacian 69
least action principle 259
Legendre differential equation 181
Legendre polynomials 180, 184, 187
Leibnitz's formula 192
Levi-Civita density 64
Lorentz force 83
L'Hopital rule 252
L'Hopital's rule 113, 212

[M]
Maclaurin series 120
magnitude 157
matrix 147
matrix multiplication 136
Maxwell stress tensor 91
minor 138
modulus 30
moment of inertia 88
momentum operator 110

[N]
Neumann function 210
Neumann functions 223
neutral equilibrium 130
non-holonomic 259
norm 157
normal 56
normalized eigenvector 153

[O]
odd function 238
operator 146
order 135
ordinary differential equation 36
ordinary differentiation 18
orthogonal 59, 146, 157
orthogonality 198, 201
orthonormal 157
orthonormal condition 204
orthonormality 204, 205
oscillator 176
outer product 158
overdamped 44

[P]
Parseval's theorem 250, 253
partial differential equation 36

partial differentiation 18
partial sum 114
particular solution 46
pendulum 264, 267
period 128
permutation symbol 64
plane polar coordinate system 53
Poisson's equation 69
polar angle 53
polygamma function 176
position vector 94, 101
power series 115
Poynting vector 92
probability density function 56
projection operator 158

[R]
radial component 267
ratio test 115
Rectangular coordinate system 52
rectangular matrix 135
recurrence relation 182, 199, 200
right-hand rule 52
right-handed coordinate system 52
Rodrigues' formula 188
rotation matrix 148
row matrix 135

[S]
saddle 131
scalar 57
scalar (or dot, inner) product 59
Scale factor 110
scale factor 95
scale factors 100
scattering theory 221
second order differential equation 40
self-adjoint 146
similarity transformation 161
simple harmonic oscillator 43
simple pendulum 127
singular 184
Snell's law 263
SO(2) group 150
solid angle 97
spherical Bessel 223
spherical Bessel function 221
Spherical coordinate system 53
spherical coordinate systems 94
Spherical Hankel functions 224
spherical harmonics 204, 205
spherical Neumann function 221
spherical wave 235
square matrix 135
stable 130
step function 252
steradians 97
Stirling's formula 178
Stokes' theorem 77, 78
SU(2) group 150

[T]
Taylor expansion 118
Taylor series 118, 120
tension 279
Tensor 86
Three or Four Vectors Operation 64
time-dependent Schrödinger equation 74
time-evolution operator 164
time-independent Schrödinger equation 73
torque 278, 279
trace 137
transformation 147
transpose matrix 138
triad 86
Trigonometric Functions 11
triple product 64
Two Vectors Operation 59

[U]
underdamped oscillator 44
undetermined multiplier 272

unit 138
unit tensor 90
unit vector 57
unitary 146
unitary operator 162
unitary transformation 160, 161
unstable 130

[V]
vector 57
vector (or cross, outer) product 59
vertical 33, 34
vertical ellipse 33

[W]
wave function 256
wave number 256
Weierstrass expression 284